THE DYER'S HANDBOOK

ANCIENT TEXTILES SERIES VOL. 26

THE DYER'S HANDBOOK

Memoirs on Dyeing by a French gentleman-clothier in the Age of Enlightenment translated and contextualised

DOMINIQUE CARDON

OXBOW | books
Oxford & Philadelphia

First published in 2016. Reprinted in paperback in 2021 and 2024 in the United Kingdom by
OXBOW BOOKS
The Old Music Hall, 106–108 Cowley Road, Oxford, OX4 1JE

and in the United States by
OXBOW BOOKS
1950 Lawrence Road, Havertown, PA 19083

Paperback Edition: ISBN 978-1-78925-549-2
Digital Edition: ISBN 978-1-78570-212-9 (epub)

A CIP record for this book is available from the British Library

Library of Congress Cataloging-in-Publication Data

Names: Cardon, Dominique, editor.
Title: The dyer's handbook : memoirs on dyeing / by a French gentleman-clothier in the Age of Enlightenment ; translated and contextualised ; edited by Dominique Cardon.
Other titles: M?emoires de teinture. English
Description: Oxford ; Philadelphia : Oxbow Books, 2016. | Series: Ancient textiles series ; vol. 26 | Translation of M?emoires de teinture. | Translation into English of an anonymous French manuscript held in a private collection consisting of four essays produced around 1763 possibly written by Paul Gout of the Royal Manufactures of Bize, France in the Langdoc Region, originally consisting of about 100 pages in which were held swaths of sample dyed woolen cloth and including recipes for their coloring. The work has been published in French under the title M?emoires de teinture. | Includes bibliographical references and index. | Description based on print version record and CIP data provided by publisher; resource not viewed.
Identifiers: LCCN 2016017651 (print) | LCCN 2016013179 (ebook) | ISBN 9781785702129 (epub) | ISBN 9781785702136 (mobi) | ISBN 9781785702143 (pdf) | ISBN 9781785702112 (hardback)
Subjects: LCSH: Dyes and dyeing--Textile fibers--France--Languedoc--History--18th century. | Dyes and dyeing--Wool--France--Languedoc--History--18th century. | Dyes and dyeing--Early works to 1800. | Manuscripts, French--France--Languedoc.
Classification: LCC TP897 (print) | LCC TP897 .M46513 2016 (ebook) | DDC 667/.2--dc23
LC record available at https://lccn.loc.gov/2016017651

Printed in the United Kingdom by CMP Digital Print Solutions

For a complete list of Oxbow titles, please contact:

UNITED KINGDOM
Oxbow Books
Telephone (0)1226 734350
Email: oxbow@oxbowbooks.com
www.oxbowbooks.com

UNITED STATES OF AMERICA
Oxbow Books
Telephone (610) 853-9131, Fax (610) 853-9146
Email: queries@casemateacademic.com
www.casemateacademic.com/oxbow

Oxbow Books is part of the Casemate Group

Front cover: The spring of Las Fons (D. Cardon); a page from the original manuscript of *Mémoires de teinture*
Back cover: Three sample cards with English cloths bought in Constantinople in 1733 and their closest French equivalents, Archives départementales de l'Hérault, C 2200 (D. Cardon)

For my dear friends and family
in the British Isles and North America

Contents

Acknowledgements

I am very grateful to:

The family who owns the manuscript of *Mémoires de teinture*, who generously allowed me to study and publish it, and to Jean-Claude Richard, who brought me into contact with its members.

To Yvonne Mathieu, Roger Monié and Benjamin Assié for their help to locate the Bouillette spring and for their warm welcome to Bize.

At the *Archives départementales de l'Hérault*, I am greatly indebted to Julien Duvaux, Monique Bourseau, Vinciane Thomas; at the *Archives départementales de l'Aude*, to Sylvie Caucanas, then Director, and Georges Delmas; and to all the staff in both archives, who have been unfailingly kind and helpful.

I received a similarly friendly and highly efficient welcome from their English colleague archivists at Gloucester Archives – Helen Timlin, Andrew Parry, and all the staff – as well as at the Wiltshire and Swindon History Centre in Chippenham – Gill Neal, Steven Hobbs, and all the staff.

I am very grateful to the following friends and colleagues who are sharing my interest in the sleeping treasures of dye books with samples and pattern books preserved in our respective countries' archives:
Anita Quye, of the University of Glasgow, and Lucy Tann and the team at the Southwark Local History Library and Archive, for sharing their first assessment of the Crutchley Archives.

Jenny Balfour-Paul, of the University of Exeter, and Mary Henderson and Anne Buchanan, at Bath Central Library, for sharing information and photos concerning the Wallbridge Mill Dye Books.

Hero Granger-Taylor, for perseveringly, albeit unsuccessfully, trying to locate "a pattern of woolen cloth made in North Wales, after the manner of that made in France" in the *Journal of the Commissioners for Trade and Plantations*.

I gratefully acknowledge my indebtness to the friends and colleagues who communicated results from their ongoing research and allowed me to use information of great relevance to some of the issues discussed in this book:

Aki Arponen, on the history of alum exploitation in Sweden;

Annie Mollard-Desfour, on the history of some colour names;

David Pybus for his information on exports of Yorkshire alum to the south of France;

Tristan Yvon for his information on exports from the French West Indies to Bordeaux.

I am very thankful to Jennifer Tann for her encouragements and answers to my questions on the history of the Gloucestershire woollen industry she knows so well.

To the staff at the inter-library loans department of the Université des Sciences et Techniques du Languedoc in Montpellier, Monique Hibade and Sandrine Beraud, go my heartfelt thanks for their extraordinary efficiency and perseverance in providing me with the greater part of the documentation consulted for this book.

I am very grateful to Witold Nowik, for his dye analyses and comments on the colorants he identified in some of the samples of cloth in the manuscript of *Memoirs on Dyeing*, and to Iris Brémaud, for her colorimetric measurements and characterisation of all the samples, and enlightening discussion of the results. It is thanks to the generous welcome of Jean Gérard, Patrick Langbour, Daniel Guibal and Marie-France Thévenon, of the Research Unit BioWooEB of CIRAD (International Centre of Agronomic Research for Development) in Montpellier and their sharing of their equipment that these colorimetric measurements could be performed.

I thank Pierre-Normann Granier for the beautiful and exact photos he has taken of all the pages with samples in the manuscript.

I thank Karyn Mercier very much for her beautiful map and diagrams.
It has been a great pleasure working with the friendly staff at Oxbow Books – especially Clare Litt, Mette Bundgaard and Hannah McAdams.

I thank the Pasold Research Fund very much for a grant which has enabled me to visit archive collections and examine dye books and pattern books of different clothiers and dyers, in Gloucester and Chippenham.

I am grateful to the Région Languedoc-Roussillon-Midi-Pyrénées for a grant which has contributed towards the expenses of the present publication.

Foreword

The core part of this book is the translation into English of the text of a French manuscript, *Mémoires de teinture* (Memoirs on Dyeing), the critical edition of which was published recently.[1]

The manuscript is privately owned, and only one copy is known to exist. It is fragile and beautiful; the descriptions of dyeing processes contained in it are illustrated by samples of fine broadcloth dyed in the corresponding colours. This single French edition was the sole medium from which to make the memoirs accessible to the public.

This volume includes a translation of the original manuscript, with the addition of a number of essays that I hope will put this exceptional document in its historical, economic, technological contexts.

For those historians who have long been fascinated by the change in scale and the amount of innovation that occurred in woollen cloth production in Europe during the 17th and 18th centuries, the *Memoirs on Dyeing* brings first-hand insight into the daily preoccupations and tasks of a key actor in the success story of the Languedocian broadcloth production specially devised for export to the Levant. Even non-specialists may be interested in understanding the clever management and technical organisation that made it possible for the author's manufacture to produce, dye, finish, pack and export up to 1,375 pieces of superfine broadcloth per year, representing nearly 51 km of cloth of an estimated total weight of about 15,000 kg. Per day, it implied dyeing a minimum of eight half-pieces of cloth, each measuring 15 to 17 ells (18 to 19 metres) in length and that's without counting any holiday off.

The *Memoirs on Dyeing* also contribute new elements to clarify important technological issues about the competition that took place between the textile centres of Venice, the Netherlands, England and France to conquer the vast markets of the Ottoman Empire and beyond. The author's and his Languedocian colleagues' best rivals at the time being the English clothiers of the West of England, I have started a research into their archives, pattern books and dye books, which will, of course, have to be continued, but has already allowed me, in this book, to propose some comparisons between the production systems in these two regions of Europe, each with a long tradition of wool weaving and dyeing.

My interest in these colourful documents is not purely historical. I hope that this book may be of use, not only to readers with an interest in economic history and in the history of techniques, but also to the growing numbers among the young generation of colourists, designers and dyers with a keen interest in the colours of the past. Some may want to revive them, as a natural and essential part of the new conception and production process emerging with the "Slow Fashion" movement. Others may simply use them as an inspiration for new colour trends.

It is with such uses in mind that the present book has been planned. Its dimensions in height and width not being too much reduced compared with the dimensions of the original document, the colour plates of all the pages of the manuscript illustrated with dyed textile samples can do full justice to the beauty of these colours miraculously preserved for us since the 18th century. Thanks to the corresponding recipes, and to the conversion of their *Ancien Régime* quantitative data into metric figures provided, they can be reproduced exactly, with the same ingredients. Alternatively, the same shades can be reproduced by dyers and colourists with any colorants – either natural of synthetic – they may wish to use, because they can match the results of their experiments with the coloristic definitions of the samples in the *Mémoires de teinture*. Colorimetric measurements have been performed by Iris Brémaud, scientific researcher at the *Laboratoire de Mécanique et Génie Civil*, CNRS (National Centre of Scientific Research), University of Montpellier 2, on all the samples preserved in this document. The results are published as CIE L*a*b* data at the end of the book.

Imagine. Colours of the past, escaping from the pages of old dye books and pattern books. Persian blue, raven, dainty blue, pomegranate flower, spiny lobster, winesoup, pale flesh, dove breast, golden wax, grass green, green sand, rotten olive, modest plum, agate, rich French gray, gunpowder of the English, finding their way to the streets of our cities, enlivening everything we wear, all allied to dissipate the bleakness of the times. "You may say I'm a dreamer" – historians do like to think that understanding the past can inspire the future.

Note

1 Cardon, D. (2013) *Mémoires de teinture – Voyage dans le temps chez un maître des couleurs*. Paris: CNRS éditions.

Part I

A new life for a mysterious manuscript

1

In Search of an Author

First encounter

My story starts a few years ago in the south of France, at a dinner party in Montpellier, a city famous for its early medical school and university with long history of research into plants and their uses. During the dinner, an eminent albeit retired professor of botany and pharmacology happened to be sitting beside a historian friend of mine. Neither remembered how their conversation induced the professor to mention an extraordinary manuscript in his possession. It had been handed down through his family and – said he – supposedly originated from an ancestor who was a clothier in the 18th century. What made the document so special was the world of colours revealed in its pages, illuminated by 177 samples of fine broadcloth, dyed in all the hues of the rainbow, glued to the paper in front of nearly all the recipes. Being quite old and having just been told he was seriously ill, the professor felt he would like to have the manuscript examined by a specialist who could assess its historical value and the usefulness of planning its publication. Ancient textiles, natural dyes – my interest in such fields being pretty widely known locally, in a matter of days I had been alerted, put into contact with the professor, and found myself ringing at his door bell in one of the beautiful medieval houses in the historical centre of the city.

What awaited me inside was a courteous and kind welcome by the professor and his wife. After some polite talk, revealing their earnest interest in historical research, they brought me the manuscript, kept in a drawer of the professor's desk: a high, thin notebook with a frail, partly torn, cream-coloured paper cover. I opened it – or rather, it opened itself – at the middle page: on both left margins, a column of colour names; just beside, petals of thin, velvety cloth of different shapes and shades of green corresponding to each colour name, compellingly named: black green, dark green, obscure green, duck green, grass green, emerald, parrot, light green, gay green, nascent green...

Facing these swatches, on the right column of either page, were the author's explanations on how he had made these hues. For each special shade of green sample, the exact degrees of woad and indigo blue ground necessary to obtain it were clearly defined. As the professor had said, all the pages of the memoirs similarly bloomed with delightful assortments of vivid or subtle colours and detailed recipes for their creation.

That summer afternoon, I fell in love with the document and resolved to publish it so that it could be shared with other lovers of colour. Our common enthusiasm for this beautiful project, I like to think, brought some light into the ensuing painful period of illness that befell the professor, ending in his death. The recent publication in French of the scholarly edition of the manuscript,[1] after several years of hard work, was a great joy for his widow.

Betrayed by a spring

Immediately after the first enchantment at discovering this extraordinary document, came an overwhelming sense of the difficulty of the task involved in trying to understand by whom it had been written, and why. The manuscript is anonymous, no date is to be found anywhere in the text, nor any mention of a known place name that could help locate precisely where it was written. True enough, in the very first page of the manuscript, the author defines both the technical and geographical limits of his work: he is only writing – he warns – about "the dyes that we make in Languedoc for the Levant". But this is not very helpful in terms of location. The wealth of historical studies that have been dedicated to the planned revival and development of the woollen industry oriented towards Eastern Mediterranean markets in 17th–18th century Languedoc, has revealed how widespread the production centres were all over this vast province, then stretching from the River Rhône westward, nearly to the outskirts of the Pyrénées mountains and the border with Spain

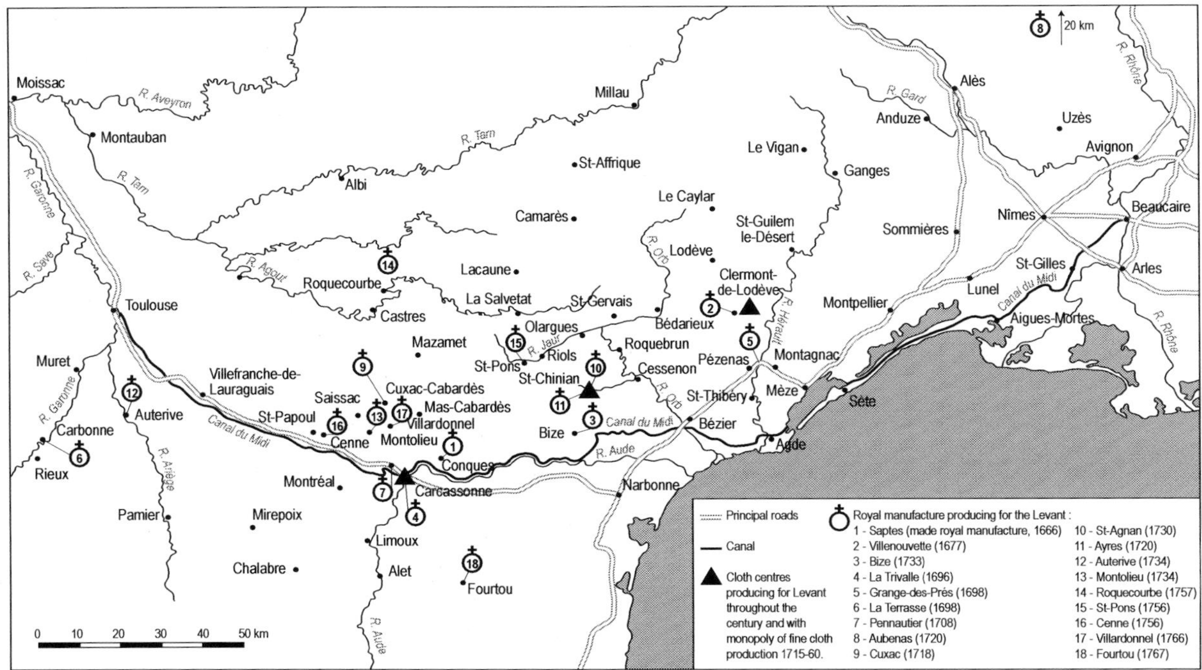

Fig. 1.1. Textile centres in Languedoc exporting broadcloth to the Levant in the 18th century. Map K. Mercier/D. Cardon, CNRS, CIHAM/UMR 5648, after J.K.J. Thomson, Clermont-de-Lodève 1633–1789 – Fluctuations in the prosperity of a Languedocian cloth-making town. *Cambridge, Cambridge University Press, 1982, frontispiece.*

(Fig. 1.1). How could one guess where to start looking in Languedoc without further clues?

Well, there is a hint, but it only appears in the last of the four memoirs. There, the author announces that he is going to complete the methodical presentation of "all the colours that are done for the Levant", which he has offered in the second memoir, by a record of the results of his personal experiments to improve some dyeing processes or to create new colour shades. This naturally induces him to give some details on the practical conditions in which he conducts these experiments, starting, as one would expect, by comments on the quality of his water resources, so vital for a dyer. This is how at last, at page 80 of the manuscript, the author reveals that the place where "his colours" are produced is a Royal Manufacture that, he explains, has two different water supplies. For dyeing and washing the cloths, a river that receives additional water from a group of springs a short distance upstream is used; another spring is directed to a fulling mill. The name of the Royal Manufacture he does not mention, nor that of the river that flows by, and the name he does mention for the springs upstream that flow down into the river, "Las Fons", is not very telling since it just means "The Springs" in the local Occitan language. Which leaves only one hope of identifying the place he is writing from: the name of the other spring, "La Bouilhete", whose water he disparagingly describes as "*molasse*" – or "softish".

Locating a spring with that name in the hydrographic network of a whole province: one might expect it to be about as easy as looking for a needle in a haystack. It actually turned out not to be as hopeless as it appeared. To begin with, the simple mention that the author worked in a Royal Manufacture allowed to significantly restrict the research: on the 26th January 1735, the *Etats du Languedoc* (the assembly of the representatives of the Estates of the Province) had "definitively" limited the number of Privileged Royal Manufactures to no more than twelve.[2] Although six more Royal Manufactures – not Privileged – were actually created after that date, it wasn't necessary to extend our search to encompass them.

Indeed, thanks to a network of volunteers, all particularly knowledgeable about the natural environments and local history of the region, within a year the place was found where a spring called La Bouillette still exists (Fig. 1.2) – although it more and more frequently dries up, like many other springs in Languedoc. In that same place, a river, the Cesse, flows right at the foot of the high walls of a former Royal Manufacture (Fig. 1.3), and a few miles upstream, a group of springs below the paleolithic caves of Las Fons mingle their tepid waters in a whirl of bubbles at the bottom of a mossy rock basin, before overflowing into the river (Fig. 1.4). The place where all these sources of water can be found is a small town, Bize-en-Minervois (in the present department of Aude).

Its broadcloth manufacture was raised to the status of Privileged Royal Manufacture in 1733.[3]

The "*molasse*" Bouillette had taken its revenge by proving crucial to identify the place where, in all probability, the author was working and writing. This marked a real breakthrough because it gave significance to the few clues scattered in the text about the approximate time when the *Memoirs* were written. Cross-referencing these pieces of data on place and time of writing of the *Memoirs* then logically led to a hypothesis about the author's identity which proved unerringly coherent with the contents of the text. Happily, it further opened a fascinating insight into the social context and historical conjuncture in which the author and his peers were striving not only to keep their manufactures afloat, but to make as much progress as possible in terms of quality as well as quantities, most of them obviously moved by a true passion for this branch of industry which they looked upon as a form of art.

On the other hand, the identification here proposed for the author has just added more mystery to the story of the manuscript. While the ancestor from whom the present owners thought it was inherited had indeed been a clothier in the 18th century, he apparently had no known connections in Bize, his factory was located in a different part of Languedoc and it never figured among the prestigious Royal Manufactures.

Because the circumstances in which the manuscript was integrated into the family's archives still are an enigma, and as a token of respect for what seems to have been the author's will to remain anonymous, since he neither signed his *Memoirs on Dyeing*, nor mentioned the name of his manufacture, I shall keep just calling him "the author" in the following chapters, in which I endeavoured to situate the translation of his text it in its technological and historical context.

In reality, however, identifying the place where the author was writing as the Royal Manufacture of Bize,

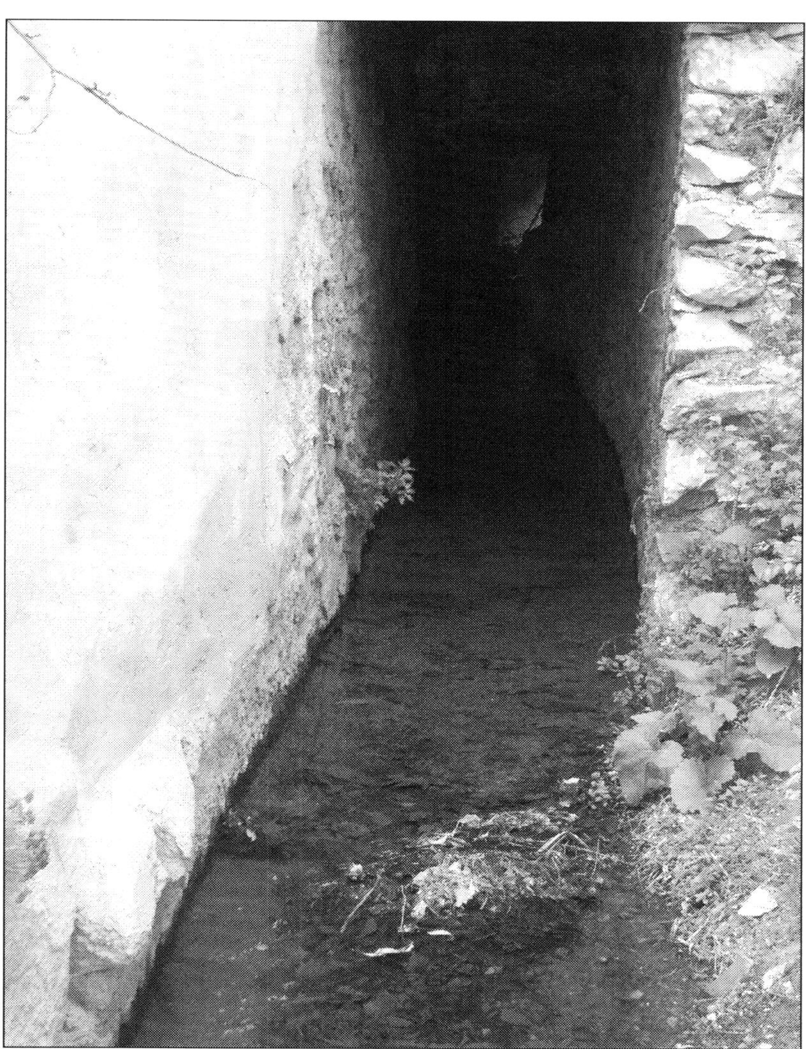

Fig. 1.2. Water course from the spring of La Bouillette to the former fulling-mill of the Royal Manufacture of Bize. Photo D. Cardon.

Fig. 1.3. The Cesse in Bize, under the walls of the ancient Royal Manufacture. Photo D. Cardon.

narrowed down the possibilities until one person stood out as the most likely, both to have possessed the capacities and to have been in a position to write such a masterpiece of technical literature.

Spreading branches, deep-rooting

Assuming that the *Memoirs on Dyeing* were written in Bize, the next, and more difficult, step towards identifying the author was trying to understand in what period of the eventful history of the manufacture the composition could have taken place.[4] Not a single date figuring in the text, it was the author's flirt with plagiarism in his first memoir, the *Memoir on dye drugs*, that helped. There, the author felt he should complete the records of his personal knowledge of the mordants and dyestuffs which he was accustomed to use by some more general information on their provenance. This he looked for in the copious popular scientific and technical literature, issued year after year in that Age of Enlightenment. Without acknowledging it, he copied whole passages from different sources including the famous *Encyclopédie, ou Dictionnaire raisonné des Sciences, des Arts et des Métiers* (Encyclopaedia, or a Systematic Dictionary of the Sciences, Arts, and Crafts) by Denis Diderot and Jean d'Alembert, and also, fortunately, from less well known books including strange stories inherited from medieval almanacs and "books of secrets", in which characteristically misspelled names betray the original source.

The two latest publications that could be identified in this way as having been "recycled" by the author in his *Memoirs* turned out to be the *Dictionnaire du citoyen, ou abrégé historique, théorique et pratique du commerce* (The Citizen's Dictionary, or abridged historical, theoretical and practical Dictionary on Commerce) by Honoré Lacombe de Prézel, published in 1761, and the *Dictionnaire domestique portatif, contenant toutes les connoissances relatives à l'oeconomie domestique et rurale* (Portable Domestic Dictionary, including all knowledge related to domestic and rural economy), by Roux, Aubert de La Chesnaye-Desbois and Goulin, published in 1762. Therefore, it became clear that the *Memoirs on dyeing*

Fig. 1.4. The spring of Las Fons. Photo D. Cardon.

could only have been written after 1762. More precisely and most probably, the author may have been working on the manuscript during the two or three ensuing years. This can be deduced from two mentions he makes in the same *Memoir* about the economic consequences of the Seven Years' War (1756–1763) on his trade; at page 4, he still complains about the considerable rise in the price of Rome alum in the previous years, while at page 10, he can already point to a beginning of decrease in the price of cochineal which had soared, obviously due to the disruption of sea trade routes. He may therefore be writing toward the end of the war or just after it, to make good use of the spare time imposed to him by the slack in business, and before the boom in broadcloth production that, in Languedoc, followed the end of the war.[5]

The *terminus post quem* of 1762 is particularly important because, from this date to the moment it had to close down, at an uncertain date during the French Revolution, the Royal Manufacture of Bize constantly remained a property of the "Pinel group" from Carcassonne, while it was all that time directed by an *entrepreneur* to whom the Pinel family had entrusted its management in 1757. There is therefore a very high degree of probability that this one man, Paul Gout, is the author of the *Memoirs*. During all these years, Paul Gout was the only person who lived permanently in the Manufacture in Bize, managing the production, organising and supervising all operations in the dye-house. He alone was in a position to complain about the water resources of the Manufacture, and above all, to feel free to cut off samples from the cloths as they were coming out of the press.

The position of *entrepreneur*, executive manager of the factory, was as crucial from a technical point of view as prestigious in terms of social status in 18th century Languedoc. There were good reasons for Paul Gout to have been chosen by Germain Pinel, then the head of a powerful family of businessmen and clothiers from Carcassonne including Germain's younger brothers and later his nephews.[6] Gout had already had the opportunity to reveal his talents and to be distinguished by the family when he was entrusted with the responsibility of replacing

François Pinel as manager of another Royal Manufacture, that of Saint-Chinian (about 17 km north-east from Bize), when Pinel decided to go back to live in Carcassonne, at the end of 1754.[7]

At Saint-Chinian, between September 1754 and 1756, Paul Gout makes a good start, easily maintaining the level of production decreed by the government for all Royal Manufactures. According to the Fixation System, intended to distribute production quotas fairly between clothiers and to prevent the glutting of the Levant markets, each Royal Manufacture was allowed a maximum of 420 pieces of the type of cloth, named *Londrins Seconds,* per year.[8] As soon as the Fixation System, unpopular among many clothiers, was abolished in the course of 1756, Gout immediately reacted and started intensifying and speeding up all operations, to reach an output of 450 pieces before the end of the year.

At the beginning of 1757, however, he had already moved to Bize where he took over after Jean Mailhol, a former independent clothier who had managed the manufacture for the Pinel family during the previous years.[9]

Paul Gout's rapid professional ascension may be explained by old links with the Pinel family. He came from a family of small clothiers and cloth-workers in Carcassonne, most of whom belong to the same parish as the Pinels, St Vincent Church, in the heart of the wool-working district of the town. In the 1720s, presumed period of Paul's birth and prime youth, several babies with fathers named Gout described as *marchand-drapier* (clothier), *pareur* (cloth-finisher), *marchand-teinturier* (master dyer) or just *teinturier* (dyer) are christened there, and so are François Pinel in 1725, and Louis in 1726.[10] Though no mention of a Paul Gout could be found in the parish registers, this is not conclusive, since many pages are too damaged to be legible. As a boy among the wool workers of Carcassonne, Paul may well have already been distinguished as bright and promising by Pierre Pinel, the founder of the dynasty, or later by his sons. He certainly worked as *commis* (assistant, clerk) for François Pinel at Saint-Chinian for several years before succeeding him as the manager of the manufacture.[11] The years he spent there were very fruitful, giving him the opportunity to become acquainted with Antoine Janot, a commission dyer of the town whose expertise was widely respected and who left several memoirs on various aspects of the art of dyeing, including a beautiful dye book illustrated with samples of *Londrins Seconds.*[12] There also, Paul met his future wife, *demoiselle* Marie Bonnefous, who happened to be the daughter of Barthélemy Bonnefous, the director of the other Royal Manufacture at Saint-Chinian, the Manufacture of St-Agnan.[13]

Paul Gout got off to an excellent start at Bize. From 785 pieces of *Londrins Seconds* produced during the last year of Jean Mailhol's management, the output rose to 860 pieces in 1757.[14] But the following year, "*les derrangements que cause la guerre*" – the disruptions caused by the Seven Years' War – hit the trade to the Levant and the clothiers and their workers in Languedoc. At Bize, production declines to 760 cloths in 1758. Reacting at once, with the commercial flair which ensured his long-term success, Gout is one of the two first entrepreneurs producing for the Levant market to face the difficult conjuncture by trying to open new markets, diversifying his production with small quantities of both superfine and middling fine cloths.

In 1759 the economic situation worsens, and the production of *Londrins Seconds* drops further to 735 pieces at Bize. However, Gout holds on and turns to the King's Intendant in Languedoc to ask for the permission to revive a type of cloth called *Nims*, made with cheaper Spanish or local wool, which had provided the Royal Manufacture at Bize with good outlets in earlier years, but had been disallowed later on. On the 8th of February, he justifies his request in a letter to the Intendant: "I intend to attempt this production in the hope of faring less badly with it than with the *Londrins Seconds* on which I have been losing very much money since I have managed this Manufacture". The Intendant flatly rejects his request. Instead of bowing down to this decision, the young *entrepreneur* immediately challenges it in an indignant answer to the Intendant: "After the immense losses I have been suffering in my business for the past years" – a slight exaggeration – "and after exhausting myself to support my workers", he cannot admit not to be "permitted to make use of the only way which is left for me to-day to provide for their subsistence". He even threatens to go over the Intendant's head and directly appeal to the social conscience of the Minister: "It appears to me that the major part of the workers of this province having no work to-day, the Minister should be grateful to those manufacturers who give them employment. This is my motivation. I dare hope, My Lord, that you will find it fair, and that you will spare me turning to Mr. the *Controlleur Général* to obtain the said permission". Gout ironically ends his letter with the flowery formulas typical of the century: "I have the honour of being with respect, My Lord, your very humble and very obedient servant" (Fig. 1.5).[15]

The case actually does go all the way up to the minister, Daniel-Charles Trudaine, but through the administrative channels. Some information on the bold protester is requested and the Inspector of Manufactures in charge of Bize admits that "this entrepreneur is intelligent" but does not conceal his doubts "that he exactly conforms to the rules". The result is that as from June 1759, Gout is allowed to produce 40 pieces of *Nims*, which will be registered in the statistics for 1760 as *Londres Larges* (a type of cloth technically close enough) to avoid openly

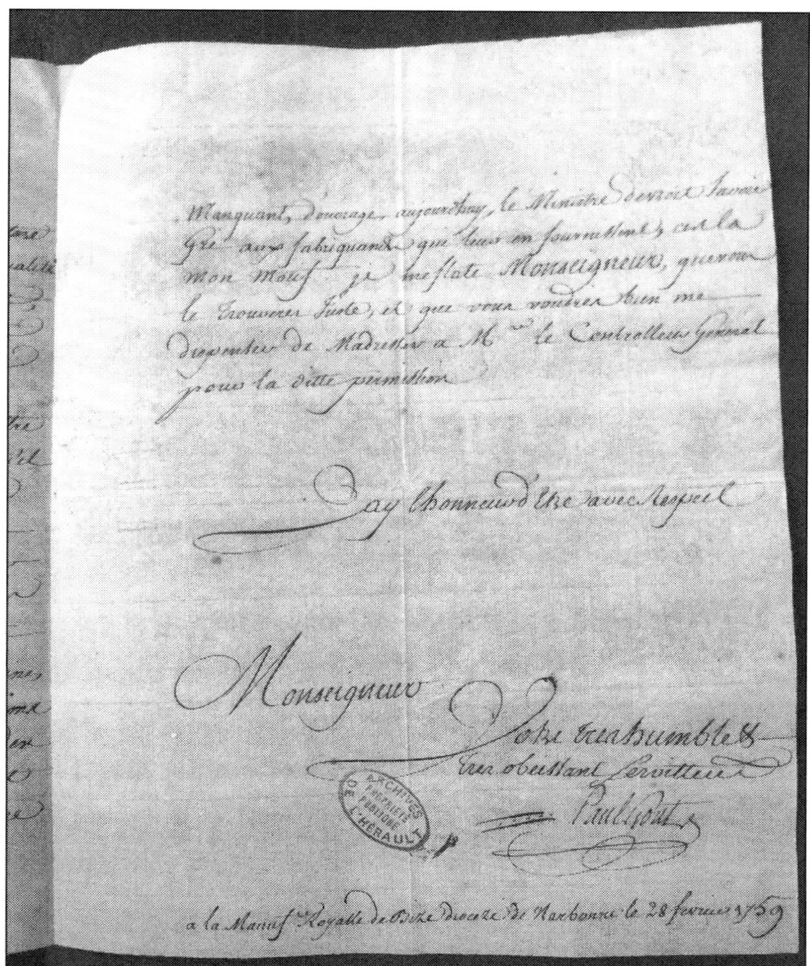
J'ay l'honneur d'être avec respect

Monseigneur

Votre tres humble & tres obeissant serviteur

Paul Gout

Fig. 1.5. Letter from Paul Gout dated 28th February 1762 to Monsieur de Saint-Priest, Intendant in Languedoc. AD34 C 5550 doc. 17. Photo D. Cardon.

admitting such breach in the rules. That year, he also regains confidence in the economic prospects for *Londrins Seconds*, producing exactly 900 pieces, and makes a new attempt at diversification by launching the production of 50 pieces of *Londrins Premiers*, the quality of cloth immediately superior to the *Londrins Seconds*.

The following years are at least as productive, in more than one respect: in September 1761 his first son, Barthélemy Paul, baptised in Bize church, is registered as "the son of Mr. Paul Gout director of the Royal Manufacture" in the parish book. The boy's godfather is none other than his maternal grandfather, Barthélemy Bonnefous, "director of the Royal Manufacture of Saint-Chinian". In May 1763, a second son, Guillaume, is born and Paul Gout is again described in the parish register as "director of the Royal Manufacture".[16] If indeed he was the author of the *Mémoires de teinture*, couldn't it be the birth of these two first sons – potential successors in the trade – that gave Gout the incentive to prepare such treasure of technical information for them? As discussed earlier, this is the time around which the writing of the manuscript may be dated.

In any case, Gout goes on working hard for the prosperity of his growing family. The Manufacture produces 825 *Londrins Seconds* and 48 "superfine" cloths in 1763, the output reaching a record of 1255 pieces of *Londrins Seconds* plus 120 superfines in 1764.

Such truly impressive figures certainly serve as a warning not to underestimate the efficiency of the woollen cloth industries that flourished in many parts of Europe long before the Industrial Revolution. In the essays that follow the translation of the *Mémoires de teinture* in the present book, the technical implications of such huge production will be examined in detail from the point of view of the management of dyeing processes. For the time being, suffice it to point out that dyeing 1375 "double" pieces of cloth in a year (hardly any were sent in the "white" or undyed state) actually meant handling a minimum of eight "half" pieces measuring 15 to 17 French ells (18 to 19 metres) in length per day.[17]

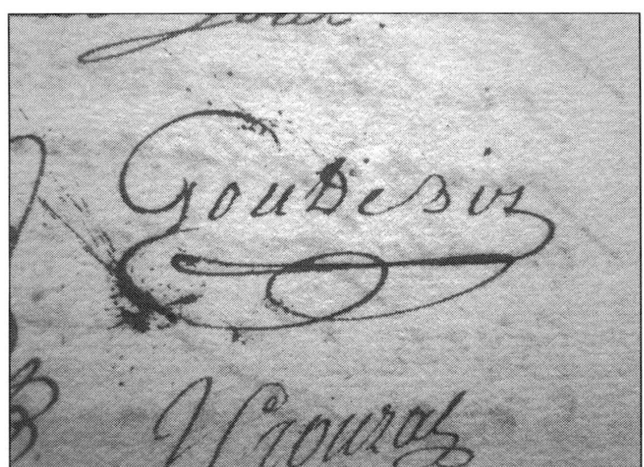

Fig. 1.6. Paul Gout's signature in the itemised inventory established after Germain Pinel's death, dated 21 March 1774. AD11 3 E 1246. Photo D. Cardon.

On the 12th of May 1766, on the baptism certificate of his third son Jean-Joseph (who will not be his last), Mr. Paul Gout is for the first time mentioned as *"coseigneur de Bize"*. He signs his newly acquired name "Gout de Biz" (Fig. 1.6).[18]

Gout had become a "gentleman-clothier",[19] one of the four joint squires of Bize, following in the steps of Germain Pinel, owner of the Manufacture, already mentioned as *coseigneur* of Bize in the *compoix* (land register) for 1761.[20]

"Gout de Biz"

Ridiculous snobbery; exhilaration in the fulfilment of a child's dream to raise above his initial condition and integrate the ranks of the gentry; proud display of a well-deserved social success based on merit: all such interpretations may be proposed, depending on the observer's viewpoint. What is certain is that the case is by no means isolated at the time, either in France or in England.[21] In both countries, the manufacture of woollen cloth for export to the Levant was one of the best ways to facilitate upward social mobility. This is particularly true of the management of a Royal Manufacture in 18th century Languedoc, and it does not only apply to owners-capitalists but also to the *entrepreneurs*, their executive managers. *"Négossients... fait icy noblesse"* ("Merchants-clothiers... here pass for gentry") – sneered a captain in the King's army visiting the broadcloth-producing diocese of Lodève in 1749.[22]

A quarter of a century earlier, Daniel Defoe offered a similar picture of the West of England society: "many of the great families, who now pass for gentry in these counties, have been originally raised from, and built up by this truly noble manufacture".[23] The external signs of such change in status did not pass unnoticed: *"ce n'étoit alors que peruqués"* ("there were but wigged people around in those days"), a former clothier from Clermont-Lodève remembered in 1782.[24] In 1751, Pierre Gout from Carcassonne (Paul's father, uncle or cousin?) mocked his fellow clothiers, *"poudrés à blanc et le commis aussy comme le maître"* ("white-powdered, the clerk just like the master") among whom *"beaucoup... qui ne sont pas connoisseurs"* ("many that are not connoisseurs") in the trade.[25]

The last criticism assuredly could not apply to Paul Gout. The addition of the name of the place where he worked to his patronym should best be viewed as the ultimate sign of his deep and sincere commitment to his task. By signing Gout de Biz, he declared himself a man who had not only found his way in life but also the place where he could best express the *supériorité de lumières* (superior enlightenment) he had obviously been born with, and which was judged indispensable by an inspector of manufactures for a successful career of cloth manufacturer.[26] The manufacture of Bize was the instrument Gout chose to create colour symphonies composed of thousands of cloths dyed in all colour hues, and he conducted it like an orchestra for more than thirty years, with unfailing inspiration.

Capitalists and entrepreneurs: a successful symbiosis

Superior intelligence, thorough technical expertise, acute feeling for textile and colours: none of this, however, would have been enough to ensure the ongoing success of the enterprise, in that second half of the 18th century, a difficult epoch of *"crise permanente larvée"* ("permanent latent crisis")[27] in the cloth trade to the Levant, which saw some going bankrupt while others prospered (Fig. 1.7). If the Manufacture of Bize managed to fare well through the recurring episodes of glutting of the markets in the Levant and maintain a level of production (Fig. 1.8) that put it in second position among the Manufactures of the Carcassonne area,[28] it definitely was thanks to the large amounts of capital invested in the Manufacture by the Pinels, particularly by Germain who, when he died in 1774, left a fortune estimated at about 1,600,000 French pounds.[29] Paul Gout could further benefit from preferential prices for buying Spanish wool, cochineal, indigo and alum, all major commodities in the very diversified international trade carried on within the Pinel group.

Too busy managing this considerable business from his headquarters in Carcassonne, Germain Pinel could not live in Bize, but he always showed a keen interest in his Manufacture. In May 1769, when the *loyer des Manufactures royales* (an annual government subsidy of 3,000 pounds per privileged manufacture) is threatened

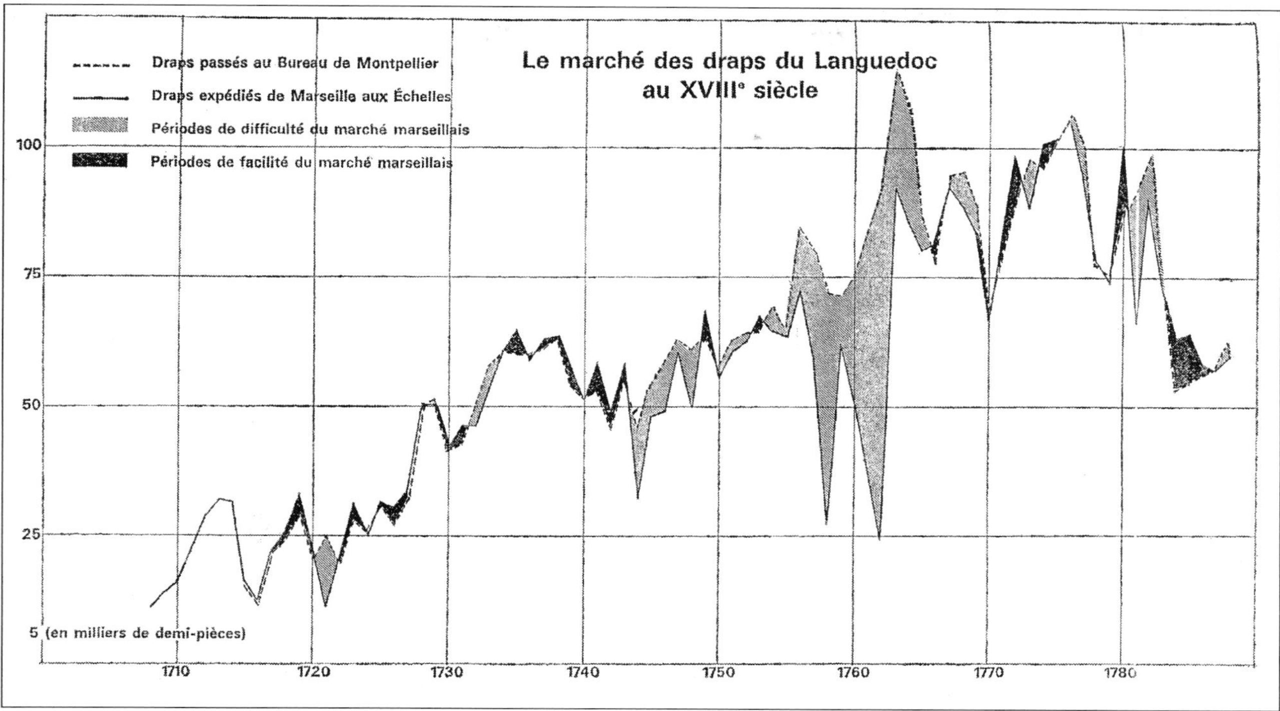

Fig. 1.7. Trends in the market of Languedocian broadcloths produced for the Levant in the 18th century. The dotted line corresponds to numbers of cloths controlled in Montpellier; the full line, to numbers of cloths actually exported from Marseilles to the Levant. (Figures in thousands of finished pieces.) The grey zones indicate periods of difficulty in finding outlets, the black zones, periods of easy selling. After M. Morineau and Ch. Carrière, "Draps du Languedoc et commerce du Levant au XVIIIè siècle", Revue d'Histoire économique et sociale, *56, 1968, p. 117.*

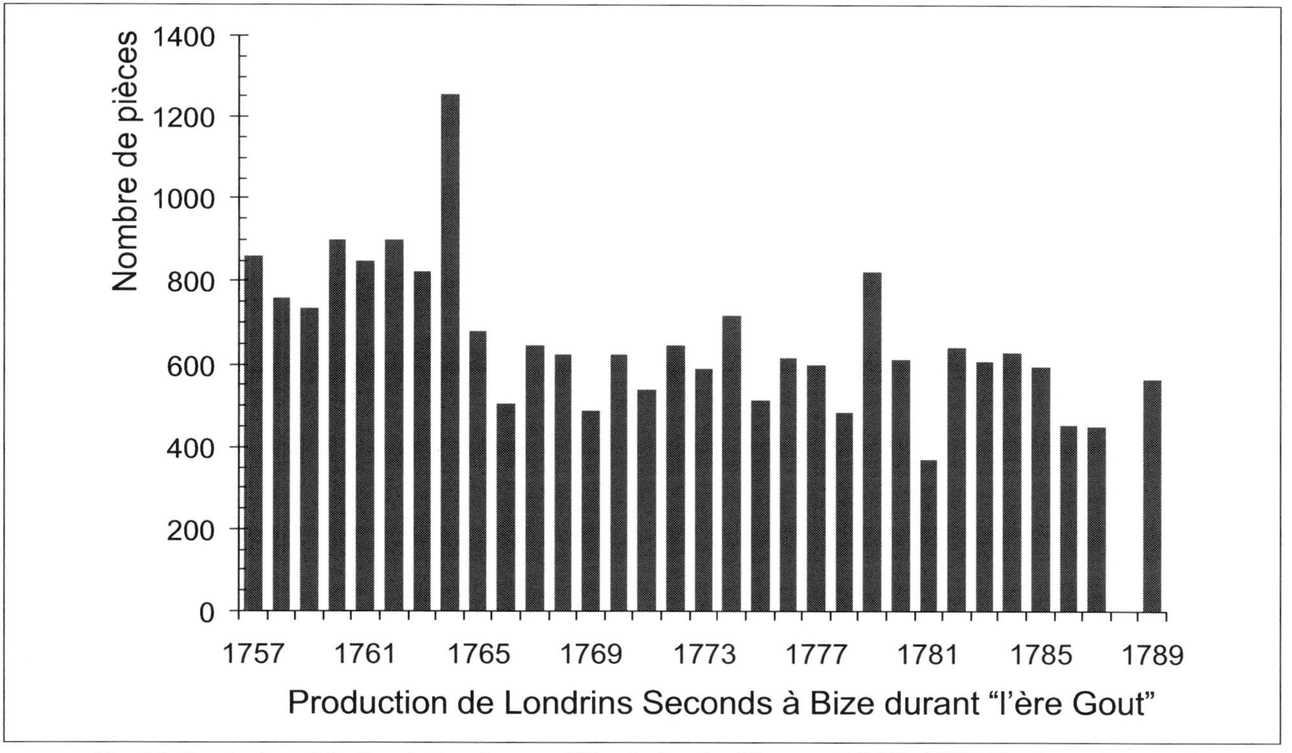

Fig. 1.8. Production of the Royal Manufacture of Bize during the "Gout era" (1757–1789). Numbers of pieces of Londrins Seconds cloths. Data for 1788 are missing in source, AD11 9 C 31. Graph D. Cardon / I. Brémaud

of suppression,[30] he personally writes to the *syndic* (representative officer) of the *Etats de Languedoc* to protest against the project, demonstrating the social impact of the Manufacture of Bize and stressing the importance of the financial investment implied:

> "It is of real usefulness, since I have very much augmented its buildings and the number of workers in the place, all the countryside around it is kept busy by the employment I provide, not only for the Levant but for many other markets. Since 1732, the year when I have become its owner, I have been maintaining it by my money, by my associates and by my clerks… it produces broadcloths of superior quality… and in spite of the present bad state of trade, this Manufacture has never been so busy… I have as many as 50 broad looms beating in Bize."[31]

His commitment must have much contributed to strengthen his links with Paul Gout.

When Germain Pinel dies in 1774, one year after his only son Pierre, Gout collaborates with Germain's brother Louis Pinel to provide all necessary information to the notary who comes to the Manufacture to proceed to the itemised inventory of the estate. He clarifies what belonged to Germain and what is his own in the house as well as in the workshops and mill. This inventory shows what degree of intimacy reigned between the two families, who did not only share the premises but even objects of daily life: in the office where Gout works every day and Germain often came, nearly all the furniture belongs to the latter. But on the upper floors, in nearly every room, after a list of furniture, curtains, bedspreads, lamps and decorative knick-knacks described as Germain Pinel's, the inventory ends in the repetitive remark: "all the rest belongs to Paul Gout".[32] He also shares financial interests with the Pinel brothers through capital he has invested in the family's business. Many years after Germain's death, the firm "Gout de Bize, père, fils et Cie" in which his sons now young adults are his partners, figures in the *Grand Livre* (main account book) of the company "Louis Pinel et Cie" in connection with various dealings.[33]

All these years of shared interests have built up such strong links between the Gouts and the Pinels that it is on an idyllic family scene, very much in the spirit of 18th century painting, that this brief biography of Paul Gout can best be ended. On the 2nd of April 1790, Zoé, Paul's first grand-daughter, is christened. She is the daughter of his eldest son Barthélemy Paul, now described in his turn as the "director of the Royal Manufacture of Bize" in the baptism certificate. The priest who baptises her is none other than one of Germain Pinel's grand-sons, François-Xavier, rector and pastor of the estate of Truilhas near Narbonne, who has come for the purpose. His brother, Messire Antoine Jacques Xavier Pinel de la Taule, "*gentilhomme ordinaire du roi*", has also come all the way from Paris to Bize to rejoice in the company of Gout de Biz and his wife Marie, who acts as her grand-daughter's godmother, and three of their sons – "Gout the elder", Zoé's father; "Gout the younger"; and another brother, Jean-Joseph – who "have all signed down together with ourself", the rector-pastor concludes at the end of the certificate (Fig. 1.9).[34]

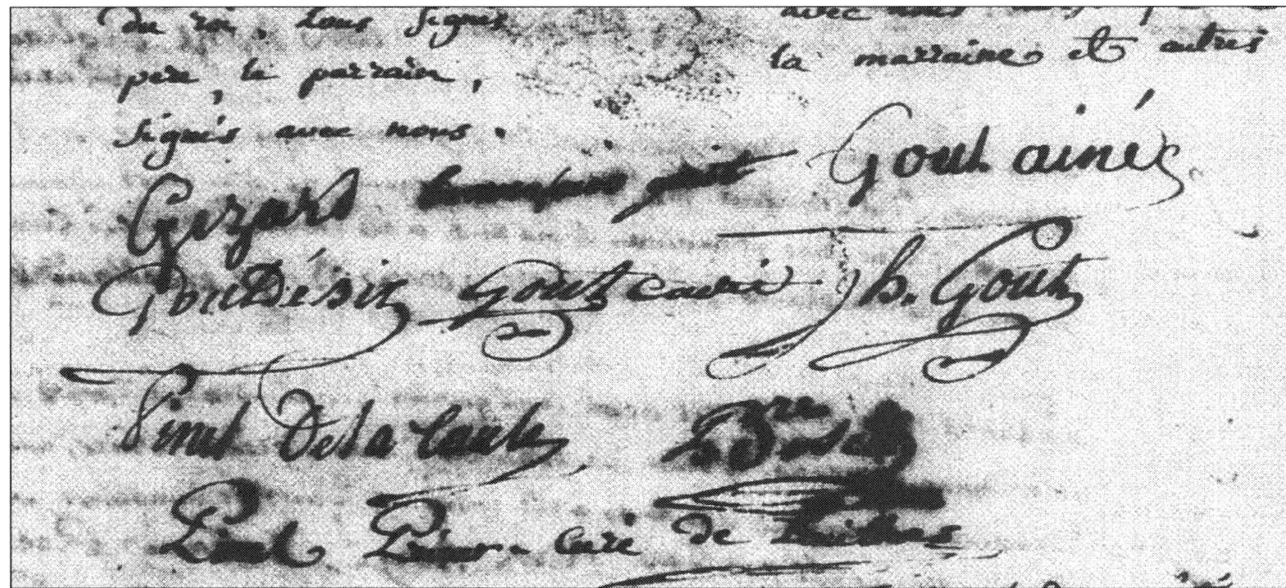

Fig. 1.9. Signatures of two of Germain Pinel's grandsons figuring with those of Paul Gout, his wife and three of their sons, on the certificate of baptism of "Gout de Biz's" first grand-daughter, Zoé, born on 25 April 1790 in Bize. AD11_5Mi0289_5E065_006_nmd_0003.

The *Mémoires de teinture* in context

One cannot but wonder how the director of one of those Royal Manufactures, busy with the to-ings and fro-ings of technical operations and the permanent flow of workers and goods, could find or take the time to write on dyeing. One could further find it surprising that he may have possessed the thorough knowledge of the art of dyeing necessary to undertake such a challenging task.

A first explanation lies in the crucial part played by the beauty and quality of the colours in the development of the Languedocian broadcloth production targeted to conquer the markets of the Eastern Mediterranean. At the time, no clothier, no inspector of manufactures, and certainly none of the factors in the Levant, has the slightest doubt about it. "The perfection of dyeing is as necessary as that of the construction of the cloth, it being the most considerable and essential finish that is given to the goods and it is the deficiency, or the perfection of this process that hinders their dispatch or makes them easy to sell," declares an expert dyer from Carcassonne in 1731.[35] In his book *Le Parfait Négociant* (The Perfect Merchant), first published in 1675, Jacques Savary already stressed that "the Turks, the Armenians and the Persians are very demanding concerning colours… It is the vividness of the colours that makes them [the cloths] sell easily."[36] Moreover, the elites of these countries of colour lovers expect the producers from the other side of the Mediterranean to follow the fluctuations of the tastes and fashions of their different social circles with the utmost reactivity: "Above all, the assortments for the bales must be strictly observed concerning the colours, which change here nearly twice a year; one must take care to send new ones every six months," advises the author of still another memoir, written in the Levant around 1680.[37]

Two merchants from Marseilles confirm the fickle nature of the market in a letter to the Intendant of Languedoc, dated 1757. They urgently request to be exceptionally permitted to export six bales of cloth bought from a clothier of Carcassonne who has already exceeded the export quota allotted him for that year by the Fixation System, otherwise these cloths "may fail to find buyers, due to the slightest delay that would cause their assortment to become old-fashioned".[38] Letters of English factors at Aleppo quoted by Gwilym Ambrose, though earlier, already show the same preoccupation for the correctness of colours: "I doe herewith send you a small muster of cloath, onely for the collour, which is here the first thinge desired," emphasizes one of them in October 1658; another in 1677 illustrates the fickleness of fashions in the Levant: "For colours I formerly sent home patterns of the choicest, what new colours may be now come or be best I cannot yet tell… if any particular colour be in fashion at Stambol you may expect 'twill be esteemed by the great ones here". "Persian colours", however, correspond to particular shades, "which I shall endeavour to sent the patterns of in my letter", writes another in 1699, adding "these Persian colours are only worn by persons of the best quality, who give a good price but wear only the finest cloth they can get".[39] The reactivity and expertise of dyers in Languedoc have been considered as key assets in the growing success of French cloth in the Levant.[40] They largely contributed to ensure outlets there that lasted longer, albeit on a reduced scale, than has been commonly assumed.[41]

To be considered as a good manager, the director of a Royal Manufacture was therefore expected to be knowledgeable about all aspects of dyeing, from technical processes and regulations to the latest technical improvements, real or claimed as such. Indeed, he needed to possess practical, technical expertise, especially since a dye-house most often was part of the factory and functioned under his direct supervision: "the entrepreneurs of Royal Manufactures that include dye-houses, employ dyers as workmen, but are expected to be their own master dyers," one inspector explains in 1745 to the Intendant of Languedoc. The problems that might arise from such direct involvement were being illustrated at that moment by a series of black cloths that had just been confiscated for their defective dyeing, the guilty dyers being further threatened of banishment from the trade. Two of them happening to be the entrepreneurs of renowned Royal Manufactures, the inspector expresses his feeling that "there is some difficulty in enforcing the exclusion penalty decreed by Mr. the Minister's letter."[42]

The issue of the quality of dyeing is indeed a technical, economic and political matter. The *Mémoires de teinture* possibly represent the first published text that demonstrates this so clearly and this is what makes this document so unique and exceptionally interesting. It is not written by a commission dyer recognised for his expertise, working for the most prominent clothiers but not having directly to worry about foreign markets, like Antoine Janot's *Mémoire sur la teinture du grand et bon teint des couleurs qui se consomment en Levant* (Memoir on the best mode of dyeing the colours that sell in the Levant), finished in Saint-Chinian on the last day of March 1744. Neither is it a report on scientific enquiries and experiments, like Jean Hellot's *L'Art de la Teinture des Laines et des Etoffes de Laine en grand et petit teint* (The art of dyeing wool and woollen cloth in the best and in the ordinary modes), published in Paris in 1750, as part of an enlightened government's scheme to encourage technical progress. The author of the *Mémoires de teinture* knows both documents, he has obviously read and used them.[43] I think that he wrote his *Memoirs* precisely because nobody better than he could realize that they did not fully answer the needs of the trade and of the epoch. Nor was anybody in a better position to gather together every piece of relevant information, taking into account all different aspects of the trade, to provide an even more useful contribution than his predecessors.

He writes his first memoir on dye and mordant drugs – which has unfortunately been only partly preserved – from the point of view of the manager of a big enterprise facing fierce international competition, with an aim to keep the costs of buying raw materials, both colorants and mordants, as low as possible. He writes the second memoir of recipes of fast dyes for the markets of the Levant as a dyer trained with good masters, convinced of the usefulness of royal regulations to ensure the quality of dyeing of broadcloths. This was a belief shared by a vast majority of French dyers and manufacturers, and even today, many historians agree with Philippe Minard's analysis of the "economy of quality" and "the usefulness of standards" that had been set in France at national level by Colbertian regulations: "Standards and specifications are in no way purely arbitrary. They are rather the results of a long process of coordination between producers and consumers, leading to an agreement," and warranting the latter's trust in the quality of the product – dye and dyeing processes in this instance.[44] With the third memoir on testing procedures for the different kinds of dyes, he begins to reveal himself as an adept of the "middle path", consisting in trying to find a balance between attempts at reducing production costs and an aim to maintain a high level of quality. Finally, he appears all at once a wonderful empirical chemist, an expert colorist and an outstanding acrobat in the art of the "economy of quality" in the last part of his manuscript.[45] This actually is a precious extract from his logbook as entrepreneur of a Royal Manufacture, in which he records some of his experiments to reproduce the colours of the best mode of dyeing at lower cost: using slightly lower quantities of the right ingredients, but counterbalancing this by additions of other permitted but cheaper colorants, or by chemical tricks; trying progressive doses of forbidden colorants… to eventually fall back onto identical effects, verified today by the colorimetric measurements of cloth-samples corresponding to the same colour names in the second and fourth memoir.[46] Hats off to the artist!

The text translated here into English is the first published 18th century treatise on woollen dyeing whose author is a clothier writing while engaged in business. The only equivalent known so far, mainly dealing with silk dyeing, is the masterly *Tratado instructivo, y practico sobre el arte de la tintura* (Instructive and practical Treatise on the Art of Dyeing) written by Luis Fernandez, Master Dyer, Director of the Royal Manufacture of Valencia, published in Madrid in 1778.[47]

The *Memoirs on dyeing* unquestionably surpassed it in one respect that turned out to be an insurmountable handicap to their publication until modern times: the presence of 177 samples of broadcloth, tangible witnesses of the wonderful colours described in this manuscript, itself a miraculous survivor of the period of turmoil that saw the end of the manufacture where it had been written.

Notes

1 Cardon, D. (2013) *Mémoires de teinture – Voyage dans le temps chez un maître des couleurs*. Paris: CNRS Éditions.

2 *Archives départementales de l'Aude*, AD11 9 C 23, *Extrait des délibérations par les Etats en faveur des Manufactures de la province*.

3 By a decree of the State Council (*Conseil d'Etat*) of the Kingdom of France dated 27 June 1733 (Marquié 1993b, p. 69).

4 Regarding the difficult beginnings of the broadcloth factory at Bize and the crucial part played in its development by industrial capitalists from Carcassonne, cf. Marquié 1993b, pp. 68–71.

5 The disruptive effects of the Seven Years' War on the "ordinary course of the outlets of their broadcloths that were consumed both in the Levant and in the East Indies" were still lamented in a *Mémoire des fabricants de draps* written by the clothiers of Carcassonne in 1765, AD11 9 C 24; cf. also Marquié 1993a, pp. 54–55; Thomson 1982, pp. 372–373.

6 On the financial power and international ramifications of this "Pinel complex": Marquié 1993a, pp. 177–196.

7 AD34 C 2076.

8 The *Londrins* were meant to compete with the types of middling-fine broadcloths that were made in the West of England for export to the Levant. These were mostly dyed in London, and assorted and packed there, hence the name *londrine,* first given in the Levant to a type of Dutch broadcloth that emerged to compete with English models in the seventeenth century, before being adopted for the French cloths which completely replaced the Dutch product by the end of the century (Davis 1967, p. 103; Sella 1961, pp. 64, 119–120). Sold at reasonable prices for a very good quality, *Londrins Seconds* had by then become the best-selling kind of broadcloth in the Levant. A discussion about this production and the English cloths it was competing with is proposed below, in chapter 3.

9 AD34 C 2238.

10 AD11 AG069/GG 245–251.

11 This was the usual way in which the brightest children of the less well-to-do clothiers could hope to reach the enviable position of *entrepreneur*: they were learning the trade on the job, working as *commis* to the wealthiest clothiers. Examples for Carcassonne in Marquié 1993a, pp. 307–308. The sons of the government's inspectors for the manufactures who chose to follow in their father's steps also frequently tried to train with the most renowned clothiers or with their entrepreneurs, cf. Minard 1998, pp. 122–123.

12 AD34 C 5569, dated 31 March 1744. Its edition by this author is forthcoming. Other essays by this great master of colours are preserved in the *Archives départementales* in Montpellier, AD 34 C 2238, C 2240.

13 AD11 AC041–1E007 and AD34 C 2238, doc. 70. The Gout family is but one example of the many families of clothiers of lesser importance who adopt the same strategy of upward social mobility as the gentlemen clothiers, owners of the most prestigious Royal Manufactures, that is through inter-marriage among clothier families, Marquié 1993a pp. 273–274; the same policy is described among clothiers around Clermont-de-Lodève by Thomson 1982, p. 333. Close parallels are to be found among their peers and rivals, the clothiers of the West of England, Tann 2012, pp. 40–43.

14 All data on the production of the Bize Manufacture come from the same document, *L'Etat des Manufactures*, AD11 9 C 31, recording the production of all the manufactures of Languedoc from 1753 to 1789, with only the data for 1788 missing. It is particularly useful in that it gives production figures both in numbers of bales and in numbers of cloths, which eliminates any possibility of confusion. It is further precised that each bale contains two *ballots* (bundles), each of those including five "double pieces" called *naucades* (from the Occitan name *nauc* for the scouring or fulling troughs) that were actually cut into two after scouring, giving two "half pieces" measuring 15 to 17 *aunes* (ells) in length, that is 17.82 to 20.20m, dyed separately, in order to offer potential buyers a wider assortment of colours in each bundle. For each year, this statistical document first gives the number of *ballots* and cloths already produced and despatched at the time of the inspector's visit, and then it adds the number of cloths being processed and due to be finished "before the end of the year". It is the total figure that is counted here as the yearly production of each manufacture.

15 The whole file, AD34 C 5550, preserved in Montpellier, offers an amusing example of the manufacturers' awareness of the awkward position in which "the predominantly liberal leanings of the high spheres of the administration of the *Bureau de Commerce*", particularly under the Trudaines, put the Intendant and the inspectors who mostly shared Colbert's views, as highlighted by Minard 1998, pp. 331–334; and Thomson 1982, p. 366.

16 AD11 AC041–1E007.

17 This is the lowest possible estimation, obtained without deducting any holidays – impossible to quantify for Ancien Regime Languedoc.

18 *Ibid.*

19 This is how those English clothiers of Gloucestershire "with a great stock and large credit", the nearest equivalents to the French directors of Royal Manufactures, were called: Moir 1955.

20 Marquié 1993b, p. 70.

21 Claude Marquié quotes many testimonies of the "strong desire" to integrate the gentry prevailing among the clothiers of Carcassonne, Marquié 1993a, pp. 231–260; the same was true in Gloucestershire, Moir 1955, pp. 239–240, 242.

22 AD34 C6763.

23 Defoe 1927, vol. 1 p. 281.

24 AD34 C5592.

25 AD34 C2514, *Traité sur la fabrique des Londrins Seconds de Carcassonne*.

26 *Mémoire sur les diverses qualitez des draps destinez pour les Echeles du Levant*, dated 18 November 1750, AD34 C2380.

27 Minard 2003, p. 78. His analysis is in agreement with many other French and English historians' views on the fluctuations of that trade during the second part of the century. For the West of England, Tann 2012, pp. 55–56, 60.

28 Marquié 1993a, pp. 94–95; 1993b, p. 71.

29 Marquié 1993a, pp. 282–284.

30 *Ibid.*, p. 89.

31 AD11 9 C 25, doc. 12. These broad looms, scattered all over the town in the weavers' houses, belonged to the owner of the manufacture. At the same period in England, both in Yorkshire and in the West Country clothing districts, the weavers owned the looms they were working with, Ponting 1971, p. 24; Moir 1955, p. 248.

32 AD11 3 E 1246.

33 AD11 3 J 277, pp. 4, 10–11, 13.

34 AD11 AC041–1E008.

35 *Memoir* by Batizat, a Carcassonne dyer, sent in 1731 to Monsieur de Bernage, Intendant of Languedoc, AD11 9 C 11.

36 *Le Parfait Négociant*, Reed. Richard 2011, pp. 402–403.

37 *Mémoire pour le commerce des draps de Levant* (Memoir on the cloth trade for the Levant), AN G7 1684.

38 Letter from Mesrs Gouffre and Gautier dated 6 November 1754, AD34 C 2076.

39 Ambrose 1931, p. 248.

40 Wood 1935, pp. 141–143.

41 My forthcoming edition of a clothier's dye book and another clothier's pattern book of the beginning of the 19th century, with samples of colours produced specially for customers in different countries of the Eastern Mediterranean, will show that the popularity of fine broadcloths from Languedoc survived the vicissitudes of the French Revolution and Napoleonic Wars.

42 AD34 C 2241; the long and intricate case of "the black cloths" is further examined in one of the essays that follow the translation of the text of the *Mémoires de teinture*; see below p. 139.

43 He both mentions Hellot's name and quotes his treatise, as will be seen in the translation that follows; Janot's influence, or at least a body of technical knowledge and culture shared by Janot and the author of the *Memoirs*, clearly appear in the dyeing recipes for similar colours in their respective dye books and they are confirmed by the colorimetric measurements of the corresponding cloth samples, published in the last part of this book.

44 Minard 2003, p. 79.

45 As defined by Minard 2003, after Karpik 1989.

46 The results of colorimetric measurements of all samples in the manuscript by Iris Brémaud can be found in Appendix 1.

47 Fernandez 1778. Fernandez introduces himself on the cover of the book as "*D. Luis Fernandez, Maestro Tintorero, natural de la Ciudad de Toledo, vecino de la de Valencia, Visitador de los Tintes de esta Ciudad y su Reyno, Director de la Real Fabrica que los cinco Gremios mayores de Madrid mantienen en dicha Ciudad...*"

2

A Maimed Manuscript

The recent critical edition of the only known manuscript of the *Mémoires de teinture* includes its complete codicological description.[1] Here, therefore, are only summed up such data on its present state of conservation as are indispensable for a full comprehension of the translation of the text and comments that follow.

Under the general title *Memoirs on dyeing*, four distinct but complementary essays are grouped together in one booklet. On its cover, under the main title, the topics of the four memoirs are described, in terms that differ slightly from the titles given at the beginning of each memoir inside the book. The first essay, announced as *Memoir on dye-drugs which serves as an introduction to dyeing of the cloths that are made in Languedoc for the Levant*, occupies page 1 to page 37 of the manuscript. The second memoir, entitled *General method – To dye broadcloth in the best mode of dyeing for the ports of the Levant*, corresponds to page 38 to the upper part of page 77. The third essay, described as *Instruction on the testing procedures for false colours*, corresponds to the bottom of page 77 to page 79. The fourth essay, entitled *Annotations on the colours made for the Levant with – Their patterns, their processes, and some observations*, corresponds to pages 80 to 96. Page numbers are written at the top left corner of each page, starting after the cover, as in a printed book.[2]

The book measures 36 cm high per 24.6 cm wide and was originally made of fifty paper sheets, folded into two and bound together at the centre fold, thus forming a note book of one hundred pages. The two last pages, not numbered, were glued together to make the back cover stronger. The front cover is not reinforced and its verso is not numbered (Fig. 2.1). The title page of the first memoir is page 1 of the manuscript. Page 2 offers a summary of the contents of this memoir indicating the page number corresponding to the beginning of each section devoted to a specific colorant or mordant (Fig. 2.2). The layout is similar to the table of contents of a printed book.

This feature, the mode of pagination of the manuscript and its layout are clearly devised to make it look like a printed book, and this also applies to most of the countless memoirs written in France at the time. This type of literature, dealing with sciences, arts and trades – like the *Encyclopaedia* – was very much *dans l'air du temps*, in keeping with the spirit of the Age of Enlightenment and this trend was reinforced in France, in 1746, with the creation of the administrative body of the *élèves des manufactures* (apprentice inspectors of manufactures), who all had to write such a memoir as part of their training.[3] Good examples of the standard presentation that came to be generally adopted are provided by two other memoirs on different aspects of dyeing, written at the same period: the *Mémoire sur les opérations de la teinture du grand et bon teint des couleurs qui se consomment en Levant* (Memoir on the best mode of dyeing the colours that are consumed in the Levant) by Antoine Janot, of Saint-Chinian, written in 1744, already mentioned in the preceding chapter;[4] and a very informative *Mémoire sur la teinture du Bleu et du Noir de Sedan* (Memoir on Dyeing into Sedan Blue and Black) by a Mr. Delo, former *élève des Manufactures*, dated 1761.[5] Like the present manuscript of *Memoirs on Dyeing*, these two documents consist of a simple paper booklet without a reinforced cover. Both present the same simple binding device: the centre fold is pierced in two places, a few centimeters from the upper and lower edges, and two loops of greenish-blue ribbon or string passing through all pages keep them tied together.

These two well preserved memoirs help us imagine the original aspect and binding of "our" manuscript. In its present state, however, pages 13 to 32 are missing (Fig. 2.3). They were neatly cut off with a very sharp blade, most probably a quill knife, in a parallel line to the central fold of the book, a few millimeters from it. This happened at an unknown date and in unknown circumstances. With ten left halves of sheets cut off and such a rudimentary binding, the book could not but start

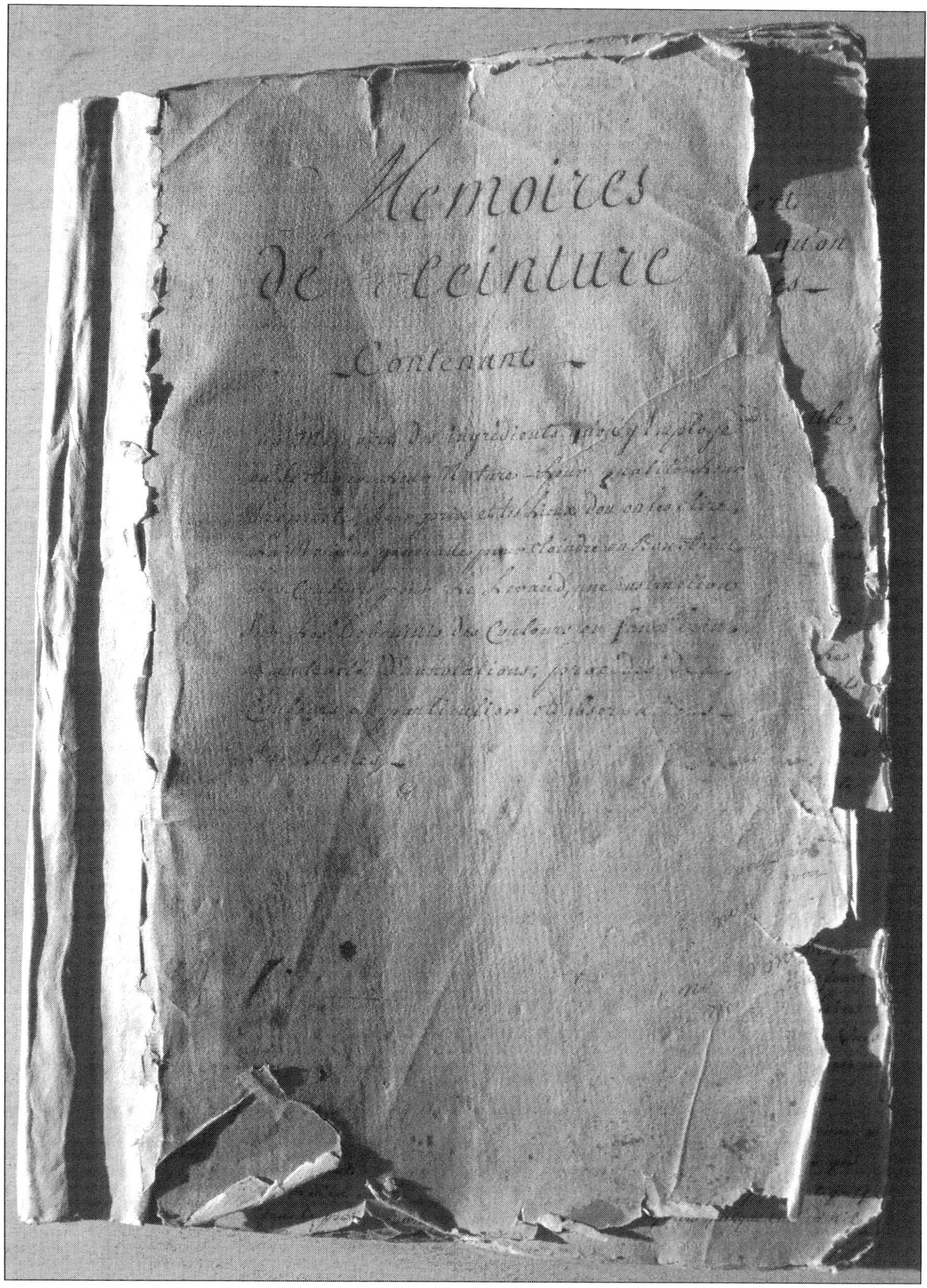

Fig. 2.1. Front cover of the manuscript of Memoirs on dyeing. Photo D. Cardon.

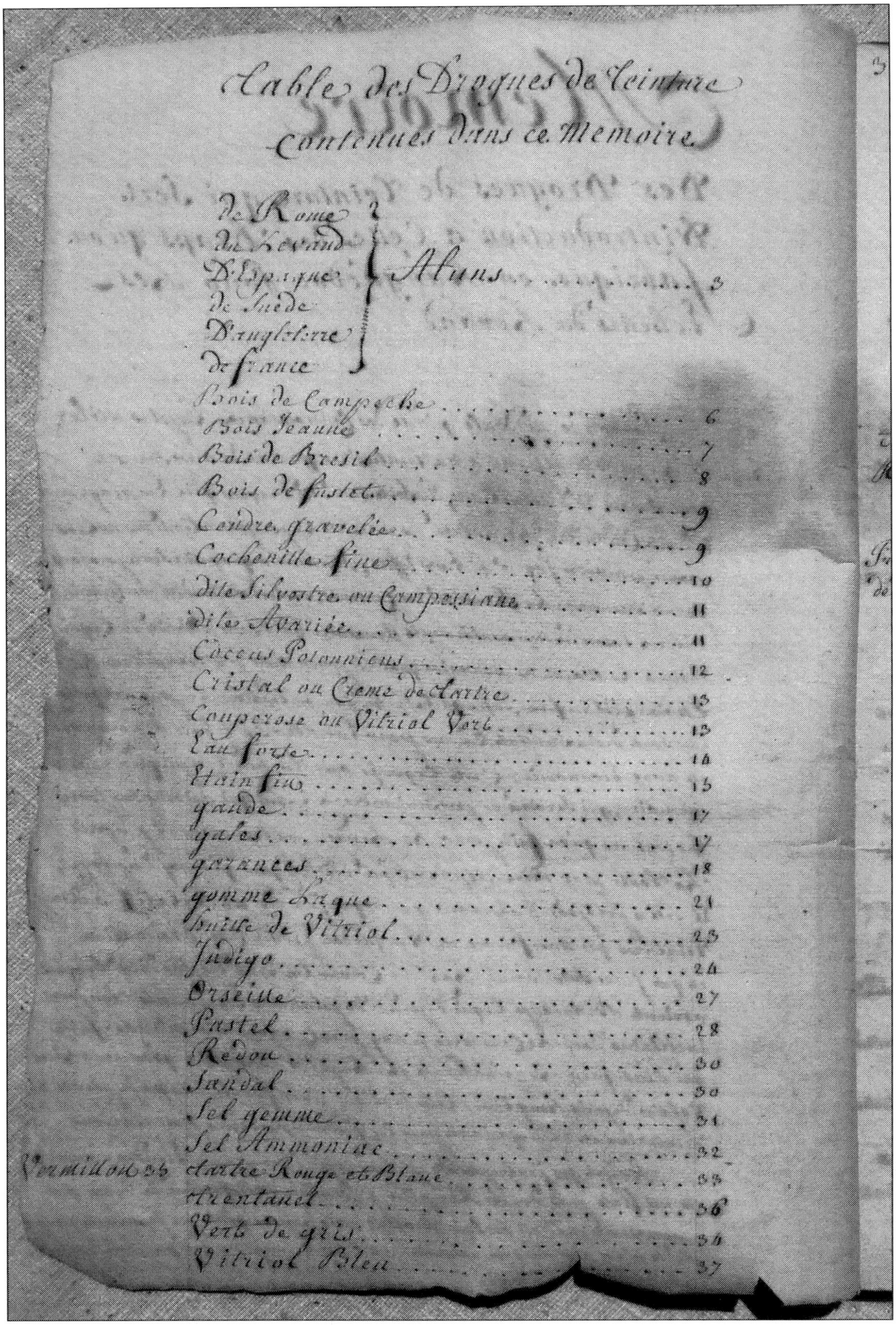

Table des Drogues de Teinture
Contenues dans ce Memoire

de Rome
du Levant
D'Espagne } Aluns … 3
de Suede
D'angleterre
de france

Bois de Campeche … 6
Bois Jeaune … 7
Bois de Bresil … 7
Bois de fustel … 8
Cendre gravelée … 9
Cochenille fine … 9
… 10
dite Silvestre ou Campesiane … 11
dite Avariée … 11
Coccus Polonnicus … 12
Cristal ou Creme de Tartre … 13
Couperose ou Vitriol Vert … 13
Eau forte … 14
Etain fin … 15
gaude … 17
gales … 17
garances … 18
gomme Laque … 21
huille de Vitriol … 23
Indigo … 24
Orseille … 27
Pastel … 28
Redon … 30
Sandal … 30
Sel gemme … 31
Sel Ammoniac … 32
Vermillon 33 Tartre Rouge et Blanc … 33
Terentanel … 36
Vert de gris … 36
Vitriol Bleu … 37

Fig. 2.2. Page 2 of the manuscript with summary of the first Memoir. Photo D. Cardon.

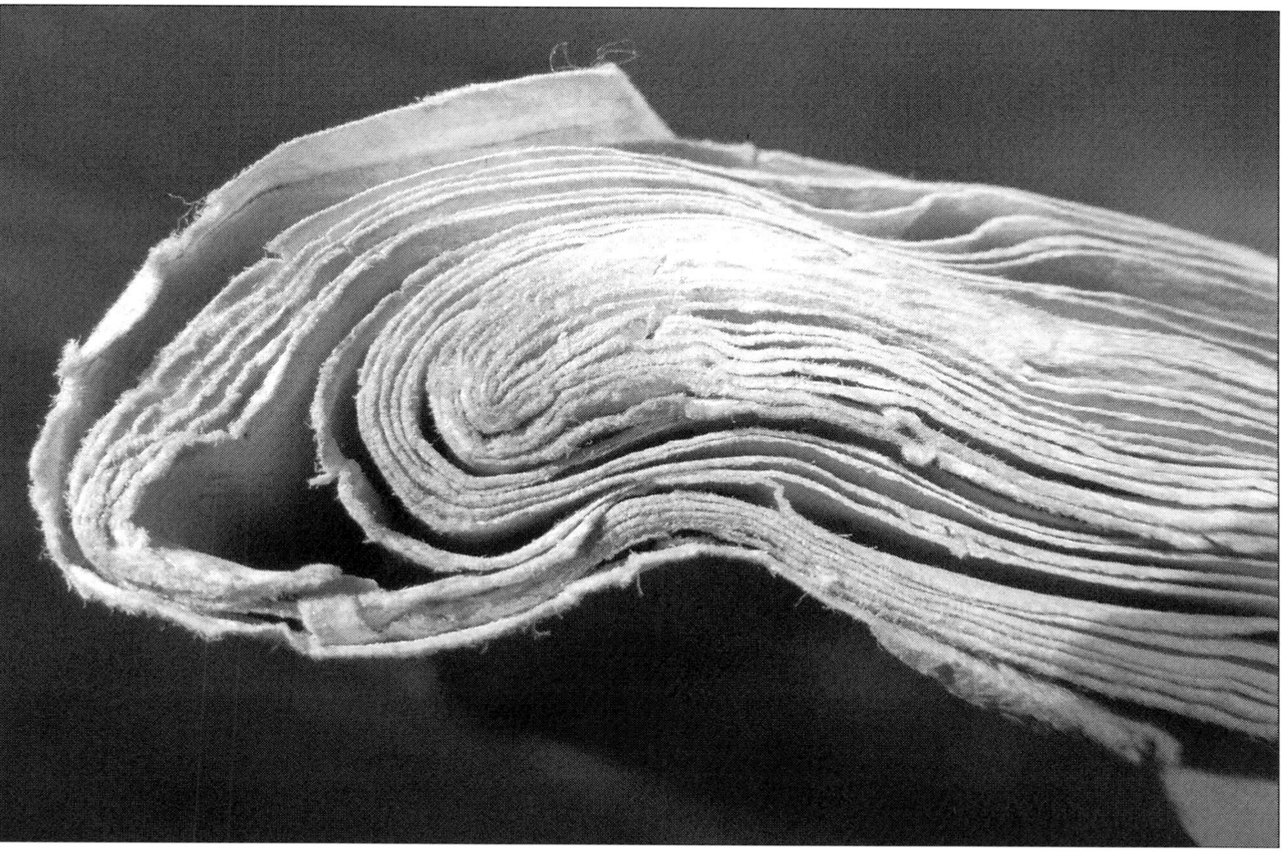

Fig. 2.3. View of the manuscript showing gap left by cut off pages. Photo D. Cardon.

dislocating. It was restored – again at an unknown date and in unknown circumstances – by carefully sewing the maimed sheets of paper and the intact ones in the new middle of the book, between page 58 and page 59 and it was reinforced by a strip of paper, glued along the back fold of the cover. We know what the contents of the missing pages were, thanks to the summary of the first memoir on page 2. The missing descriptions of dye or mordant drugs include cream of tartar, copperas, aqua-fortis, tin, weld, gallnuts, madders, lac-dye, oil of vitriol, indigo, orchil, woad, *redoul* (a tannin-rich plant typical of damp road sides in Languedoc), sandalwood and rock salt. The only type of information now provided about these products, in the present state of the manuscript, comes from the detailed recipes of the second and fourth memoirs, indicating to what use they were put, in what quantities, and what colours were obtained, as illustrated by the corresponding cloth samples.

Another puzzling feature of the manuscript is that none of these cloth samples has survived intact. From the two parallel lines, lightly drawn with a pencil, that run from top to bottom of the pages with samples, it can plainly be seen that all samples were originally of the same rectangular shape and size and were glued onto the page in a regular vertical column, in the same way as on the beautiful pattern sheets commonly sent to factors in the Levant, many of which are still preserved in public archives in Languedoc and in Marseilles (Fig. 2.4). The same layout is also found in Antoine Janot's *Mémoire*, which the author of the present *Memoirs on Dyeing* had certainly seen and had probably had the opportunity to consult. In his own manuscript, however, most samples have had a bigger or smaller portion neatly cut off with a sharp blade that has cut through the paper in places, while other samples have been entirely peeled off the page, only leaving a few coloured fibres stuck in remnants of glue, or even tearing a hole in the page (Fig. 2.5). Explanations are proposed further down, in the light of a study of some English clothiers' pattern books, contemporary with the *Memoirs* and presenting similar features.

Regrettable as such damages are, they at least serve to prove that the manuscript is the original document conceived by the author, and not a copy of an already incomplete text. In spite of being the original manuscript of the *Memoirs on Dyeing*, however, it was not written by the author in his own hand. Although he writes in the first person, declaring, for instance, "what I am saying only applies to the dyes we are making in Languedoc for the Levant", some mistakes in the text are characteristic of someone who either is writing under dictation and has not heard or understood well what was dictated, or is trying to copy draft pages hastily written in a handwriting hard to decipher. At page 9, *les luthiers* (stringed instrument makers, mentioned there because they use the wood of young fustic) have been transformed into *lithurliers*, which does not mean anything in French; *agaric*, the larch agaric used as a mordant, has become *alaric*, another meaningless word. Another clue is the inverted order of pages, in two places in the manuscript. In the first *Memoir* where the colorants and mordants are treated in the alphabetical order of their names, *tartre* (argol or tartar) which should come before *trentanel* (flax-leaved daphne) at page 33, occurs on the next page, p. 34; and the title of the second memoir, page 39, comes after the introduction to that memoir on page 38. Obviously, the draft version of the manuscript was dictated or written on separate sheets of paper and put into a pile. In the pile, two sheets of paper were put upside down, the verso before the recto, and they were copied in the same wrong order in which they had been placed. When the author considered that he had completed the draft, he must have passed it on to a secretary or clerk, entrusting him with the task of preparing the notebook, presenting it with a clear and elegant layout and copying the text in as legible a handwriting as possible. Only one person did the writing, with the exception of a few words of comment, added in a different hand, at pages 53 and 70 of the manuscript. The forensic examination of the document by an expert in handwriting has confirmed that neither hand is Paul Gout's. This was established by comparison with Gout's letters and signatures, many examples of which are still preserved in public archives in Languedoc.

Such a result did not come as a surprise. At the time, important clothiers, particularly the directors of the Royal Manufactures, had to face such complex tasks and responsibilities, they were so frantically busy, that they all had got into the habit of employing at least one *commis* (clerk, secretary). It was, moreover, "a sign of social distinction".[6] In various letters from clothiers of Clermont-de-Lodève, an important clothing centre situated some 80 km to the north-east of Bize, the manufacturers exchange complaints about the expenses they are "obliged to face for their *commis*' wages". In 1742, one of them, Jean-Pierre Desalasc, at the head of a less important manufacture than Bize, mentions that he employs as many as three clerks "who cost [him] some 1,200 pounds annually".[7] At a later period, in 1787, Louis Pinel, Paul Gout's associate in Carcassonne, has two clerks, paid 450 pounds per year each.[8] No document has been preserved on this aspect of management in the Royal Manufacture at Bize, but whoever was in charge of giving the document its definitive form did a beautiful job, fully answering the author's desire that "the dyer and merchant" may "find something useful in this booklet". The layout of the two memoirs with dye recipes and cloth samples is much better designed than Antoine Janot's *Mémoire*, which probably served as model: colour names, samples and corresponding recipes are placed side by side on each page. Recipes are presented in a standard form, ingredients mentioned in the sequence of use, with quantities written in Arabic numerals and standard abbreviations, allowing a quick reading as if the document was meant to be consulted as a guide, while the dyeing processes described are under way. The rearrangement of the third memoir, based on Jean Hellot's *Instruction sur le déboüilli des laines, et etoffes de laine* (Instruction on the testing of wool and woollens dyes),[9] is also a great improvement on its source, being much more practical and user-friendly for manufacturers wanting to avoid problems at the stage of the quality controls of their cloths.

All this effort to render the *Memoirs on dyeing* as clear and easy to read as possible, flows from the author's purpose. More than a manufacturer and dyer's private notebook, they are the expression of the philanthropic ideals of a practitioner eager to generously share every bit of knowledge he has accumulated through his training and professional experience. Sixty years later, William Partridge similarly expresses his philanthropic motivation for publishing his *Treatise on Dying*, declaring that he, "by birth and education an Englishman", "will be amply repaid in the satisfaction of having done his duty as an adopted citizen" by contributing to "establish a branch of manufacture that will provide employment for his fellow-citizens of the United States of America".[10]

Not having had the opportunity or time to have it printed, or being unable to solve the technical problem, or face the cost, of including dyed cloth samples into a printed edition of the *Memoirs on Dyeing*, the author must have been willing, not only to share his beautiful manuscript with his sons and successors, but to show it among a wider circle of relatives, friends, associates and colleagues. Did he maybe allow some recipes to be copied, going as far as adding presents of snippets from the corresponding cloth samples in his book? This would indeed be a nice explanation for the present maimed state of the manuscript.

Fig. 2.4. "Assortment" of colours of Londrins Seconds cloths in a bale joined by Pierre Gout (probably a relative of "Gout de Biz's") to his "Treatise on the fabrication of Londrins Seconds of Carcassonne" sent to His Lordship, Monseigneur de Saint-Pries, Intendant in Languedoc, in 1751. It shows how a bale is composed of two bundles of ten cloth pieces dyed different colours. Note the inner part of the list, preserved in two samples in the upper part of the assortment. AD34 C 2514. Photo D. Cardon.

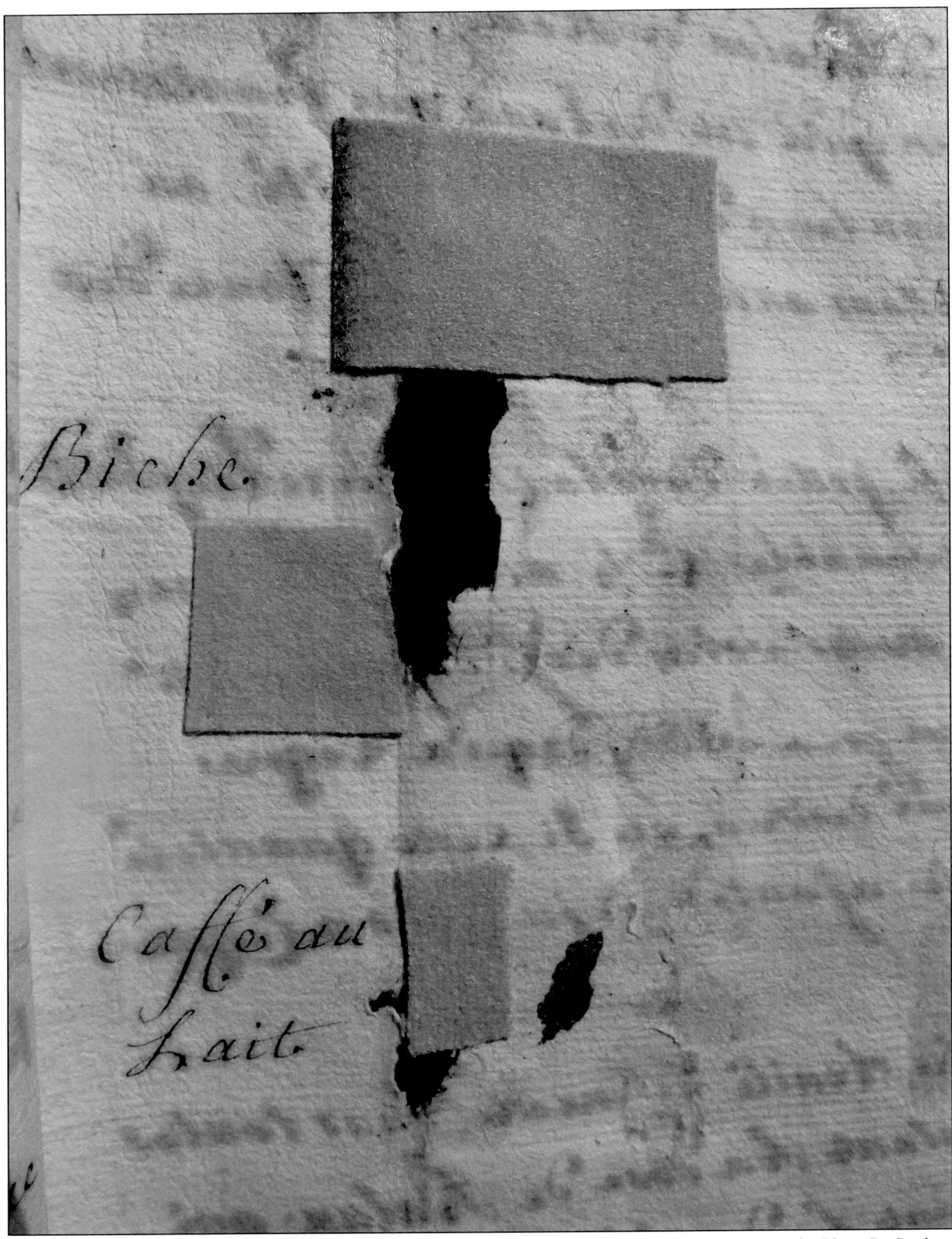

Fig. 2.5. Page 47, torn where sample on verso has been stripped off; part of the list is included in top sample. Photo D. Cardon.

Another one is suggested by a comparison with pattern books of English clothiers, contemporary with the *Memoirs*. A wealth of such documents is preserved in public archives in the west of England. In some of them, offcuts from the preparation of the patterns have remained trapped between the pages of the book. They show that patterns of the same standard size were cut from the piece of cloth freshly dyed, probably with a quill knife, using rectangles of paper cut from drafts or obsolete documents as stencils (Fig. 2.6). In many pages, snippets have been cut off from patterns, or a whole pattern has been stripped from the page and replaced by the name of its colour, while the snippet or pattern was sent to factors in London (Fig. 2.7).[11]

Similarly, the author of the *Memoirs* may have had to face urgent requests to send a pattern of a particular shade to factors in Marseille or in the Levant, and not having any cloth dyed in that particular colour at hand at the time, he may have used some of the samples in his manuscript.

Notes

1 Cardon 2013 pp. 39–66.
2 To prevent any confusion, all page numbers mentioned in this book refer to the page numbers in the manuscript.
3 Minard 1998, pp. 128–129.
4 AD34 C 5569.
5 AN F/12/737. Sedan here refers to the prestigious Royal Manufacture of broadcloth that had been created in this city close to the northern border of France. The quality of the black dye of Sedan cloths was famous and dyers from other regions tried to emulate it, as witnessed in the second Memoir.
6 As pointed out by Thomson 1982, p. 311.
7 AD34 C 2039, C 2040.
8 AD11 3J277; Marquié 1993a, p. 185.
9 It forms the last part of Hellot's famous treatise on the *Art of dyeing wool*, Hellot 1750, pp. 617–631.
10 Partridge re-ed 1973, p. xii.
11 Pattern book of Usher and Jeffries, clothiers in Trowbridge, Wilts., Wiltshire & Swindon History Centre 927/4, last period of use, from 1760 to 1768.

Fig. 2.6. Pattern book of Usher and Jeffries, clothiers in Trowbridge, Wilts., offcuts from cloth pieces used for the patterns and paper stencils. Wiltshire & Swindon History Centre 927/4, last period of use, from 1760 to 1768. Photo D. Cardon.

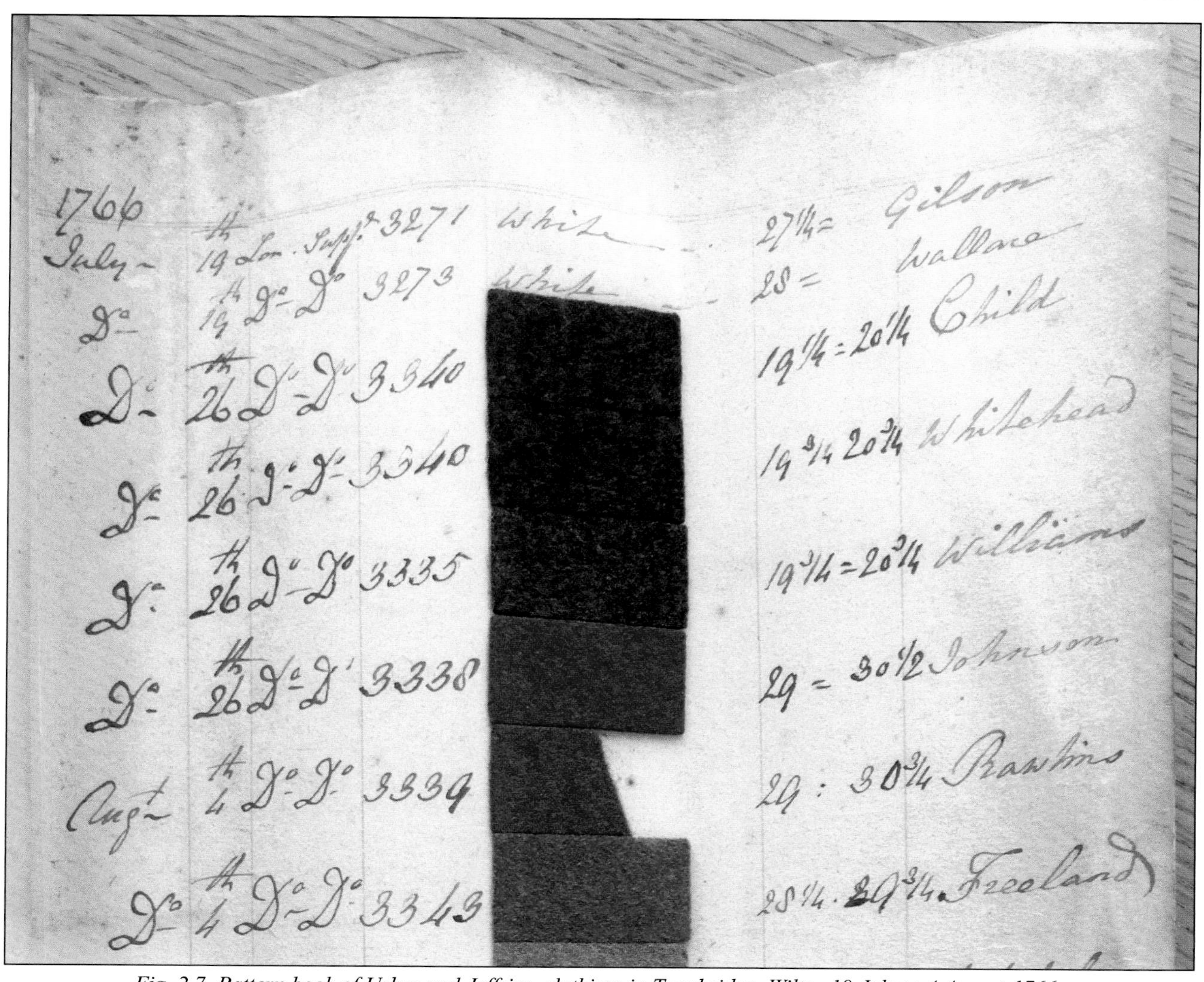
1766
July – 19th Lon. Supp.r 3271 White — 27 1/4 = Gilson
D.o – 19th D.o – D.o 3273 White — 28 = Wallace
D.o – 26th D.o – D.o 3340 19 1/4 = 20 1/4 Child
D.o 26th D.o – D.o 3340 19 3/4 20 3/4 Whitehead
D.o 26th D.o – D.o 3335 19 3/4 = 20 3/4 Williams
D.o 26th D.o – D.o 3338 29 = 30 1/2 Johnson
Aug.t 4th D.o – D.o 3339 29 : 30 3/4 Rawlins
D.o 4th D.o – D.o 3343 28 1/4 . 29 3/4 Freeland

Fig. 2.7. Pattern book of Usher and Jeffries, clothiers in Trowbridge, Wilts., 19 July to 4 August 1766, Wiltshire & Swindon History Centre 927/4. Photo D. Cardon.

3

Cloth and Context

L. S., L. L., L. P. or Mahoux?

Nowhere in the text does the author explicitly state from what kind of cloth the samples still glued in the manuscript of the *Memoirs on Dyeing* have been taken. That they all be of the same fine, lightweight quality is easy to recognize at first glance. But it is not as easy to identify it precisely. The range of types of broadcloth produced in Languedoc for export to the Levant, as defined by regulations (Table 3.1), included subdivisions corresponding to subtle technical differences within each type, which resulted in making the best made cloths of an officially lower type look pretty much like the lowest grade of the superior quality, in the eyes of any but the professional experts – clothiers, inspectors, factors in the Levant – or their most refined and wealthy Oriental customers.[1]

The issue is of particular relevance in the present case, since at the beginning of his second memoir, at page 42, the author gives, not one name only, but a list of the types of cloth he commonly dyes and to which his dye recipes will apply. This he has to do before he starts giving the proportions of ingredients to be used – in this first instance, to prepare the tin mordant for scarlets and other cochineal-based colours – because in dyeing processes, the quantities of mordants and colorants are commonly calculated in proportion to the dry weight of fibre or of woven textile to be dyed.[2] Now, the weight of each piece of cloth will depend on its technical characteristics, namely: the length of the warp which will determine the final length of each piece; the number of warp threads ("ends") and the width across which they will be spread on the loom through the reed, these warp threads being more or less finely spun and more or less densely set. This, in turn, will affect the amount of weft threads – usually less fine and less tightly spun than the warp – that can be beaten into the warp during weaving. As a result, the author warns, the weight of the cloth pieces he is accustomed to deal with can vary from 25 to 30 pounds (10.350 to 12.420 kg),[3] depending on which type(s) of cloth he has to dye in a particular batch.

These cloths he mentions in abbreviated form, by the initials of their names, so familiar they are to him: "L. S.; L. L.; L. P." – only making an exception for the highest quality, the so-called Mahoux. Evidently, "L. S." stands for Londrins Seconds, "L. L." for Londres Larges and "L. P." for Londrins Premiers. A first remark to be made is that this list – covering the whole range of Languedocian cloths commonly exported to the Levant during the second half of the 18th century – adds much strength to the hypothesis on the author's identity proposed earlier in this book. While only Royal Manufactures were allowed to make and export the best quality mentioned here – Mahoux – not all of them did it on a regular basis – or did it at all – and moreover, very few are actually known to have also diversified their production to occasionally include the less prestigious and slightly coarser Londres Larges, more commonly produced by independent, smaller clothiers from all parts of Languedoc. Not only was the Royal Manufacture of Bize one of the places where Mahoux (of the lower category, Mahoux Seconds) were regularly produced, but it has been seen in the first chapter that readiness to react to the fluctuations of the trade by prompt qualitative and quantitative adjustments in production was, from the start, one of the strong features of Paul Gout's managerial style. He will go on distinguishing himself and his Manufacture by an important and diversified output, combining a permanent great bulk of Londrins Seconds with a non-negligible production of Mahoux Seconds and Londrins Premiers until 1785, when 120 pieces of Mahoux Seconds and 45 pieces of Londrins Premiers are recorded for the last time, in addition to the 595 pieces of Londrins Seconds produced that same year.[4] After this date, the production figures recorded for Bize only concern Londrins Seconds, until the statistics themselves stop in 1789.

A second remark is that the mention of Londrins Seconds in first position, in the list of cloths commonly dyed by the author of the *Memoirs*, makes it likely that it is from this type of cloth that the dyed samples in the manuscript have been taken. This is all the more probable as it is the main production of all royal manufactures and the absolute best seller cloth in the Levant, representing 84% of the imports of French cloths into Constantinople in the 18th century and between 88 and 92% of the exports of cloths to the Levant, all ports considered.[5]

The silent messages of samples

In point of fact, better than any hypothesis, conclusive evidence can be found in the manuscript that the samples have indeed been cut from pieces of Londrins Seconds. Such evidence does not come from the text though, but from a small group of samples – and this is a brilliant demonstration of the exceptional usefulness of this kind of document where texts and textiles complete each other. While seven different samples were being taken from freshly dyed pieces to illustrate the corresponding dye recipes, the clothier's or his assistant's scissors cut into the side selvedge of the piece of cloth, including the innermost coarse thread of the list. This can be observed in a sample of azure blue on page 41, of *chamois* (buff) on page 47 (Fig. 2.5, top sample), of *noisette doux* (soft hazelnut) on page 70, in two samples of *agathe* (agate) on page 82, a sample of *cire doré* (golden wax) on page 89 and one of *cerise* (cherry) on page 95. Since regulations had defined characteristic lists for each type of cloth produced for the Levant, each having a special pattern of longitudinal stripes of coarser threads (*liteaux*) and different colours, the coarse dark blue threads visible on all seven samples unmistakably identify them as belonging to Londrins Seconds, whose lists must be "2 *pouces* (inches) wide (5.4 cm), white, with three or four *liteaux* of dark blue thread", depending on the subcategory of Londrin Second.[6] There were to be three *liteaux* for the lower kind with 2,400 ends in the warp, and four *liteaux* for the upper sort, with 2,600 ends in the warp; in the present case, it is impossible to know which of the two sorts of Londrins Seconds has been used, since only the innermost *liteau* is preserved in the samples. Two pattern sheets of Londrins Seconds preserved in the *Archives départementales de l'Hérault* in Montpellier confirm this identification: there, too, some samples include the inner portion of the list, showing from one to three dark blue *liteaux* similar to those in the samples of the *Memoirs on dyeing.*[7]

Much has been written on Londrins Seconds. Regulations had imposed a strict frame of technical specifications for this type of cloth. Many different technical and economical explanations for its popularity in eastern Mediterranean markets were proposed at the time, both by French and by foreign – particularly English – observers. But never, before this identification of the type of cloth used for the samples in the manuscript, had it been possible to publish a technical analysis of a Londrin Second, in the finished state in which the pieces, unwrapped, were displayed to prospective buyers in Constantinople, Smyrna, Aleppo, Sidon or Cairo a quarter of a millennium ago. The following analysis was only made possible because one of the samples in the manuscript was cut from a damaged part of a cloth, where the nap had been destroyed in spots (maybe by splashes of corrosive tin mordant?), exposing the threadbare ground of the cloth. Well preserved broadcloth is normally impossible to analyse, the fulling and finishing processes completely hiding the structures of threads and weave under the uniform, velvety and opaque surface of the nap.

Technical analysis of a sample of Londrin Second cloth

Based on the sample of scarlet page 82 of the *Memoirs on dyeing*, the technical data revealed by this analysis are proposed as representative for all the other samples in the manuscript, quite obviously taken from the same type and quality of broadcloth. Most likely, it is equally representative of the production of Londrins Seconds in the Royal Manufacture of Bize during the whole period of Paul Gout's management.

Dimensions, preservation

c. 3.7 x 2.5 cm; the four edges are slightly indented, the velvety scarlet nap of the cloth is corroded in very small spots in several places, revealing the ground weave of the cloth, faintly tainted of a pale pink colour; the rest of the sample is very well preserved; the scarlet red dye of the cloth's nap has retained great vividness.

Technical characteristics

Wool, tabby weave, almost balanced, regular and close texture.

Warp, woven undyed, looking very pale pink in the parts where the nap of the cloth has been destroyed; tight S-spin; c. 25.5 ends per cm.

Weft, woven undyed, same colour as warp in the parts of the sample without nap; slightly thicker, medium Z-spin; c. 18 picks per cm.

Discussion of the results

The most immediately apparent – and very useful – information derived from the examination of the sample concerns the stage at which dyeing took place. Since only the raised fibres of the nap are red, the core of both warp and weft threads not having been impregnated by the colorant, it means that the cloth piece – or "end", in William Partridge's terms[8] – must have already been fulled, and its nap raised and cropped, when it was plunged into the dye bath. Characteristically, this is an important technical choice about which regulations, dyers' books and memoirs are too often silent. It was so obviously part of common knowledge in the trade that only the authors of the *Dictionnaire universel de commerce* cared to precise that "the cloth having been well woven, fulled, and its nap raised and shorn, its lists are girt-webbed and it is sent to be dyed."[9] A scientist like Jean Hellot also feels it necessary to explain that concerning broadcloths, "reds, blues, yellows, greens and all the other colours that are wanted perfectly uniform, are only dyed [...] after they have been completely fulled and finished".[10] About scarlet dyeing, he further mentions that a good test to check that it has been done according to regulations is "to cut a small sample with scissors and look at the *tranche* (edge), it will be of a beautiful white [...] One calls *tranche* the inside, the core, or the densest part of the ground of the cloth. When this dense ground is dyed like the face, in any colour, it is said that this colour or dye *tranche* (seeps through) and the contrary is said when the central part of the cloth has remained white. Legitimate scarlet never *tranche*, never seeps through."[11] In the west of England too, traditional broadcloth was dyed in the piece after fulling and gigging, while by contrast, Spanish or medley cloth was dyed in the wool or, in some imitations, "in the say, that is before it is thickened in the mill".[12] One important consequence and advantage of dyeing after fulling and raising that Hellot does not mention is that only the outward surfaces of the cloth being dyed, it allows a substantial economy on the quantities of mordants and colorants to be used, which is of particular significance when such a costly dyestuff as cochineal is involved.

A less apparent, unexpected and important result of this analysis is the high warp count of the sample: with more than twenty-five warp threads per centimetre, it is much denser than the count of, at most, nineteen ends per centimetre that would have resulted from the strict respect of regulations concerning Londrins Seconds (Table 3.1). Indeed, it is even higher than the warp count that can be calculated from the data specified by regulations for the top quality Mahoux Chalys. It is higher also than the warp counts in samples of top quality English cloths bought in Constantinople in 1733 (Tables 3.2 and 3.3); higher too than the warp counts that can be calculated from the data mentioned by William Partridge as recently obtained from a clothier in Gloucestershire, in the part of his *Practical Treatise* dedicated to "the Manufacture of Broadcloth" (Table 3.4).[13]

The high warp count of the sample of Londrin Second in the manuscript is typical of a cloth that has been deliberately conceived to be especially thin and fine, using very finely spun yarn and warping more threads than stipulated by regulations. The explanation, simple and based on common sense, will also serve for some of the English cloths that will be examined further down: the manager of the Manufacture was a competent clothier who wanted to ensure large and durable outlets for his production through a good repute. He obviously knew exactly what were the optimal characteristics of the best possible cloth he could afford to propose his customers, at a competitive price that would, at the same time, prove profitable enough to keep his Manufacture and its work force busy, and himself and his family well off. This implied mental gymnastics "which distinguish the good *Fabricants* (manufacturers), excites their ingenuity, stimulates emulation among the others, hence the good repute that several of our Manufactures have acquired", analyzed Henri Louis Duhamel du Monceau in his *L'Art de la draperie, principalement pour ce qui regarde les draps fins* (The art of broadcloth manufacture, mainly as regards fine broadcloths), published in 1765.[14] It also brings new evidence that common practice in a royal manufacture allowed for some degree of initiative, within the frame of detailed norms established by regulations.

Another unusual technical feature of the sample is the spinning direction of the warp and weft yarns. It is not the fact that the two groups of threads are spun in opposite directions, which is surprising; this is a technical trick which had been generally adopted since the Medieval ages in all European woollen cloth producing regions. Its rationale is that in the woven cloth, the spires and wool fibres of warp and weft threads that have been spun with opposite twists are now slanting in the same direction (Fig. 3.1), which will make it easier to obtain an even, smooth surface during the raising and fulling processes.[15]

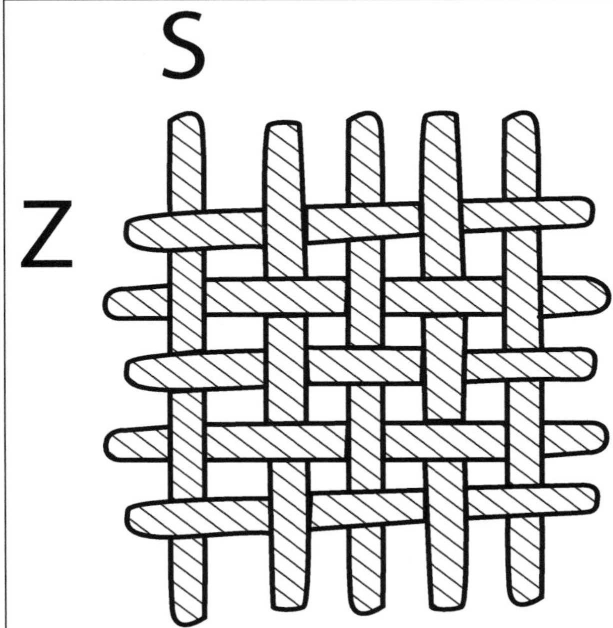

Fig. 3.1. In a woven cloth, spires and wool fibres of warp and weft threads spun with opposite twists are slanting in the same direction. Diagram D. Cardon.

William Partridge, in his *Practical Treatise*, explains that "the twist for warp and filling should be different, when one is twisted with an open band, the other should be done with a cross band" and a little further on, he adds that "the yarn for filling, when intended for broadcloth [...] will have to be spun different from the chain, with a twist the reverse of this; it must not be twisted so hard and somewhat coarser,"[16] which closely corresponds with the differences in the warp and weft yarns of the sample described above. What is intriguing here is that the twists of the warp and weft in this sample are the reverse of the usual spinning directions in broadcloths made in other parts of France, as described, again, by Duhamel du Monceau, with their warp threads "always spun on the spinning wheel *à corde ouverte* (with an open band)", that is with a Z-twist, and the weft threads "always *à corde croisée* (with a closed band or cross band)", resulting in a S-twist (Fig. 3.2).[17] Nevertheless,

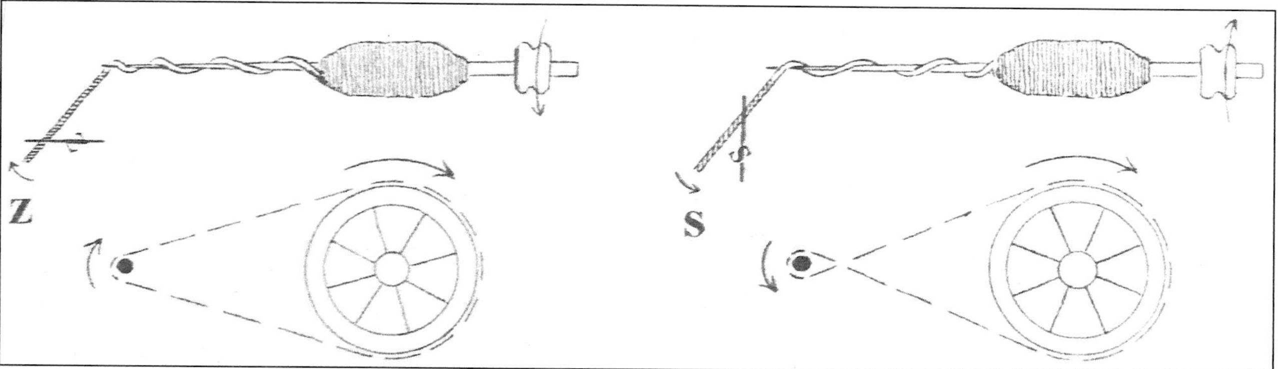

Fig. 3.2. Threads spun on the spinning wheel with an open band receive a Z-twist, while threads spun with a closed band or "cross band" receive a S-twist. Diagram D. Cardon.

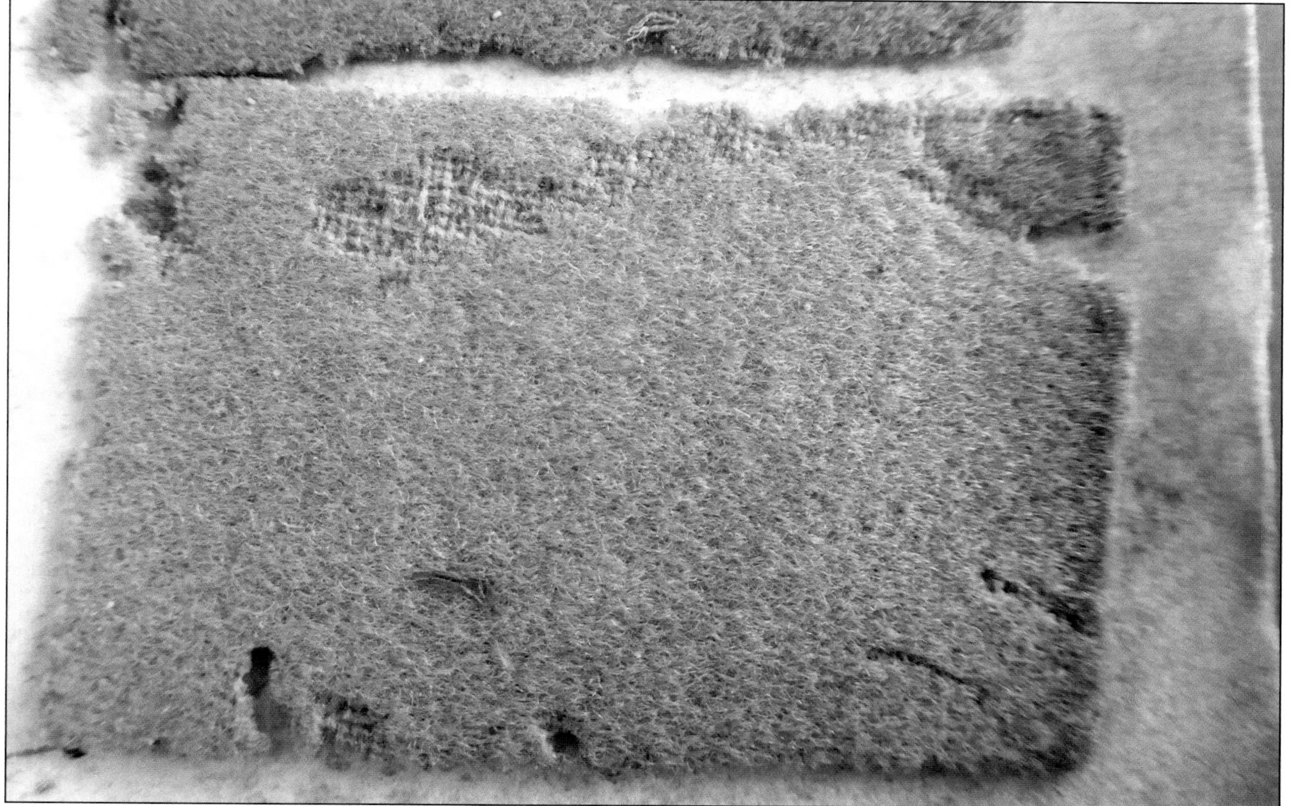

Fig. 3.3. S-twist warp and Z-twist weft of pattern dated c. 1725, in the book of an unidentified clothier. Archives of John and Thomas Clark Ltd., records of firms preceding the establishment of their firm. Wiltshire & Swindon History Centre, Chippenham 927/14.

looking at the successive regulations elaborated for the types of broadcloths that were produced in Languedoc for export to the Levant, one finds that from the start, the *draps à la façon d'Angleterre et de Hollande*, that were initiated with the aim of imitating English and Dutch broadcloths, were to have their warp "*filée à rebours*" (contrary to normal, obviously meaning with closed band) resulting in a S-twist, and the weft *à droit* (in the normal way, with open band), giving a Z-twist.[18] Each of the other samples of securely identified Londrins Seconds preserved in the Archives départementales of Hérault, also have a S-twist warp and Z-twist weft, and the same is true of samples of Londres Larges and of Nims "*façon d'Angleterre*", showing that although the articles defining the spinning directions of warp and weft were not repeated in the successive later regulations on cloths produced for the Levant, the rule of spinning them in opposite directions to the other cloth types was maintained.[19] On the other hand, their conformity with their alleged English models is confirmed by the technical analyses of samples of English cloths bought in Constantinople in 1733, preserved in the same archives and presented in more detail below; of three different categories and in a wide range of qualities, they all have a S-twist warp and a Z-twist weft. The same is true of "Best Super" cloths made in Trowbridge around 1725, preserved in the pattern book of an unidentified clothier: part of the nap on some samples having been moth-eaten, the ground is clearly visible (Fig. 3.3).[20]

Game of names

French-made "*Londrins*" and "*Londres*", spun "*à la façon d'Angleterre et de Hollande*"; Venetian-made "*londrine all'uso d'Olanda*"; English-made "Nims" and "French Colours"... it is time now to try and disentangle the threads of the intricate story of cloths' names, a game played for two centuries by European factors in the Levant, while commercial war was raging between their countries and technical adaptations inspired by industrial espionage constantly took place in the concerned manufacturing regions.

The game starts at the beginning of the 17th century. The Venetians, until then exclusive providers of high quality broadcloth to the Turkish Court, do not feel much threatened when English-made cloths appear on the markets of the Levant. These *londre*, as they call them in letters, are cheaper, not made for the same consumers, they think.[21] In the years 1640–1660, a new wave of broadcloths, the *lakens*, arrive from the Netherlands. Being "lighter and thinner" than both the traditional Venetian broadcloths and the English *londre*, they come to be called *londrine*, the diminutive form of *londre*. The Cretan War (1645–1669) between Venice and the Ottoman empire offers the Dutch merchants a golden opportunity to win the favour of the Turkish elites who will get to like the fine new type of cloth and never turn back to wearing heavy Venetian cloth.[22] This induces the Senate of Venice, in 1679, to allow "the manufacture of broadcloth in the manner of Holland", the so-called "*londrine all'uso d'Olanda*". They could never compete with the genuine Dutch *londrine*, however, and from 454 pieces in 1679, their production did not stop declining until 64 pieces only were eventually produced in 1698.[23] Nevertheless, the Venetian name for this type of cloth will survive, like the former *londre*, and the two most important categories of broadcloths in competition for the Levant markets will go on being designated with names derived from *londre* and *londrine*.

In France, the first regulation mentioning these names and giving detailed specifications for "*Londrines Premières larges*", "*Londrines Secondes*", "*Londres larges*" and "*Londres*" was issued in 1697.[24] However, *Londrine* sounds like a feminine name in French and it must have been hard to conciliate with the concept of *drap*, broadcloth, a masculine name, in French clothiers' subconscious. As soon as a new regulation was deemed necessary to precise some details for the fabrication of the new types of cloth, the name was changed to the masculine, definitive form: *Londrins*.[25]

All these cloth names – *londre*/*Londres*, *londrine*/*Londrins* – clearly deriving from the name of the city of London (*Londres* in French), demonstrate a general understanding among European merchants that it was there only, in London, that the English cloths supposed to have served as models for these imitations started being considered as "cloths for the Levant".

In England, effectively, they were not made specifically to be exported to the eastern Mediterranean. They were sent to London by clothiers from Gloucestershire, Wiltshire and increasingly Yorkshire for the cheaper sorts, already dyed or more often in the undyed state. In Gloucestershire and Wiltshire clothiers' books, they are classified as "superfine, 4 arrows, 3 arrows, 2 arrows" or, a little later, as "superfine, best super, super, super gold, gold".[26] In London, they were entrusted to Blackwell Hall factors and it was through them or through packers, that the Levant merchants selected and bought the assortments of qualities and colours answering the current demands of the Levant markets. The cloths that were bought undyed were entrusted to commission dyers in London who conformed to patterns or orders sent by the merchants' factors in the Levant and they were finally packed and shipped to the ports of the eastern Mediterranean.[27] There, they were given commercial names which naturally depended on their quality but appear to have also varied over time, and from one trading station in the Levant to another.

It is for Aleppo that these names and categories are best known, thanks to Ralph Davis' book on English traders in Aleppo in the 18th century, largely based on the archives of the Radcliffes, an important family of Levant merchants. Davis was able to highlight that "the English broadcloths sold in Aleppo fell into three fairly well defined categories, according to their quality, and these categories were defined by terms little used outside the Levant."[28] According to his classification, the cheapest category of cloths sold in Aleppo consisted of the so-called "londras or reds [...]; nearly half the bales sold were of this kind. A similar number of bales of middling quality, known as mezzo-londrines, fangot, or French Colours, were normally sold" at double the price of the londras. "Finally, a much smaller quantity of better cloths, described as astracans, or later as half-drabs" was sold at still higher prices, about 1 ½ times superior to those of the middling qualities.[29] "Londra", Davis goes on to explain, "was the word Italian merchants had used at the beginning of the seventeenth century as a general name for the broadcloths from London [...] and it is impossible to say how the name lost its general meaning and came to be used only for the cheapest grade of English broadcloth. There was a considerable variety within the londra grade, perhaps reflecting the survival of some Gloucestershire cloth alongside cheaper but inferior Yorkshires [...]. There is ample identification of coarse broadcloth with londra, apart from the evidence of the Radcliffe records."

These cheap cloths did, in fact, find their chief market in Persia; "for Arzeroon and Bagdad, Londra and very thin cloth of Persia colours are most esteemed, being for the Persia market", wrote a factor in Aleppo in 1738. Though there was a fairly steady demand for londra, it was not popular with the Levant factors, because it could not be sold at a profitable enough price: "Londra makes a terrible account", wrote the same factor in 1747, "but I must have some by me to please some buyers."[30] An important share of the londra trade consisted of whole bales containing only red cloth, dyed cheaply with madder rather than with expensive cochineal, which sold easily to the fez-makers of Aleppo and other cities of the Ottoman empire. "Moreover", Davis concludes, "londra stood up well to French competition." This was obviously represented by the "Londres" and "Londres larges" cloths from Languedoc.

Still following Davis' classification, "the second grouping in quality was of middling cloth – mezzo-londrine, fangot and French colours [...]. Nearly half the bales of cloth sold by the Radcliffes were in this range, and they accounted for considerably more than half the value of sales [...]. They evidently came from Gloucestershire or Worcestershire clothiers. One of the factors wrote of 'Worcesters, called Mezzo Londrine'. The names applied to them – mezzo-londrine, fangot, French Colours – were used only in the Levant trade; these three varieties sold at similar prices and it is impossible now to say exactly what distinguished them. But there were real differences, for fangot was regarded as a good selling line at a time when mezzo-londrine was being almost overwhelmed by French competition,"[31] – certainly from the Languedocian Londrins Seconds, steadily gaining ground during the second quarter of the 18th century. Davis is at a loss to precisely define what type of cloth the term "fangot" – the Persian word for a bale – could refer to. On the other hand, he gives precious, unique evidence about the third type of middling cloth: "French Colours were English cloths of the londrine type, dyed in light colours in imitation of French cloths. The term only appeared after 1741, coinciding with a period in which English mezzo-londrines, so described, found a poor sale in the Levant. Large quantities of French Colours were included in the cloth consignments of the next few years. After the French became involved in the war with Britain in 1744 French Colours rapidly disappeared from the Radcliffe consignments, the last ones being shipped in 1746. The name was never used after that, and mezzo-londrines were sold again in large quantities. The term appears to have been a temporary one, adopted in the early phases of the use of new colours, and dropped when they had become a regular part of the English colour range for middling fabrics. All these middling-quality cloths suffered severely from French competition in the thirties and forties."[32] This was the beginning of the triumphal progress of the Londrins Seconds.

"Finally", Davis continues, "an altogether higher grade of cloth was sold, in much smaller quantities than the first two [...]. There was a growth in the demand for this better cloth in the fifties [...]. Cloths of this kind were generally described as astracan, or later when they were much more popular at Aleppo, as drabs, half-drabs, or mock-drabs. The term "drabs" was again simply a description of a type of cloth, and has no relation to the present-day meaning of the word. Drabs were dyed in every colour, and in fact the expensive red dyes producing rich colours – scarlet, crimson, deep wine – were used for a higher proportion of these than of other cloths. The Radcliffes secured most of this cloth from Worcester clothiers."[33]

Davis further sums up such scantier insight into the other markets in the Levant as can be gained from his documentation. "The requirements of the other trading centres differed from those of Aleppo. At Stamboul demand was concentrated on the higher qualities of cloth [...]. Smyrna, on the other hand, had no sale at all for the better quality cloths, but handled londra and the cheaper mezzo-londrines, largely for the trade with Persia. The demand from Cairo was small and intermittent, and no generalization can safely be made about it."[34]

In Constantinople, the distinction between "three well defined categories" of English cloths also prevailed, similar to that identified by Davis in the documents concerning Aleppo, but with different names for the middling and top qualities. This is shown by an exceptional series of documents preserved in the *Archives départementales de l'Hérault* in Montpellier,[35] collecting the results of an extravagant operation of industrial espionage, planned in Paris at the beginning of 1733 by the *Contrôleur général des Finances*, Philibert Orry, and executed in Constantinople under the direction of the French ambassador, Louis Sauveur de Villeneuve. With an aim to allow a more precise comparison between the English and French cloths in competition in the Levant, the scheme was to send envoys around wholesale and retail shops in the various ports of the Levant and buy a two ells-length (2.37 m) of English cloths of every available quality within each category, together with the head of each piece, tie a label with a number onto it in decreasing order of quality and price, and send it back to France with all relevant details of prices, colour assortments, quantities commonly sold, etc. Only one complete range of 39 samples was actually collected, in Constantinople. It was sent back to Marseilles "on 25th April 1733, on Captain Argeme's *pinque* (pink)". The information was later completed by the answers to a detailed enquiry sent around the other *Echelles* (ports) of the Levant, namely Smyrna, Aleppo and Cairo. Luckily for historical research, although the lengths of cloth were distributed among various experts for examination, comments and propositions for future actions, a sample of each of the 39 original samples has been preserved, with its original number and all the relevant documentation (see back cover).

Samples and documentation both confirm Davis' statement that "there was a considerable variety within the cheapest grade", the londras – called "Londres" in the French documents (Tables 3.2 and 3.3). Seven different qualities are represented among the ten samples of English "Londres" of various colours collected; the name "reds" mentioned by Davis for a kind of londras regularly bought by fez-makers, is illustrated by a sample representing the third quality, which is said to be always sold in whole bales containing only cloth dyed red with madder.[36] A technical report on 38 of the lengths of cloth by a French clothier, consultant for the project, assimilates the two best qualities of these English Londres to French "Londres Larges", the others being closer to the French "Londres".[37]

The middling category is neither designated as mezzo-londrines, nor as French Colours, as in Aleppo. It is called "Nims", referring to the city of Nîmes, in Languedoc; it must have been the English riposte to the use of the names "Londres" and "Londrins" by the French to pass their cloths off for English products. These English Nims also encompass a wide range of qualities: the 20 samples collected in Constantinople are classified in six different qualities. In the French clothier's opinion, the lowest is only slightly better than the French "Londres Larges" while the highest is equivalent to some Londrins Seconds. He therefore proposes to create a new category of cloths in France, to fill the gap he has perceived and imitate the intermediary qualities of English Nims, and he adds some technical recommendations to that effect. As a result, the new category of French "Nims" is created and its technical characteristics defined by an ordinance dated 31 June 1741.[38] As seen above, this is precisely the time when the English "French Colours" appear in Aleppo factors' letters. This new name, I think, may have been created in an attempt to distinguish the grade of English cloths previously called Nims, from their French imitations with the same name. This is all the more likely as the name "French Colours" disappears in the Levant at about the same time as the new category of cloth is disallowed in France, ten years later, because of "their bad quality and uselessness".[39] Nims will actually go on being made occasionally, in Languedoc, as a cheaper alternative to Londrins Seconds, each time depending on a special permission that had to be obtained from the Intendant.[40] This is why they still figure during the second half of the century in the synthetic tables presenting the technical data imposed by regulations for all categories of cloth produced for the Levant. Amusingly, this is the time when English clothiers of the West of England start designating some of their cloths as "Lon. Sup", the initials of which, L. S., coincide with those of the French Londrins Seconds.[41]

In the same series of French documents, the highest grade of English cloth found in Constantinople is not designated as "astracans", nor as "drabs" as in Aleppo, but is called "Mahouts". It includes three different qualities, the top one being also called "Chaly" or "Mahou de Chaly". The *Memoir on the broadcloths sent by the English nation to Constantinople* corroborates Davis' statement that the highest quality of cloth was mostly sold in Constantinople. It adds that it fetched such high prices that only three or four bales were imported each year, most of the consignment being bought "for the use of the Sultan himself, his Grand Vizier and a few grandees of the empire who often have this commodity collected on board the ship or at the custom" before it had time to be offered to any other customers.[42] Only one length of it, of a beautiful light yellowish green, was bought in Constantinople, but apparently not communicated to the French clothier in charge of studying the different grades of English cloths, for he does not comment on this special quality. My technical analysis (Table 3.3, top line) of the preserved sample (made possible by the fact that in a tiny patch the nap has disappeared, probably eaten by an insect) showed that it is more balanced, with more weft threads in proportion to warp ends, but, surprisingly, not finer than the sample of Londrin Second in the manuscript of the *Memoirs on Dyeing* analysed earlier (see back cover: right, bottom right).[43] It came from a considerably wider piece of cloth, however, if the comments accompanying these Mahout cloths are to be trusted; they mention that the top quality English Mahouts were 2 ½ pikes (169.5 cm) wide without counting the width of the selvedges on either side.[44]

It actually was the second quality of English Mahouts that it was planned to study and copy as exactly as possible. This was firstly because it was the best selling one, with 40 to 50 bales sold each year in Constantinople.[45] Another reason was that the top quality could not so easily be reproduced economically in Languedoc. Experiments made in Royal Manufactures run by a manager recognised for his expertise, had shown that it took "as much time to make one hundred pieces of Mahoux cloth as to make eight hundred pieces of Londrins Seconds".[46] The third and lowest quality of English Mahouts, on the other hand, was found to already have equivalents in the best made French Londrins Premiers and in any case it was of lesser economic importance, having a smaller outlet in Constantinople, in the order of "ten to fifteen bales at most in a normal year".[47]

This review of the commercial names given to the cloths imported from Europe to the Levant, and the insight it offers into the technical characteristics of the grades of cloths designated by these names, mainly serve to show how attentive and reactive the English and French competitors in this momentous tournament were to each other's progress. This game of names and long-lasting war of industrial espionage highlight their mutual recognition of the technical and commercial strengths of the products they planned to imitate or supersede. It implied a thorough knowledge and keen understanding of the technical processes and economic and social conditions that allowed the production of each particular type of broadcloth. Such knowledge and understanding were the cornerstones of the clothiers' culture, in the West of England as in Languedoc. This common expertise was like a common language, echoes of which are preserved in their private papers and in the mountains of pamphlets and memoirs they were addressing to their respective governments all along the 18th century, very often with similar queries and in similar terms. The comments that follow, on each other's "cloths for the Levant", help us figure out what thoughts and preoccupations the author of the *Memoirs on dyeing* may have had in his mind in the mornings, at the start of a new day of managing "his" Royal Manufacture.

Visions on broadcloth by contemporaries – French and English points of view

Very good analyses have been written about the successes and declines of both English and French cloths in the Levant at different periods of the 18th century, taking into account the complex international context and the social and economic evolutions that were taking place, in the producers' as well as the consumers' countries.[48] What is proposed here is much less ambitious. It is only a small selection of texts whose authors try to define and explain the respective technical qualities and weaknesses they have actually seen, felt, measured, in English and French cloths competing for the same fluctuant Eastern Mediterranean markets.

Comments on English cloths sold in Constantinople in 1733, by a French clothier

The report written by an anonymous French clothier on the English cloths bought in Constantinople in 1733 has already been mentioned.[49] The exploitable technical data resulting from his detailed examination of the samples are collected together in Table 3.2. His comments offer a shrewd appraisal of the similarities and differences between the various grades and qualities of the English cloths and their French counterparts. The report being quite long, only excerpts can be presented here.

> "It is good to make an exact analysis of the English cloths that have arrived from Constantinople and of the way in which they are made, which being understood, all that will have to be done will be to make cloths that will be similar in wool as well as in finish [...] The two most essential points that must therefore be examined are the quality of the wool and the number of threads in all the sorts of cloths that they send to Turkey. This is what has been done as will be reported, starting with the Londres and so on, from one quality to another successively, up to the superior ones [...]. The Londres [of the sixth to third qualities] are consistent in number of threads as well as in widths – although there are some little differences of one finger's breadth more or less from one to another, which only proceed from the lack of regulations, which are unknown in England [...]. The body, goodness, stoutness of these Londres is also the same, as well as their finish. For the quality of wool there is some diversity, it being impossible that a perfect similarity could ever be found in that respect, since in France, where there are regulations which prescribe qualities of wool, more disproportions are found in that respect, between one manufacturer's Londrins Seconds and those from another, than there are between all these Londres. [...] The Nims of the sixth quality are a little superior to our Londres Larges from the point of view of their wool. Composed of 1,800, 2,100, 2,300 and 2,400 threads, these differences are a little surprising in a same grade of cloths, all the more so as in the higher grades are found some cloths with a lower number of threads, as will be seen in what follows. The freedom which the English manufacturers have, to work with whatever number of threads in the warps as they deem appropriate (not having regulations in their country, as has already been said), and the

differences in spinning, some being thinner and others fuller, explain these variations. [...] the finer or coarser one spins, the higher or lower the length of thread, but always the same weight of wool. This no doubt is the rule the English must follow and the reason for the differences in numbers of threads that have just been mentioned".

About the four lower qualities of English Nims, he estimates that they could all be imitated following technical data prescribed by only one regulation, "because from one regulation, at least three qualities of Nims will result, depending on the manufacturers' genius, some will be less beautiful, others will be ordinary, others more beautiful [...] since it is not possible that all the wools that are used, and the skills of those who use them, may be uniform in all respects; and to this reason must be attributed such diversity in the qualities of the English Londres, Nims and Mahous of the second sort joined to the Memoirs sent from Constantinople." The end of his report summarises the lessons learnt from the careful examination of the samples: for the wefts of their cloths, the English use short wools, which cost less than the Spanish wools used in Languedoc, and "allow using narrower reeds and having the warps and wefts spun thicker; hence the cloths do not need to be fulled for so long and nevertheless they have more body and are firmer." Another important difference he highlights occurs at the finishing stages: "in Languedoc, as much nap is raised as possible, so that the cloths are weakened and become soft and deprived of springiness; the English, on the contrary, only draw out as much wool as necessary to cover the ground, and crop it very little, as is easily ascertained: hence they are good, firm, dry, springy, and long lasting."

My own visual comparison between the samples of English and French cloths stuck together on the three sample cards preserved confirms that the nap in the former is generally less dense and closed, and that the threads – where the ground of the cloth is visible – also tend to be less thin than in the corresponding quality of French cloth. Whether these differences made them better or not really depends on one's point of view. It does look as if the English manufacturers were making their cloths with an aim to ensure a long life of use, while the French were more concerned about their beauty at first sight. This is apparently what most impressed William Partridge, whose comments on this point are quoted below. Another remark made in this report is confirmed by the examination of the samples: there are indeed more variations in the warp and weft counts of English samples within the same grade (as appears from Table 3.3), while the French samples tend to be more homogeneous within each category, which is probably due to the technical frame imposed by regulations – however, it has been seen earlier that this allowed for exceptions, such as the particularly fine type of Londrin Second that was chosen by the author of the *Memoirs* to illustrate his dye recipes.

Comments on French cloths by Thomas Hale, around 1756[50]

The only mention of French cloths made by Thomas Hale in his book, *A Compleat Body of Husbandry*, is – as could be expected – in connection with sheep-breeding and wool, the production and trade of which was of such major importance to the British economy. Still, it is quite precise from a technical point of view:

> "We hear much of late of the French cloths; people praise them for their pliantness and easy wear; and though they are in reality greatly inferior to the English, many prefer them for this reason. The difference is altogether owing to their spinning two threads out of every parcel of wool, and keeping one for the warp and one for the woofe, the latter of which they make so loose that 'tis hard to work it. A very little beating brings the cloth into a body, and when finished it is soft and pliant. One would think these French cloths, by their look and feeling, could do little service, but the countrary is found upon trial. They never crack, and the looseness of the woofe thread rising from time to time, keeps them from wearing threadbare. Our hard cloths are liable to grow bare at the seams; these never do; because they are less harsh. They are fitter for gentlemen's service than for laboring people; but it would be easy to follow that practice more moderately".

Hale proves well informed about French spinning practices and what he says about the distinction between warp and weft (woofe) is particularly relevant for Languedoc. In some important Royal Manufactures, such as Les Saptes, near Carcassonne, the warp and weft of the different grades of cloths were even spun in different specialised workshops, respectively the *grand filoir de chaîne* (big warp-spinning room), equipped with 93 spinning wheels and five reels, next to which the *commis de la chaîne*, a clerk specially in charge of watching and registrating warp production and warping, had his small office; and the *filoir de trame*, the weft-spinning room, with 86 spinning wheels.[51]

It must be pointed that Hale's idea that it might be interesting to find a compromise between the English and French conceptions of broadcloth had already been implemented in 1741 by a Mr. Lowndes who sent a report on his experiment to the *Journal of the Commissioners for Trade and Plantations*, joining "a pattern of woollen

cloth made in North Wales, after the manner of that made in France".[52]

Comments on French cloths by English diplomats in the Ottoman Empire, c. 1760–1765

These texts are of particular interest in that they refer to the very period when the author of the *Memoirs on dyeing* was writing his masterpiece, or had just finished writing it. At the same time as the English consul and the factors of the Levant merchants in Aleppo were working on their report to Lord Halifax, Paul Gout's record production of Londrins Seconds of 1764 was on its way to the factories and retail shops of the cities of the Levant, on board a pink or on camels' backs – some of it, maybe, bound for Persia and beyond.

From the English Consul and factors in Aleppo, 1765:[53]
Copy of the representation of the British Factory at Aleppo to the Right Honourable the Earl of Halifax, Aleppo the 30th July 1765.

In pursuance of the King's Royal Pleasure signified to Mr. Grenville, His Majesty's Ambassadour at the Ottoman Porte by your Lordship's letter of 18 january 1765 [...] and communicated by His Excellency to the British Consul and Factory residing in this City in his letter of the 4 April 1765. Whereby they are commanded to lay before your Lordship an account of the increase or decrease of the British trade to the Levant, the causes or occasions of such increase or decrease, and the methods which may be deemed proper to remove such obstructions as from time to time it may have met with: the said Consul and Factory having seriously considered the state of the British trade at present, and for about twenty years last past, have the honour to inform your Lordship: that during the said space of twenty years there has been a continual decrease of the British trade, at this place, as very evidently appears by a retrospect of our annual manifests of our imports here, for from 1748 to 1756 we imported at Aleppo one thousand bales of English cloth (*communibus annis*) whereas from 1756 to 1764 the annual imports are reduced to five hundred bales of cloth, and so in the same proportion of tin, lead, indigo, cochineal and other goods. The causes of this decrease are various [they mention the "depopulation of Persia" and the competition from the East India Company in Bassora, and go on...] a great obstacle we have had to struggle with has been the decline of our cloth fabricks with respect to quality, and the improvements made therein by the French nation who, perceiving that the chief consumption of cloth hath of late years been confined to the mere southern parts of Turkey, have wisely adapted their fabricks to the climate and by an introduction of cloths of a very fair appearance, of thin, and light quality, with a continual variation of colours, as the taste of the Turks have required, and from a vicinity to this market, and the cheapness of their labour at home, they have acquired a manifest superiority so that whilst our nation import annually about 500 bales of cloth, they, for these five years past have found vent for 1 500 bales of their fabricks, whereas formerly they imported only 500 bales whilst the English imported 1500. A piece of French cloth in these times weighed about 25 pounds and sold at 2 ½ dollars per pike, whereas of late they weigh only 17 ½ pounds and sell at 1 ½ dollars per pike on an average for ready money, altho' we are of opinion that this price does not turn them to account, and we are informed that proper methods are taking in France to restore their fabricks to their former quality when we imagine the price will rise [...] As to the methods of removing these obstructions, they appear to us in general very difficult, and in some respects insurmountable, as is that of the government of the country, in which we cannot interfere. The reestablishment of Persia must also be left to the effects of time. To cope with the French, and outrival them in their fabricks, will require great art and attention, and still greater frugality; some attempts of this kind have indeed been made by some members of the Levant Company, who amidst several trials have now and then succeeded in imitating the quality and colours of the French cloth fabricks, but even they could not come near them in price by about 10 % per yard."

Signed: William Kinloch Consul, David Haÿs, Charles Smith, Colvit Bridger, Henry Preston and Isaac Hughes Sund

From Sir James Porter, British Ambassador in Constantinople from 1747 to 1762, in the second edition of his Observations on the Religion, Law, Government and Manners of the Turks, *1771:*[54]

The French cloth was undoubtedly of a finer wool and finer spun, but thinner, of little substance, a looser woof, yet with a short nap, and a variety of bright fancied colours: and indeed, it has been so thin of late, that they greatly undersell us, though perhaps at a loss. [...] we may in part ascribe the decline [of our trade] to the quality and cheapness of French cloth.

Like Thomas Hale, the authors of both texts highlight the thinness and lightness of the French cloths, and Porter rightly relates the fineness of the threads to that of the wool used – the qualities of which were specified for each grade by the French regulations. There had actually existed, since the Middle Ages, a nearly non-interrupted tradition of broadcloth production based on the use of short, comparatively fine wools in Languedoc and the neighbouring coastal Mediterranean regions of Roussillon, Catalonia and Valencia, which exported part of their cloths to the Levant.[55] This long expertise has too seldom been taken into account to explain the successful development of a range of light broadcloths in 18th century Languedoc. Moreover, the spinning wheel had been adopted very early in these regions, through the influence of the cotton-working regions of Andalusia under Moorish control.[56] The Languedocian spinners' skill at the wheel was typically a *savoir de la main,* transmitted from mother to daughter for many generations, and it certainly was an asset for the fabrication of a "soft and pliant" cloth. Nevertheless, the English diplomats and factors were right in pointing that some of the French cloths – probably of the lesser qualities, usually made by independent clothiers – were getting excessively thin: the weight of 17½ British pounds (7.937 kg) they mention for some of the cloths they had examined, marks a debasement in the order of 30 % as compared with the average weight of 25 pounds (11.339 kg) that had been normal until then, deriving from the technical characteristics defined by regulations. This – it has been seen – was almost exactly the average weight of the cloth pieces commonly dyed by the author of the *Memoirs*. Indeed, such excessively low weights and selling prices were a form of dumping which actually did not adversely affect the English trade only: the clothiers who produced such cloths "of little substance" were "gaining such bad reputation that they hardly dared to go on inscribing the name of their guild at the head of the pieces" and were getting lower and lower prices for their cloths until "some among them could not sell theirs at any price, either in Marseilles or in the Levant," as noted some French inspectors.[57]

A right balance between "quality and cheapness", however, was kept by many manufacturers and particularly by the *entrepreneurs* of the Royal Manufactures, who strived to keep up the "very fair appearance" of their cloths' finish, the technical basis of which is explained below by William Partridge, as well as the "continual variation" of "bright fancied colours" whose recipes are given in the most minute details in the *Memoirs on Dyeing*, the translation of which follows.

Comparison of the finish of French and English cloths by William Partridge in his **Practical Treatise on dying of woollen, cotton and skein silk, with the Manufacture of broadcloth and cassimere including the most improved methods in the West of England,** *1823*[58]

> When a perfect nap is raised on a piece of thick, fine, firm cloth, the face is well and closely covered, and equally so in every part. When the nap is thin on stout cloth, it proves either that it has not been sufficiently raised, or that much of it has been taken off by defective workmanship. The latter will be the case [...] when the teazles used in the first courses have been too strong. Under these circumstances the wool will be dragged off the face instead of being drawn out and laid on it. [...] When a cloth comes from the fulling-mill, the wool is always closely matted together in the ground, and if in the first courses, the work be too strong, it will tear it out with a force sufficient to break the staple of the wool [...]. The French, in order to avoid this, give more time in the working, and use more of dead work than the English. [...] The English are about eight hours in raising the nap of a fine stout felt, and the French are from twelve to sixteen. A French cloth, therefore, of any given fine quality, has a better nap on it, and a much finer face than the English. I am aware that in making this assertion, I am treading on very ticklish ground; but the fact is well known in London by every respectable trader in the article, and I can see no reason, excepting sheer national prejudice, why it should not be candidly acknowledged in a work of this description, having for its object to instruct manufacturers in the best mode of working.

Notes

1 A *Mémoire sur la draperie que la nation angloise envoye a Constantinople* (Memoir on the broadcloths sent by the English nation to Constantinople) sent in 1733 by the French Ambassador in Constantinople in answer to a request from the *Contrôleur Général des Finances* (the equivalent of a Minister of Economy and Finance for 18th century France) convincingly illustrates this point. It reports that the subtle gradations in qualities within the three main grades of English broadcloths exported there around 1733, encourage the cloth traders and retailing shopkeepers to sell the finest English Nims (the name given to the second grade of English cloth) for Mahouts (the name designating the top English quality), and the French best Londrins Premiers for English fine Nims, because there is "little difference from one quality to another, which makes it easy for them to deceive the natives, who in general are not good enough connoisseurs to distinguish between them", AD34 C 2200, folio 1. This statement is illustrated by three sample cards, joined to the memoir, on which English cloths of the whole range of qualities, freshly bought in Constantinople, figure next to their closest French equivalents (see back cover).

2 Cardon 2007, pp. 8–19.

3 The pound used as weight unit in the text of the *Memoirs* is the *livre poids de table* of Montpellier, as demonstrated in Cardon 2013, p. 401. It weighs 0.414kg.

4 Table of production statistics concerning royal manufactures for the years 1750 to 1789, AD11 9C 31.

5 Eldem 1999, p. 38; Carrière 1980, p. 94.

6 AD11, doc. 9C 22, c. 1765.

7 AD34 C2514, doc. 15, dated 18 November 1751, illustrating a treatise on the manufacture of Londrins Seconds in Carcassonne by Pierre Gout; and AD34 C5550, doc. 44, pattern sheet of Londrins Seconds made in the royal manufactures of La Terrasse and Auterive (south of Toulouse), dated 10 April 1759.

8 Partridge re-ed 1973, p. 135.

9 Savary des Bruslons and Savary 1748, vol. 2, p. 931.

10 Hellot 1750, pp. 307–309.

11 *Ibid.* pp. 299–300.

12 Tann 1967, pp. 17–19, 23.

13 Partridge re-ed. 1973, p. 51; in his "technical notes" to this edition, Kenneth Ponting confirms that these data "probably represented West of England normal practice"and that the broadcloths he had himself examined were "around this norm", p. 232, n. 51. About the English cloths bought in Constantinople preserved in the Archives départementales de l'Hérault in Montpellier, see below p. 32.

14 Duhamel du Monceau 1765, p. 42.

15 Cardon 1999, pp. 252–253.

16 Partridge re-ed. 1973, pp. 43, 45.

17 Duhamel du Monceau 1765, p. 38.

18 *Statuts et Reglement pour la Manufacture des draps de Carcassonne, Cité, Saptes et Conques* of 26th October 1666, *Recueil des Reglemens* 1730, t. 3, p. 220, art. IX and X.

19 Sample illustrating a treatise on the manufacture of Londrins Seconds in Carcassonne by Pierre Gout, dated 18th Novembre 1751, AD34 C2514, doc. 15; samples from the Royal Manufactures of La Terrasse and Hauterrive, made in April 1759, AD34 C5550, doc. 44; samples from Antoine Janot's *Mémoire*, identified as Londrins Seconds from the thick blue *liteaux* included in some of the samples in the same book, AD34, C5569, p. 4, olive green; samples of Londres Larges from the manufacture of Viguier de Carbonne, Esqu., AD34, C5550, doc. 42; sample of Turkish blue, Nim *"façon d'Angleterre"*, AD34, C5550, file with *matrices* (standards) for various qualities and dyes; sample cards with English cloths, AD34, C2200.

20 Wiltshire & Swindon History Centre, Chippenham, Archives of John and Thomas Clark Ltd. Records of firms preceding the establishment of their firm in 1801: ref 927/14.

21 Sella 1961, p. 119.

22 Sella 1961, p. 64, letter from Venetian merchants in Constantinople, 5 January 1673: "the Turks and mainly the elites, accustomed to the comfort and to the lightness, thinness and other advantages of the Dutch cloths, do not want to consider turning back to suffer the weight of Venetian broadcloth".

23 *Ibid.*, p. 120, table 2.

24 *Arrest portant Reglement pour les Draps dont le commerce se fait en Levant* of 22 October 1697, Anon. 1730, *Recueil des Reglemens*, pp. 131–133.

25 *Ibid.*, *Arrest* of 20 November 1708, p. 138.

26 Wiltshire and Swindon History Centre, 947/1802, pattern book of Thomas Long of Melksham, dated 1698–1728; 927/4, pattern book of Usher and Jeffries, 1727–1768.

27 Davis 1967, p. 107; Moir 1955, pp. 232–235. Blackwell Hall, near the Guildhall in London, was the ancient cloth market.

28 Davis 1967, p. 100.

29 *Ibid.*

30 *Ibid.*, pp. 100–102.

31 *Ibid.*, pp. 102–103.

32 *Ibid.*, p. 104.

33 *Ibid.*, p. 105.

34 *Ibid.*, pp. 105–106.

35 AD34 C 2200; see note 1 above.

36 AD34 C 2200, 1st sample card, n° 7, and comments in a document entitled *"Explication sur la qualité des 39 coupons de draps envoyez par la Pinque du Capitaine Argeme"*, folio 3 verso.

37 AD34 C 2200, Anon., *Comparaison des draps anglois, hollandois et françois avec le plan des mesures qu'il conviendroit de prendre pour faire imiter les premiers en Languedoc* (Comparison between English, Dutch and French cloths with a plan of the measures to be taken to have the former imitated in Languedoc).

38 *Ordonnance portant reglement pour la fabrique des draps Mahoux Seconds, Nims et Londres ordinaires, façon d'Angleterre, destinés pour les Echelles du Levant* (Ordinance constituting regulation on the fabrication of the cloths Mahoux Seconds, Nims and ordinary Londres in the English manner, for export to the Levant), AD34 C5550, doc. 23.

39 AD34 C2058.

40 It has been seen above, pp. 8–9, that Paul Gout had had some trouble obtaining it in 1759.

41 First occurrence of "Lon. Sup." in the pattern book of Usher and Jeffries, clothiers in Trowbridge, Wiltshire, in 1762, Wiltshire and Swindon History Centre, 927/4.

42 AD34 C 2200, *Mémoire…*, folio 2.

43 AD34 C 2200, sample card n° 3 (see back cover: right), sample n° 39 (bottom right).

44 *Ibid.*, *Mémoire…*, folio 1 verso. The same document gives the equivalence 1 ¾ pike = 1 *aune* (French ell); therefore 1 pike = 67.8 cm.

45 AD34 C 2200, *Mémoire…*, folio 2.

46 AD11 9C30: experiments by Jean Marcassus, director of the Royal Manufactures of La Terrasse and Auterive, in 1728; a sample of one of his superfine Mahous is placed next to the English sample of Mahou de Chaly, AD34 C 2200, sample card n° 3.

47 AD34 C 2200, *Mémoire…*, folio 2 verso and *Comparaison…*, folio 8 verso.

48 Masson 1911, pp. 412–416, 476–478, 489–490; Ambrose 1831; Wood 1935; Moir 1955; Davis 1967; Tann 1967, pp. 38–45; Morineau and Carrière 1968; Mann 1971; Carrière 1978; Thomson 1982; Marquié 1993; Tann 2012, pp. 54–62.

49 AD34 C 2200, Anon., *Comparaison des draps anglois, hollandois et françois avec le plan des mesures qu'il conviendroit de prendre pour faire imiter les premiers en Languedoc*.

50 Hale 1758, vol. 3, p. 307.

51 Inventory of the equipment, dated 3rd March 1700, when the Manufacture was sold to a new owner, AD11 9C19.

52 Thursday July 23, letter from Mr. Lowndes 22 july, *Journal of the Commissioners for Trade and Plantations from January 1734–5 to December 1741*. London: His Majesty's Sationery Office, p. 395.
53 The National Archives, SP 105/184, folio 128–129.
54 Porter, re-ed. 1771, pp. 367–368, 377.
55 Cardon 1999, pp. 23, 50–54; Calderan-Giacchetti 1962, pp. 173–174; Cazals and Valentin 1984, pp. 13–14.
56 Cardon 1999, p. 236.
57 AD11 9C20 and AD34 C2589; Marquié 1993, pp. 63–64, 73.
58 Partridge, re-ed 1973, pp. 78–79.

Table 3.1: Technical data of Languedocian cloths produced for the Levant based on and calculated from regulations[1]

Names of cloths	*Number of warp ends*	*Width on loom (m)*	*Width finished (m)*	*Resulting warp count*[2] *(ends per cm)*
Chalys and Mahoux Premiers	3,400 to 3,600	2.47	1.58	21.5–23
Mahoux Seconds	3,000 to 3,200	2.37	1.58	19–20
Londrins Premiers	2,800 to 3,000	2.37	1.48	19–20
Londrins Seconds	2,400 to 2,600	2.30	1.38	17.5–19
Nims	2,400 to 2,500	2.45	1.42	17–17.5
Londres Larges	2,000 to 2,200	2.53	1.48	13.5–15

1 All measurements and technical data specified by the Regulations were summed up around 1765 in tables preserved in most administrative and clothing centres in Languedoc. The copy used for this table is now preserved in the *Archives départementales de l'Aude* (AD11), doc. 9C 22; for Nims, the data were obtained in the original ordinance fixing their technical characteristics, AD34 C5550, doc. 23.
2 Theoretical result, based on the number of warp ends and width after fulling and finishing specified by the regulations.

Table 3.2: Technical data of English cloths bought in Constantinople in 1733 based on and calculated from their technical analysis by a Languedocian clothier[1]

Names of cloths	*Number of warp ends*	*Width finished (m)*	*Resulting warp count (ends per cm)*
Mahouts 2nd quality	2,800–3,050	1.58	17.5–19
Mahouts 3rd quality	2,650	1.58	17
Nims 1st quality	2,500–2,800	1.48	17–19
Nims 2nd quality	2,500–2,660	1.48	17–18
Nims 3rd quality	2,250–2,700	1.38	16–19.5
Nims 4th quality	1,900–2,500	1.38	14–18
Nims 5th quality	2,200–2,250	1.38	16
Nims 6th quality	1,800–2,400	1.38	12–16
Londres 1st and 2nd qualities	1,600–1,900	1.48	11–13

1 AD34 C 2200, Anon., *Comparaison des draps anglois, hollandois et françois avec le plan des mesures qu'il conviendroit de prendre pour faire imiter les premiers en Languedoc*. The expert had access to the whole cutting of each quality of cloth and could therefore count the exact number of ends in the width.

Table 3.3: Technical data of some English cloths bought in Constantinople in 1728 and 1733, and of French cloths glued on the same samples cards, or preserved in other documents in the Archives départementales *of Hérault, from the author's technical analyses*[1]

Names of cloths	*Warp, twist and count (ends per cm)*	*Weft, twist and count (picks per cm)*
English Mahou de Chaly, n° 39	S, 22–24	Z, 22
English Mahout 2nd quality n° 36	S,20	Z, 20
English Mahout 3rd quality n° 31	S, 22	Z, 22–23
English Nim 1st quality bought in 1728	S, 16	Z, 16
English Nim 3rd quality n° 27	S, 16–17.5	Z, 15
English Nim 3rd quality n° 22	S, 16	Z, 16
English Nim 3rd quality n° 21	S, 16	Z, 17
English Nim 5th quality n° 15	S, 12	Z, 13–14
English Nim 5th quality n° 14	S, 12–13	Z, 12
English Nim 6th quality n° 12	S, 13	Z, 12
English Londres 2nd quality n° 8	S, 14	Z, 14–15
English Londres 7th quality n° 1	S, 10–11	Z, 9
English Londres bought in 1728	S, 12	Z, 10–11
French Londrin Second n° 848	S, 18	Z, 18–20
French Londrin Second from the Royal Manufactures of La Terrasse and Auterive[2]	S, 18	Z, 18
French Londrin Second by Pierre Gout[3]	S, 17–18	Z, 17–18
French Londrin Second from St-Chinian[4]	S, 18	Z, 16–17
French Nim, standard[5]	S, 18–20	Z, 18
French Londres n° 764	S, 16	Z, 15–16

1 Only samples presenting at least a tiny part where the nap has been destroyed could be analysed; the ground weave is totally hidden in the other samples.
2 Pattern sheet, AD34 C5550, doc. 44.
3 Sample in Pierre Gout's *Traité sur la fabrique des Londrins Seconds de Carcassonne*, dated 1751, AD34 C2514, doc. 15.
4 Sample in Antoine Janot's *Mémoire – on trouvera les opérations de la teinture du grand et bon teint des couleurs qui se consomment en Levant*, dated 1744, AD34 C5569.
5 Standard for French Nims *façon d'Angleterre* dyed in Turkish blue, AD34 C5550.

Table 3.4: Technical data of Gloucestershire broadcloth produced at the beginning of the 19th century, based on, and calculated from, figures provided by W. Partridge in his Practical Treatise on dying of woollen, cotton and skein silk, with the Manufacture of broadcloth and cassimere including the most improved methods in the West of England[1]

Number of warp ends	*Width on loom*	*Width after fulling*	*Width after raising and tentering*	*Resulting warp count (ends per cm) finished*
84 biers of 38 threads = 3191	11 ½ quarters = 2.57 m	6 ¼ quarters = 1.43 m	7 quarters = 1.60 m	c. 20

1 Partridge re-ed. 1973, p. 51.

Part II

Memoirs on dyeing

English translation of the original French text

Memoirs on dyeing

–Including–

A memoir on the ingredients employed, in which will be found their nature, their quality, their property, their price and the places from which they are obtained; the general method to dye the colours for the Levant in the best mode of dyeing; an instruction on the tests for colours in false dyes; and a treatise of annotations: particular colouring processes and observations on same.[1]

[p. 1] **(I)**

Memoir on dye-drugs, that serves as an introduction to the dyeing of broadcloth manufactured in Languedoc for the ports of the Levant

Nobody doubts that a most useful and necessary knowledge for a dyer who wants to practise his art with success and distinction is that of the drugs he uses; he must be acquainted with the nature of some, and the preparation of others, in order to be able to judge their effects; and know their properties in order to ensure the successful outcome of the colours and undertake their dyeing with confidence.

It is moreover indispensably necessary that he should be acquainted with what indicates their good or bad quality, to avoid being deceived in the choices he will make and, lastly, that he should know from where they are obtained and the price at which they are commonly sold in order to make his purchases seasonably and economically.

This is what I am going to try and explain in this memoir, which will serve as a preliminary to the next one, dealing on all the colours that are dyed for the Levant; hereafter, one will find a table in alphabetical order of the drugs that are used for the latter. It has to be observed that what I say here concerns the dyes that we make in Languedoc for that market and remains within these limits. As the prices of the aforesaid drugs are very variable, depending on several events that, in business, are always uncertain, I thought I could do no better than indicate the highest and lowest prices they have for a long time been known as costing: to my mind, such notion should suffice to guide the dyer and the merchant in their speculations.

Beyond doubt, the object of my enterprise is of a broad scope, therefore I do not pretend to fulfill it to the entire satisfaction of those who could draw some benefit from it. However, I hope they will be somewhat grateful to me for having undertaken it and that in any case they will find something useful in this booklet.

[p. 2] **Table of the Dye Drugs included in this memoir**

Alums [p. 3]

Alum is a fossile and mineral salt formed by the combination of vitriolic acid and a kind of earth the nature of which is still unknown. It is one of the ingredients most useful to dyers and a non colouring drug that is used to prepare the cloth to receive an infinite variety of colours that could not be produced without it and which this preparation renders faster at the same time as giving them brightness and vividness. It is a kind of mordant that, applied onto the cloth, fixes the colours into it, binds them and prevents the finest particles from evaporating.

Rome alum

There are very abundant alum mines in Italy, in the region of Trivoly[2] and Civita Vechia,[3] in a place named La Topha.[4] These last mines belong to the Pope who leases them. The alum coming from there is called Rome alum, which is a stone of a pale white colour.

Alum preparation

After calcining these stones for twelve to fourteen hours, they are taken out from the kiln, exposed to the air, and watered three or four times a day to make them fall into efflorescence; then they are washed, the liquor is evaporated in the sun and very beautiful crystals are obtained. This alum which is also called roche alum, always has a reddish tinge that is given to it by the colour of the mine earth.

Great quantities of alum also come from the burnt countries near Mount Vesuvius, between Naples and Pourrouzole;[5] it retains all the traces of the volcano that once existed there. It is a ridge of small hills that form a kind of basin whose ground shakes and echoes under the horses' feet. In summer, a kind of earth that is found in the middle of this basin is collected and stored in a dry place. To extract alum from it, it is put into big copper cauldrons and lixiviated with rainwater already containing this salt, since it is taken from cisterns dug in this aluminous earth. When the water is evaporated enough, it is decanted above the many natural hollows in the ground, so that no expense is needed for fuel.

When the liquor is evaporated enough, it is carried into wooden tubs where alum cristallizes as the solution cools [p. 4] down. What has just been reported on the way Rome alum has been made must suffice to give an idea of the way it is exploited in other countries, where the stages of the work are more or less multiplied, depending on the mines' contents in different substances.

It sometimes happens that the miners who work in an alum mine set it on fire. When it happens, it is not possible to put it out. There are some mines in England that have been burning for a very long time. The mines calcined in this manner still give some alum.

Calcination of alum

Our dyers refine the alum they need for the saddening of fine colors by calcining it on a violent fire in a cast-iron cucurbit; after it has liquefied, it is left to cool down and ground to powder to use it as needed.

Rome alum which is the most valued of all alums, comes to Marseilles in bales of 300 to 350 pounds.[6] The price of this commodity, which used not to change much, has much risen for some years. It could be obtained at 15 £ to 16 £ per hundredweight but nowadays it costs at the said Marseilles up to 40 £.[7] It pays customs duties for import into France at the rate of 3 £ 12 *sous* per hundredweight which, adding the weight loss, transport costs and brokerage fees in Cette[8] or Agde, adds 6 £ per hundredweight to its price, arrived in Languedoc. When choosing alum, one must observe that it should consist of transparent stones, without powder.

Alum from the Levant

There also comes to Marseilles some alum from the Levant, particularly from Smyrna and Constantinople, less pure and containing more earth than that of Rome, consequently less proper for dyers' use. The French import very little of it because this product is not rare.[9] They are of two sorts, a bigger and a smaller one. The former is better and in the Levant they give 3 hundredweights of the small one for two of the big. This alum comes in bags, it has been worth from 10 £ to 27 £ per hundredweight and is submitted to the same additional costs as that from Rome.

[p. 5] *Alum from Spain*

Some alum is obtained from Spain, that has been less appreciated at first than that from the Levant but nowadays our dyers rightly prefer it and use it even for fine colours, probably because of the high price of that from Rome. Alum from Spain has a whitish and very transparent aspect, it comes to Marseilles where it is worth from 17 £ to 34 £ per hundredweight. It is also bought in Cette and directly from Spain. When it is used in big quantities, it is convenient to choose the latter solution where much cost can be spared. The Spaniard, who come for this trade as far as Carcassonne, smuggle it in on small flat boats.

Alum from Sweden

There also comes some alum from Sweden to Marseilles, of a greenish tinge, in big transparent stones that weigh up to 200 lbs.[10] Some of it is worth more than that from Spain and some is worth less, there even is little difference in price and the additional costs are the same.

Alum from England

Alum from England, which is found in the provinces of Iorck of Lancaster,[11] is in the form of stones of a bluish tinge rather similar to slate. This alum is called roche alum because the crystals come as big lumps, clear, transparent, or because it is extracted from a stone like Rome alum.

In the same quarries some alum is found whose crystals are white; this is the sort that our dyers use, like the alums from Spain or Sweden.

Alum from France

Some alum is prepared in France near the mountains of the Pirenées.[12] There is a vein of it running along the surface of the ground in the *viguerie* of Roussillon,[13] which measures from one up to four *toises* in width, per nearly four *lieues* in length and which is very abundant;[14] this alum turns out of better value for dyers because of the vicinity of the mine.

Alum will clarify liquors and, because of this property, is employed in sugar factories.

Alum, which is regarded as one of the most powerful astringents, is of great help in pharmacy, it is used in all haemorrhages, etc.

[p. 6] **Logwood**

Logwood is the heart of a tree that grows in plenty in several islands of America, particularly in those of Campeche, Jamaica and St Croix.[15] This wood is sorted out according to its cutting, the best is the Spanish cut, that is to say that the ends are cut with an axe, which shows that it is true Campeachy wood, the English of Jamaica being accustomed to saw their logwood, which is the same.

Logwood comes in big blackish logs. The inside is red, of a nearly cinnamon shade, the action of the air causes it to darken. One must choose it high in colour and take care when one buys it that it should neither be rotten nor worm-eaten.

The price of logwood varies a lot, which depends on the greater or lesser quantities that are brought from America. In Bordeaux, it has been found to cost from 12 £ up to 50 £ per hundredweight, mark weight, which amounts to 118^{lbs}, Languedoc weight.[16] Freight charges via the Garonne and then the canal, are not considerable, they come up to 45 *sols* at most per hundredweight, toll in Toulouse included.

It must be observed that what I say here about the weight and costs for logwood bought in Bordeaux must serve for all the other drugs that come from there with which I shall have the opportunity to deal upon in this memoir; I shall further remark that when one orders them to be bought during the time of the March and October Fairs, they are free of duty. Lastly, logwood is also brought from Marseilles. It is a matter of combination to see from what place it is most convenient to obtain it, in order to get it at the best value.

Logwood gives colours ranging from *gris-de-lin* (linen grey) to dark purple but they are fugitive and it is not permitted by the Regulations to use this ingredient, except for common colours employed on low-priced goods. When one applies copperas on logwood, very beautiful greys verging to slate blues are obtained, in countless shades in proportion with the quantities employed but these colours loose much of their vividness once exposed to the air; they are more lightfast when one adds nutgall. This is the usual practice for blacks, for which logwood is very good; our dyers also use this wood to intensify some colours that are then to be saddened. Provided that they have received the necessary woad ground, this practice is all the more tolerable as it would be very difficult to make some of the requested shades without the help of this ingredient. In Bordeaux, logwood is taxed 12% entrance tax and 8 *sols* to let it out of the city; I was about to forget to say that an inferior grade of it is known, that costs from 6 £ to 40 £ per hundredweight. [p. 7]

Old fustic

Yellow wood or *fustok* grows in all the islands of the West Indies and particularly in that of Tobago where it grows very high. The English and the Dutch do most of this trade. Some of it comes to Bordeaux in logs, and some comes from Holland in barrels as sawdust, both give a beautiful golden yellow, quite fast; it turns out of better value to use the one in logs, I have always noticed that it produces a stronger and more vivid effect. Also, one does not run the risk to be fooled by the mixture that can be done in the process of reducing it to sawdust.

To use it, the dyers split it into chips and have it reduced with a plane into shavings as small as possible; in this way the substance is best extracted. The best one is that which is of a beautiful yellow inside the log, for the outside is always blackish. One must also take care that it should not be worm-eaten, too old or rotten. In Bordeaux, old fustic costs from 6 £ to 17 £ per hundredweight, it is taxed 7 £ per hundredweight, entrance tax, and 5 £ to let it out.

Brazilwood [p. 8]

Brazilwood grows in Mexico, in the part that belongs to the Portuguese, in the island of Santa Marta which belongs to the Spanish.[17] There are five qualities of it, brazilwood from Fernambuco, brazil from Japan, brazil from Lamon, brazil from Santa Marta and lastly Brasiletto or Jamaïca wood. The brazilwood from Fernambuco is considered the best of all, its name is derived from a town of the province of Brazil, and those from Japan and Lamon are not known by our dyers, who only use fernambuco, Santa Marta and brasiletto.

This wood is extremely hard and dry, it crackles a lot when burning and produces nearly no smoke because of its great dryness. To choose it well, take care that it should be high in color, heavy, compact and neat, that is without bark and without rotten parts. Moreover, after being split, its colour should intensify. It comes from Bordeaux as logs; it is worth there: brasiletto, from 10 £ to 16 £ per hundredweight, Santa Marta wood, from 16 £ to 33 £ and true Fernambuco, from 48 £ to 85 £ per hundredweight. Nowadays it also comes from Marseilles. It is a matter of combination.

These woods pay customs duty and tax when going out of the province, viz.

Small Brazilletto	3 £ per hundredweight import...	2 pounds export
Santa Marta	1 pound – 15 *sols*...	– 17 *sols* ditto
Fernambuco	3 pounds – ...	2 pounds ditto

Brazilwood gives shades ranging from onion skin to dark red; but these colours are of as poor fastness as those of logwood and, just as the latter, it is prohibited by the Regulations on fast dyes. We use it in our dyes to brighten, intensify and modify some colours for which this wood is appropriate.

Young fustic [p. 9]

Fustick or young fustic, is a small tree which grows in Italy, and in the southern provinces of France. That from Provence is the best. Young fustic is also found in cold countries, but it grows less well there. The leaves are oval-shaped with a rounded end. Its flowers, small and of a dark green, grow at the end of the branches among big tufts of much-branched, shaggy filaments. The sap wood of this small tree is white, but its heart is of a mixed colour, from a rather bright yellow to a pale green; it is its trunk or its root stripped of its bark that is sold by druggists. Cabinet makers, wood-turners and stringed-instruments makers use it for various purposes.

Young fustic gives a light yellow that is very vivid but that the air soon causes to fade; it is prohibited by Regulations. However, its use is tolerated in our dye-works because it would be impossible to produce certain colours without this ingredient, and because those that could be used instead would not have better fastness.

The wood of young fustic commonly goes to Marseilles at a price of around 11 £ per hundredweight, it comes in bundles covered with an old floor cloth.

Lees ash

Lees ash is nothing else but calcined wine lees. Here is how it is prepared: after the wine lees have been distilled or after vinegar makers have made vinegar with it, they let the marc dry up and stack it in cakes, then they burn and let calcine this substance in big hollows that they dig into the ground, in the open air and in the countryside, because of the foul smell, and this calcined substance is lees ash, that must be kept in a dry place, otherwise it looses its virtue which consists of alkaline salts.

Such ash, mixed with quicklime, is used to make the caustic stones that cauterise flesh.

[p. 10] Lees ash does not give any colour but is used in our dyes to assist the body of crimson reds, it is included into the composition of saddening mixtures for amaranth and it makes a very beautiful one for the winesoup shades.

In Languedoc, lees ash comes from Nimes, where its price is about 40 £ per hundredweight; it is also produced in Burgundy, in Lyons, and in Paris, the latter is the least valued.

Choose the lees ash that comes as stones, quite dry, newly made, of a greenish-white colour.

Fine or mesteck cochineal

Fine cochineal is an insect which feeds on a bush, a species of Opuntia, that is called Nopal, which grows in Mexico. The natives of the country, or the Spaniard who collect such precious grain, take care to take it out from the plant before the rain season, they kill and dry what they intend to sell, to propagate it.[18] There are two kinds of it, one grey, and the other reddish; both are equally good. When buying it, one must observe that the grain is beautiful, shiny, without any mixture, and with as little dust as possible.

Cochineal is brought from the West Indies to Cadix on galleons or on register ships; from there it passes on to England, to Holland, and to Marseilles. In Cadix it is worth from 65 to 75 *ducats* per *arroba* of 25 pounds, which must give in Languedoc from 27 to 28 pounds.[19] It is sold in the aforesaid Cadix without garbling while in Marseilles, it is the custom not to sell it but garbled; garbling commonly gives 8 % *garbeau* or dust, and one % of *balotes*, or big grains. Two pounds of good residue from sifting produce the same effect as one pound of fine cochineal, and the ratio is of 1 1/3 pound of *balotes* for one pound of fine cochineal. In Marseilles, this grain has been worth from 14 £ to 24 £ per pound; for some time now it is at about 20 £, and at this price, the residue from sifting is sold for from 8 £ to 10 £ and *balotes* at 13 £ to 14 £. Some residue from sifting comes from Holland that is of much less value than that
[p. 11] which is bought in Marseilles. The colour that is given naturally by the fine cochineal employed without acid ranges from lilac grey to crimson. There is an infinity of shades between these two, that it is useless to develop here. By using acids one makes them vivid, bright and orangey. *Balotes* and the residue from sifting always give less fine a colour, it is only convenient to use them for wine shades.

Mesteck cochineal pays import duties into France. By an arrest of the King's Council of 3rd October 1712, the King has granted a duty free to the manufacturers of Languedoc for a certain quantity of it, that is distributed per number of bales, on the total number of broadcloths that are made there.[20]

Sylvester cochineal or "campessiane"

Some sylvester or campessian cochineal is also brought from the Vera Cruz to Europe, it is in the woods of New Mexico that the Indian go and collect it. The insect feeds, grows and propagates itself on non cultivated opuntias, it is exposed in the rain season to all the humidity of the air, and it dies there naturally. This cochineal is always smaller than the fine one and has a more rusty aspect. The colour it gives is always faster, but it never has as much brightness. One counts 3 lb of sylvester cochineal for 1 lb of fine cochineal.

There also is some common sylvester cochineal that is of less value, it gives very little colour, and there is nearly no market for it in Languedoc.

Granille

Granille, which we only have known since a very short time, is none other than broken cochineal, very much loaded with particles of earth. It comes from Cadix in ready-made leather sacks, and it is sold like that in Marseilles. One should be aware that there is some *granille* that is absolutely nothing but earth, and gives nearly no colorant, but when it is of good quality, 1 1/3 lb of it is worth as much as 1 lb of fine cochineal for wine colours, and even for scarlet, when one knows how to employ it. I have personally experienced that there is much profit to gain by using it. It has been worth from 8 £ to 13 £ per pound in Marseilles.

Spoilt Cochineal

Spoilt cochineal is nothing else but fine cochineal that has been submerged or got wet from seawater. The salt that is deposited in it darkens its colour which becomes blackish. It is of nearly as much value as fine cochineal when using it, provided certain precautions are taken. Its degree of dampness determines its price.

Sometimes it is opportune to buy cochineal in Bayonne, where it is sold at mark weight.[21] It is a matter of calculation that everyone must make for oneself.

Coccus Polonicus, colouring insect[22] [p. 12]

Coccus Polonicus is a small round insect, a little smaller than a coriander seed; it is found adhering to the roots of *poligonum cocciferum*, it is very abundant in the palatinate of Kiovia next to Ukrainia, around the towns of Ludnow, Piatka Slobdyszczes and in other uninhabited places near to Ukrainia, Podolia, Volhynia, the Grand Duchy of Lithuania and even in Prussia around Thorun. Those who collect it know that it is immediately after the summer solstice, that the Coccus is ripe, and full of its purpurine juice. In their hand, they hold a little hollow spade made in the shape of a spud and having a short handle; with one hand they hold the plant and they uproot it with their other hand, armed with this tool, and they pick out these little berries or round insects and put the plant back into the same hole not to destroy it, which they do with wonderful deftness. After having sorted out the coccus from its earth by means of a sieve made especially, they take care not to let it transform itself into a little worm and to prevent this, they pour vinegar onto it or sometimes also the coldest water. Then they put it in a warm place, but very carefully, or else they put it in the sun to let it dry, and kill it, otherwise its insects would be destroyed and if they were dried too precipitately, they would loose their beautiful colour.

The Dutch are the only people that used to employ the coccus, half and half with cochineal. Mr Hellot in his *Art de la Teinture* says that, using this grain in the same way as cochineal, lac-dye or kermes, he could never obtain from it anything but crimsons, more of less vivid, lilacs and flesh colours and that he could never manage to make scarlets. He adds that the one he used cost much more than some beautiful cochineal since it does not give as much colour as the fifth part of this insect from Mexico. No doubt it is for this reason that trade in this drug has extremely declined.

[Pages 13 to 32 are missing in the manuscript]

Formerly a natural sal ammoniac was preferred, which was nothing else but camel urine, cristallysed in the burning sands of Arabia or Libya. This salt is so rare nowadays that it is not traded at all any longer. Good sal ammoniac must be of a grey colour on the outside, white inside, neat, dry, crystalline, without dirt, of an acrid, pungent taste, and, when broken, showing kinds of needles. It is sudorific and incorruptible. [p. 33]

It is prescribed against quartan fever.

This salt pays a tax of 3 £ per hundredweight on importation into France and 20 *sols* of its value per hundredweight when it does not enter the Kingdom.[23]

Tartar

Tartar is a salt that rises from fuming wines that form a greyish crust that crystallises and gets fixed onto the inside of barrels. This crust forms a deposit there and gets hard, taking the consistency of a stone. Tartar has the sap of the grapes for father, fermentation for mother, and the barrel for womb. Tartar is white or red according to the colour of the wine. The best is the white one that we get from Italy and Germany. This is the finest, all the more so because it is thicker, having stayed longer in the *foudres* – which are barrels containing up to one thousand pipes. Tartar from Languedoc and the one that is obtained from Montpellier is not inferior to it. For the quality of tartar depends less on the quality of the wine than on the repeated fermentations that various new wines have made successively during several years, which has a decisive influence on its goodness. So that tartar effectively is a substance embodied and like petrified from the saps of the grapes, which, having united in itself as many volatile salts as it could embrace, together they form a compact crystalline matter that settles onto the sides and bottoms of the vessels and gets separated from the lees and from wine through fermentation.

Very good red tartar is found in Languedoc. The men who go around the towns and villages to collect old rags for paper mills also buy red tartar from individuals to resell it. Others collect it on a bigger scale. For a long time, [p. 34] in Languedoc, it used to be worth from 10 £ to 15 £ per hundredweight. Today, when bought in Marseilles, it is sold around 25 £, and the refuse from its garbling, 15 £. The white one is at around 32 £, its refuse at 16 £. White tartar and red tartar have the same properties in our dyes, where they are of great use to prepare the cloth to take up the colour. The white one, which is the cleanest, is used for fine colours and the red one for common colours. This distinction will be more clearly seen in the memoir on dyeing that will follow this one.

Choose tartar, either white or red, in the form of thick plates, without dust, breaking easily and shiny inside, you will be sure to have some good one.

All tartars are aperitive, and slightly laxative, they remove obstructions and calm fevers down; for internal use, only white tartar and tartar crystal are used.

Verdigris

Verdigris, also called *verdet*, is a rust of copper or a kind of superficial calcination of it that is produced by dampening it with corrosive salts. In a crucible are placed thin copper plates, covered with sulphur salt in powder and tartar. They are left to cool down in the air and all the matter is converted into a nice verdigris. There is another way to make it by pouring vinegar or wine lees onto copper plates that are then placed in cool cellars. The acid of the vinegar corrodes the copper, a fermentation ensues that produces verdigris, which is then scraped off from the plates.

In our dyes, verdigris is only used to make a certain shade of celadon green and for blacks.

This drug is much in demand for the North and its trade is of considerable importance in Montpellier where it is worth about 20 s per lb.

[p. 35] Vermilion

Vermilion, kermes, or scarlet grain is an excrescence that is formed on a kind of knee holly, or holm oak.[24] A small worm stings the leaf and the result of the ensuing spilling of the sap is a round shell of the size of a big pea, green on the outside at first, then red and that eventually takes a grey colour when it is in its mature state; the pellicle that covers it is extremely thin, underneath it there is a second envelope that contains a very fine red dust which is called *pousset* or *pastel* of scarlet and it is this powder which is used in dyeing and in medicine.

There are great numbers of knee hollies that bear vermilion in the *garrigues* and scrub land of the provinces of Roussillon, Languedoc, of Portugal and of Spain; that which is collected in Languedoc, is valued as the best and that from Spain, as the least good.

It is a matter of importance to collect this grain at the right stage of matureness; if it is not mature enough, it does not give as much colorant and if it is left on the plant for too long, the insect that is contained in the shell eats the pulp or *pousset* on which it feeds, which much diminishes its quality.

When vermilion is collected, the pulp is taken out, and the grain is sprinkled with vinegar to kill the insect that, should this precaution not be taken, would leave the shell empty. Those who buy this grain must choose it full of *pousset* and be careful because it often happens that apothecaries sell it after having emptied it of this; if the sellers practice this trade honestly, for each lot, and for the same price, they also give the *pousset* that may have got spilt out of it.

Vermilion gives a full-bodied red and much less vivid that that of cochineal, but it is faster and is not susceptible to turn rosy. Venetian scarlet is made with vermilion. This colour has always been much valued. In Languedoc vermilion is worth from 2 £ to 4 £ per pound, it pays importation duties into France at the rate of 10 £ per hundredweight. Vermilion from kermes should be distinguished from that which is used in painting. There are two kinds of the latter, the natural
[p. 36] and the artificial. The former is usually found in some silver mines, in the shape of a red sand that is prepared by repeated washing and lixiviations. The artificial one is made with mineral cinnabar, ground on porphyry for a long time with spirit and urine, and then dried. Some is also made with lead, burnt and washed, or with calcined ceruse; some also used to be made with a red sand from near Ephesus, after it had been washed several times. But the most natural vermilion of this last kind used to be found in Spain, on inaccessible rocks; they were stones that were brought down by arrow shooting.[25] All this vermilion, both the artificial and the natural one, is used by oil and miniature painters; nowadays it is also used for the rouge which all women like to disguise themselves with.

To use kermes vermilion in medicine, the sap or pulp is extracted to make a syrup by adding a sufficient quantity of sugar to it. Kermes is cardiac, desiccative, astringent and fortifies the stomach.

The name of kermes is given to a certain preparation of antimony produced by a red powder, also used in medicine and called Carthusian powder.

Flax-leaved daphne

Flax-leaved daphne is a plant that grows naturally and commonly in the *garrigues* in Provence and Languedoc. It has a strong smell and some kind of milk comes out of its stem when it is cut. Flax-leaved daphne is used while still sappy, freshly cut, for it is not worth anything any longer after eight days and that is, provided that it is kept in a cool place and that caution has been taken not to collect it damp which would cause it to ferment, get heated and putrefy at once.

This plant makes a very dark yellow, fast enough, but it is prohibited by the Regulations; to save on expenses, our dyers mix flax-leaved daphne with weld, especially for shades of dark green for which it gives more successful results than for the others.

They cannot and must not use it except when it is in its full strength, from the month of May until the month of October.

Blue vitriol [p. 37]

Blue vitriol is a fossile or metallic salt that is formed in the bosoms of the earth by some calcination of copper or iron caused by the acid spirit of sulphur. Copper vitriol is blue, iron vitriol is green, which is the same as copperas; vitriol salt participates in these two colours if it is formed by the calcination of these two metals.

Roman vitriol is obtained by exposing pyrites or marcassites to the air until they get calcined and convert themselves into a lime of a blueish or greenish colour, according to their nature.

When they are in this state they are thrown into water, then by heating on fire, the water in excess is evaporated, the metallic salt then coagulates into beautiful crystals, long, transparent, of a light blue. Vitriols from Pisa, Germany, Hungary and Cyprus are all made in the same way, the latter is the best, we get it via Marseilles where it is worth from [ ] to [ ] per hundredweight.[26] It must be chosen of a beautiful transparent blue, with very few sulphurous parts; this drug is used to make celadon greens, of a shade close to that which is made with verdigris but much fresher and brighter.

Blue vitriol is among the goods from the Levant on which a 20 % tax is taken, it pays 7 £ 10 s import duty into France.

(II) [p. 38]

Foreword

The object I intended to treat in the memoir that follows being of a great scope – i.e. to generally deal with the colours for the Levant that are made in the dye-works of Languedoc – I must not boast to have fulfilled it, although I gave it my utmost application and I followed the gradation of all the colours susceptible to appear in different shades, from the highest to the lowest. There is an infinite number of others that vary according to chance and to fashions in the Levant. Everyday, some colours are requested that are unknown to the most experienced dyer; and some used to be much in demand twenty or even ten years ago as do not get any credit any longer today.

Nevertheless, through the order I have established in this memoir, I think it easy to manage to produce any colour that may be required, even in a different shade, by guiding oneself on the instructions for the nearest one, and increasing or diminishing the doses of ingredients.

Moreover, to make this memoir ampler, more useful, more reliable and informative, after it, as a kind of supplement, I have added special notes with patterns of the colours obtained by such processes. There, will be found many colours (some of which experimented for the purpose) which could not find their place in the order of this memoir, the means to make bright stuff and the way to economy.

Some dyers will perhaps find that I have employed the drugs with too much profusion in my colours, especially in mordant baths. They will be those who do not recognise perfection. Besides, I considered that I should fix the quantities according to the *Règlements*, to follow the great way to our art. Such persons should however be sure that economy is my main concern; let them be convinced of this by my "Annotations" to which they should refer.

General method [p. 39]

To dye broadcloth in the best mode of dyeing for the Ports of the Levant

On Blue

Blue being a primary colour, and the ground for the greatest part of fast colours, it is necessary to start with it in order to establish some order in this memoir.

This dye is done (without any other preparation than to thoroughly dampen the cloth in stale water, and to let it drain afterwards) in a large wood vat, after the latter has been well cleaned. This is the way to set the vat.

Setting of the vat

Commonly, river water is used, with which a copper vessel is filled, the one nearest to the vat; the bath is set to boil for three or four hours, after having thrown a bundle of weld into it of about 8 to 10 ^{lb}, half a *setier* of wheat bran, and 8 ^{lb} of madder.[27] The finest is the best. If one can have a used madder bath it will save some madder and it will even produce a better effect. The water is then poured into this vat where two or three bales of woad have been put. Some put it whole, as it is in the bale, others crush it with a sledgehammer; this latter way is the safest. One goes on stirring continually while the hot liquor is being poured into the vat, which is then well covered down and allowed to rest for a good four hours.

Four hours after setting, it is stirred again, and this operation must be done every four hours at most until the colour has started to come, and the vat is ready to receive the footing, that is two shovelfuls of lime for each bale of woad, which is given to feed the vat, and then more or less of it, according to the quality of the woad, and how much lime one finds the vat uses.

[p. 40] Lime is only strewed in after the vat has been stirred. One knows that the vat is ready when it does not give a bubbling sound any longer, and that it bears some blue, which is seen by striking the surface with the flat side of the plank of the rake. It is also known from the bath becoming clearer, getting of a lighter brown than before. Imperceptible veins can then be seen: in such state, the vat can receive such quantity of indigo as one judges appropriate to put into it.

Having added the indigo into the vat, and having filled this up to within six fingers' breadth from the top, one gives it a good stirring as before.

One hour and a half after this operation, it is stirred again, and given some lime if it needs some, and one other hour and a half later, it is stirred once again to dispose it for the first dip.

One checks that the vat is in working condition by dipping a sample into it, leaving it submerged during close to one hour. If it is, the sample should come out looking green, and take up a blue colour, after being exposed to the air for one minute. The experienced dyer does not make this test, he securely judges when the vat is in working state and starts working it.

If during the different times when one uncovers the vat to stir it, one notices that the head has lowered, it means it is not in fine condition; the dyer must then know whether it is engorged, or if it lacks feeding and how to remedy it.

As I have already said, cloths are dipped into the vat without any other preparation than dampening them thoroughly in common, lukewarm water, and then to heave them out and let them drain. This preparation is necessary in order to allow the colour to penetrate more easily into the body of the cloth.

Over and above all the shades of blue that are made in the woad vat, an infinity of other shades are also made in it to serve as grounds for different colours, and this first dyeing is indiscriminately described as woading.

Blues are sometimes intensified or darkened with logwood and brazilwood, putting a little alum into the bath to fix them, but this is a false dye that proves very prejudicial to the colour; it is better to pass them into a bath of cochineal already used.

[p. 41] *Bleu pers en faux*

Persian Blue, false	Woad ground of a dark Royal Blue 2 ^{lb} logwood 2 ^{lb} brazilwood per piece, in a new bath Boiled for one hour before adding ½ ^{lb} alum, heave in when cooled down. Keep it in until the shade you want is reached.
Bleu pers	
Persian blue	Passed into a cochineal bath already used for purples
Bleu turquin	
Turkish blue	Only scoured in the fulling mill and washed in hot water

Bleu de Roy King's blue	Only scoured in the fulling mill and washed in hot water
Bleu tané Tanned blue	Tanned in a reused mordant bath for a flame or a jujube colour; a little young fustic is added when the bath is too exhausted.
Bleu d'azur Azure blue	Scoured in the fulling mill and washed in hot water
Bleu céleste Celestial blue	Scoured in the fulling mill and washed in hot water
Bleu mignon Dainty blue	Ditto
Bleu de lait Milky blue	Ditto
Bleu deblanchy Off-white blue	Ditto

On Scarlet (*Ecarlatte*) [p. 42]

Liquor for scarlet, for 5 pieces broadcloth for the Levant. L. S., L. L., L. P., or *mahoux*,

30 lb soft water from the river in winter, from the well in summer
15 lb aqua-fortis of assured quality
2 lb pure tin from the mines of Cornwall, granulated
¼ lb ammonia salt, powdered

I put all this into a glazed stoneware pot, I let the boiling take place, taking care that the liquor may not be injured, which can happen by the lack of activity of the aqua-fortis, or by its excessive violence. One instantly remedies to these two inconveniences by increasing the dose at once, either of aqua-fortis or of soft water.

The liquor being successfully prepared with the doses prescribed as above, half an hour after this first operation, I graft it with one pound of aqua-fortis, stirring it with a rod; I come back to the same operation half an hour later, and I finally let it rest until the next day.

This mode of preparing liquor for scarlet has always given me very good liquors which I have been satisfied with; however there are seasons in the year, when it is necessary to change the proceeding. One will find my observations on this point further down.

After resting for twelve hours, if the liquor is of a beautiful lemon colour, and clear, it is ready for use.

The boiling for scarlet[28]

Fill up the vessel, made either of tin or of brass, with clear water to within one *palm*[29] from the top, make a brisk fire until it is about to boil, put in:

1 ½ oz cochineal
5 ½ lb tin liquor as above
2 lb white tartar, or tartar crystal — per piece of Londrin Second, Londrin Premier or Londres Large
Bran or starch if you are afraid of your water

This boiling is usually composed for 5 or 6 ends of L. S., L. P., L. L. or mahoux, which weigh from 25 to 30 lb each. One must increase the quantity of drugs in proportion to the cloth weight.

[p. 43] Stir the bath in all directions tirelessly before heaving in, and when heaving the cloth in, let the reel be turned rapidly until the goods have had two or three turns, after which let the reel be turned slowly to facilitate the boiling, which you must keep up for two hours, or at least one and a half continuously; this operation is in great part essential to the colouring. Lift the boiled cloth onto a horse, air it well, and wash in the river before the finishing.

The finishing bath for scarlet[30]

Empty the vessel in which the boiling has been done, let it be filled with clear water and put on a brisk fire. Use starch or bran again if you have bad waters, or better, sour broth, and, as soon as the bath starts boiling, let the bran be scummed clean off, and put in 1 ¾ ^{lb} mesteck cochineal in very fine powder, sifted through the sieve, being understood that you will deduce 1 ounce and a half that has been put into the boiling. The cochineal will at once form a kind of brown crust on the surface of the water, and as soon as you notice that it gets cracked, you will throw in 4 ^{lb} liquor per end, and will stir very well, to allow everything to be well blended together.

In this way you will avoid uneven dyeing, heat-wrinkles, and other inconvenience that may occur. Take care to have a good fire going, enter the cloth and let it be turned rapidly, until you judge the colour to be uniform, then turn gradually more and more slowly, and let it boil for about a quarter of an hour. Heave out, air, and let it be streamed before taking the girtweb off the lists[31].

[p. 44] *Reuse of the finishing bath*

To draw profit from the already used finishing bath for scarlet, it can be used as the boiling for flame colours, and even for other scarlets or crimsons, after which it can be reused also as boiling for the jujube, spiny lobster, pomegranate flower colours, and for oranges. Instead of using it as the boiling bath for the colours just mentioned, it can also be used as the finishing bath for the winesoups, vetch flower, morello cherry colours, the purples, violets, linen grey, mauves, etc.; after these colours, one can reuse it to make peach flower, rose, flesh colour; and eventually, after these, plum colours, pearl greys, and other light colours, the making of which I shall explain here after.

The boiling for flame colour

Choose some good young fustic, cut very fine, take 4 ^{lb} of it that you will put into a cloth bag, and let it boil for one hour and a half or two in the bath described above. Lift out the said bag and put in

5 ^{lb} tin liquor
2 ^{lb} white tartar, or tartar crystal per end

Stir well, turn the cloth rapidly and keep boiling constantly for two hours, heave out and proceed to the same operation as for the boiling for scarlet.

Also reuse this boiling for the colours mentioned above if you have the opportunity, excepting the crimsons, wine soups, and other similar colours.

I shall not deal any more on how to profit from the runs of the boiling and finishing liquors for scarlets, what I have just said equally applying to similar occasions, with only minor differences in circumstances.

The finishing bath for flame colour

1 ¾ ^{lb} fine cochineal
4 ^{lb} tin liquor per end; when the finishing bath you have used for the boiling is good, you can use ¼ of cochineal less in the finishing; otherwise, follow all that I have said about the finishing bath for scarlet.

[p. 45] There are some dyers who, before putting in the ingredients for the finishing liquor, boil a little young fustic in a bag in it, for about one ¼ of an hour. This must be observed for all finishing baths in which these ingredients are used.

Moreover, one must never undertake the making of these colours, of whatever shade, in a brass vessel, without having scrubbed it perfectly clean; this precaution is most essential to avoid staining and getting a dull colour.

Ecarlatte
Scarlet — Boiling as in the above described process and finishing the same

Ecarlatte claire
Light scarlet — Boiling as in the above described process
1 ½ ^{lb} fine cochineal in the finishing

Couleur de feu
Flame colour — Boiling as in the above described process
1 ½ lb fine cochineal in the finishing

Fleur de grenade
Pomegranate flower — 6 lb young fustic in the boiling, simmered for 1 h ½
5 lb tin liquor
2 lb tartar — per end
1 ¼ lb fine cochineal in the finishing
4 lb tin liquor

Jujube — 6 lb young fustic in the boiling, simmered for 1 h ½
3 lb tin liquor
2 lb white tartar — per end, for 1 ½ hour
1 lb fine cochineal in the finishing
2 lb tin liquor

Langouste [p. 46]
Spiny lobster — Same boiling as for jujubes here above
½ lb fine cochineal in the finishing
2 lb tin liquor

Orange — Same boiling as for jujubes, deduct 1 lb tartar
2 oz fine cochineal in the finishing
2 lb tin liquor

Abricot
Pale orange or apricot — 5 lb young fustic in the boiling
1 oz cochineal in the finishing

Gold Colours (*couleurs d'or*), Cassie,[32] Daffodils (*jonquilles*) etc.

Gold, *cassie*, daffodil, and chamois colours are made after the boiling of scarlets, by putting the necessary ingredients into the same bath.

After the boiling of scarlets, wet the undyed cloth.

Couleur d'or
Gold colour — 7 lb young fustic simmered for 2 hours
The bath cooled down
5 lb tin liquor
a trace of madder — per end

Cassie
Mimosa — After a boiling of scarlet, the cloths thoroughly wet
6 lb young fustic simmered for 3 hours
The bath cooled down
6 lb tin liquor — per end
2 oz cochineal for the whole lot

Jonquille [p. 47]
Daffodil — After a boiling of scarlet, the cloths wet
6 lb young fustic simmered for 2 hours
The bath cooled down
4 lb tin liquor
1 lb white tartar — per end

Chamois	After any boiling with young fustic, the cloths wet, the bath cooled down 1 ½ ^{lb} tin liquor per end. Observe to use a boiling made reusing a finishing bath for scarlets
Biche Doe	Wet the cloths, and pass them in an already used boiling liquor for scarlets. Let them boil for ½ hour
Caffé au lait Coffee with milk	After a boiling for scarlet made from a finishing bath, the cloths wet 1 ^{lb} old young fustic simmered for ¾ of an hour ¼ ^{lb} madder extracted separately 1 ^{lb} tin liquor per end The bath cooled down Move rapidly over the reel and heave out when on the verge of boiling One can pass a second lot in the same bath if one wishes, following the same procedure[33]
Chocolat au lait Chocolate with milk	After the above described 1 ^{lb} young fustic simmered for ½ an hour ½ ^{lb} tartar 1/8 ^{lb} cochineal per end ¼ ^{lb} madder

[p. 48] **On Crimsons (*cramoisis*) and Wine Soups (*soupevins*)**

The same operations are done for crimsons, and wine soups as for scarlets – in the boiling as well as in the finishing baths – with the exception of the wood of young fustic which is not used for these two colours.

Immediately after the finishing, and washing, the webbing has to be taken off to avoid staining during the saddening. One usually does the saddening of these colours in a vessel filled with a fresh bath heated until one cannot put one's hand in but not boiling yet, with Rome alum. Before giving the quantity, one must remark that some waters have a more saddening effect than others, according to their degree of strength or heat. I know some waters that, without adding any ingredient, give dark winesoups.

To dye crimson, as soon as the bath is at the right heat, put 1 ½ lb alum of Rome into the vessel, and after stirring well, heave in your cloth already dyed scarlet, move fairly rapidly until you get the shade you want, and if this quantity of alum is not enough, add some until you are satisfied.

Wine soup must be treated in the same way in all operations, increasing the dose of alum that goes up to about 5 lb according to the degree of saddening that is desired. Wine soup is washed clean in the stocks.

This colour can also be saddened with orchil, and lees ash. These saddening ingredients always make a more beautiful effect.

One will see very beautiful wine soups in the "Annotations" that follow this Memoir; they were boiled with alum, finished without tin liquor and saddened with these ingredients.

Cramoisi Crimson	Boiled in the boiling for scarlet 1 ¾ ^{lb} fine cochineal in the finishing 4 ^{lb} tin liquor 1 ½ ^{lb} alum of Rome in the saddening
[p. 49] *Soupevin* Wine soup	Boiled in the boiling for scarlet, also finished as for scarlet 1 ¾ ^{lb} cochineal 5 ^{lb} calcined alum in the saddening, in several turns.

The finishing baths of winesoups are commonly reused as boiling baths for vetch flower (*vessinats*) and morello cherry (*griotte*) colours, purples, violets etc. with:

5 ^{lb} alum of Rome
2 ^{lb} red tartar — per end, well boiled for 3 hours

Observe, after heaving the cloths out of the boiling, cooling and airing them, to let them be spread on a platform, where you leave them for six or seven days in winter, and two days in summer, unless you are in a hurry to do the finishing, and before proceeding to this operation let them be scoured in the fulling mill, or at least let them be streamed carefully twice.

Vessinat Vetch flower	1 ^{lb} fine cochineal in the finishing 1 ^{lb} red tartar. Turn rapidly as in the finishing for scarlet Let boil for one hour
Griotte Morello cherry	¾ ^{lb} fine cochineal in the finishing 1 ^{lb} red tartar

Cherries (*cerises*)

Cherries are made in an already used finishing bath for scarlet by boiling them with 3 ^{lb} tin liquor and 1 ^{lb} white tartar per end.
They are then finished with 2 ^{lb} tin liquor and ½ ^{lb} cochineal per end. They are also made with a freshly made bath.

[p. 50]

Cerise Cherry	Finishing for cherry in a fresh bath 5 ^{lb} tin liquor 1 ^{lb} white tartar ½ ^{lb} cochineal	per end, let boil for a ½ hour
Rose	After a finishing bath for cherries or other cochineal dye The undyed cloth soaked 1 ^{lb} tin liquor ¼ ^{lb} cochineal 1 ^{lb} tartar	per end, let boil for ¾ of an hour in the finishing
Chair Flesh	This colour is made after a finishing, by throwing out some of the bath, and letting boil only for a very short time It can also be done after violets, with a little tin liquor	
Chair pâle Pale flesh	After the above, adding 2 ^{lb} tin liquor Or after a bath of cochineal, adding enough water	
Gris de prince Prince's grey	After any wine colour The bath cooled down, then gall with one ounce gallnuts and sadden very slightly	

Purples (*pourpres*), Violets, Mauves, Lilacs (*lillas*) etc.

I have said that purples, and violets, together with vetch flower and morello cherry colours, could be boiled either in an already used finishing bath for scarlet or in a fresh bath, with 5 ^{lb} alum of Rome, and 2 ^{lb} red tartar during three hours; I add that purples and violets must have previously received the woad ground necessary to match the requested shade. These colours having been let to rest, the finishing is usually done in an already used bath of cochineal.

[p. 51] *Pourpre*
Purple
Woad ground of a light azure blue
1 ¼ lb cochineal in the finishing
1 lb red tartar
Turn the cloth briskly like in the finishing bath for scarlet, let boil for 1 hour

Violet
Woad ground of an azure blue
1 lb cochineal in the finishing
1 lb red tartar
Same as above

Mauves, lilacs, dove breast, linen greys are boiled after violets, with
3 ½ lb alum of Rome
1 lb red tartar per end during one hour, wash at the fulling mill

Lilla
Lilac
Woad ground of a sky blue
½ lb cochineal in the finishing, can be lessened if the finishing bath reused is good
1 lb tartar, let boil ½ an hour in the finishing

Gorge de pigeon
Dove breast
Woad ground of a sky blue
¼ lb cochineal in the finishing, same remark
1 lb tartar. Let boil

Mauve
Woad ground of a dainty blue
½ lb cochineal in the finishing, same
1 lb tartar. Let boil

[p. 52] *Gris de lin*
Linen grey
Woad ground of a milky blue
2 oz cochineal or suppress it altogether if the finishing bath reused is still rich in dye
1 lb red tartar. Let boil

Fleur de pêcher
Peach blossom
2 oz cochineal
1 lb tartar crystal
1 lb tin liquor in a finishing bath reused for purples, vetch flower, morello cherry colours, etc.
Let boil for ½ an hour

These small colours have more or less body, depending on the opportunities one has to make them, and on the woad ground.

Madder Reds (*rouges de garance*)

When one has the opportunity of boiling madders in a finishing bath for purples or violets, or even another dye with cochineal, they assuredly are more beautiful. Sometimes they are also boiled together with these colours.
5 lb Rome alum
2 lb red tartar per end during 3 hours. Let rest for 3 days

Let the red copper vessel be filled up to one span ½ from the top,[34] and as soon as the bath is lukewarm, put in at least 8 lb fine, or unstripped, madder, that you will leave to brew for some time, and as soon as it is fairly hot without however boiling, put in one pound of black gallnuts pounded and sieved, let the liquor be well stirred. Your cloths having been scoured in the fulling mill, heave them in, moving fairly rapidly over the reel, then turn gently until it starts boiling. Then heave out, air, and let them be washed.

Garance
Madder

Mr. Albert in his method follows the same procedure, with the difference that he puts 4 lb vinegar instead of gallnuts;[35] others put 1 lb tin liquor for scarlet. This acid gives a much brighter effect, but it makes it more difficult to get a uniform colour [p. 53]

Garance cramoisillé
Crimsoned madder

Same procedure as for madders
Simmer 1 lb fernambuco wood in a fresh bath during 1 hour. Let the bath cool down and pass the goods for madders through it until they reach the shade you want; do this operation in the madder dye bath itself, you will be all the better for it

Rouge bruni
Saddened red

The operations are the same with the difference that you save 2 lb madder
Strew a little copperas into the same bath, stir well and turn the cloth in it; one increases the dose until the desired shade is reached, observing to shut the door of the furnace; the same procedure must be followed for all saddenings
If worked when too hot, the colour is not uniform

Dark reds for blacks are saddened with logwood in several dippings[36]

Couleur de Roy
King's colour

Woad ground of a sky blue
Boiled like the colours above and let to rest
6 lb madder in the finishing
1 lb gallnuts
Saddened with copperas

Couleur de Roy sans bruniture
King's Colour without saddening

Woad ground of an azure blue
6 lb common madder
1 ½ lb fernambuco
¼ lb gallnuts

per end. Let boil for ½ an hour in the finishing bath. Wash in the river without saddening

Maure doré
Golden Moor[37]

Boiled in a strong boiling bath like madders [p. 54]
Rested and washed in the stocks
6 lb madder in the finishing
etc.
After boiling and before dyeing in madder, these colours are given a weld ground of golden yellow

Maure doré
Golden Moor

Same operation as for the above colours
Saddened with copperas in several dips

Cinnamons (*cannelles*) are usually made in the dye bath of madder reds already used, by boiling them as follows:

4 lb alum from the Levant
2 lb red tartar — after boiling for two hours they are washed at the fulling mill, and dyed with weld as for lemons or golden yellow.

Cannelle doré
Golden cinnamon — 3 lb fine madder in the finishing bath
2 lb common ditto
½ lb powdered gallnut

Cannelle rougeâtre
Reddish cinnamon — Same process as here above. Only add one pound fine madder and give a less strong welding

Canelle brûlé
Burnt Cinnamon — Same process as here above. Deduct one pound madder and sadden as for saddened reds

[p. 55] *Cire doré*
Golden wax — Boiled and welded like cinnamons and reusing a bath of these colours for broadcloth
2 lb madder, pass the cloths in it without letting boil

Yellows (*jaunes*)

The boiling for all kinds of yellows from weld:
4 lb alum from the Levant
2 lb red tartar — per end for 2 hours

About 16 lb of weld per end are put into a vessel, observing to put a small amount of quick lime, with ashes, in the bottom, on the first bundle of weld; this is what makes it give out its sap more easily. Having put the sufficient quantity of weld into the vessel for the cloths that have to be dyed, it is pressed down tightly with iron bars, specifically used to prevent the weld from floating up, which would make it impossible to move the cloth easily. The vessel is then filled up to the top, and as soon as it starts boiling, the lemons (*limons*), yellows (*jaunes*), cinnamons (*cannelles*), *festikis* etc. are passed in it. These cloths are passed one after the other, giving each one 8 to 10 turns over the reel. One should observe to add water for each cloth that is heaved in and to keep the fire going briskly, so that the bath never stops boiling.

All kinds of yellows are also made in a weld bath for greens, when one has to make both colours.

It is the same operation, as it will be seen in the section on greens.

[p. 56] *Limon*
Lemon — This colour is made with 12 lb weld, and it is always better made, when the welding bath is prepared with flax-leaved daphne

Jaune
Yellow — 16 or 17 lb weld, no flax-leaved daphne

Jaune doré
Golden yellow — After welding, pass the cloth during one hour through a fresh bath made with ¾ lb madder, without boiling however

Jaune tané
Tanned yellow — Same procedure as here above; dull it somewhat, with a little soot or pass it through a bath where you have made olives, if you have the opportunity

Feuille morte
Dead leaf — 12 or 15 lb weld are enough; they can also be made in a reused bath for cinnamons by strewing 3 or 4 lb soot into the bath. Pass them during ½ an hour without boiling

Festiqui[38]	12 lb weld in a reused bath of cinnamon or olives, strew in 7 to 8 lb soot that must boil for a good ½ hour. Pass the cloths in it for ¾ of an hour without boiling

Greens (*verts*) with weld [p. 57]

After the cloths have received the necessary woad ground, and have been well washed at the stocks, they are boiled as follows.

4 lb alum from the Levant or from Spain
2 lb red tartar per end for 2 hours

The same operation is done to prepare the dye bath with weld, as for the yellows, in all respects; with the difference that here, about 25 lb weld are put per end, and that half that quantity of flax-leaved daphne can be used, deducting as much weld, observing however to do it only when the flax-leaved daphne is in its full strength, and when its being available locally makes it cheaper than weld.

The bath being ready and beginning to boil, the duck greens are heaved in first, giving each one from 8 to 10 turns over the reel; then the dark grass greens; then the grass greens; emerald; parrots; yellows, etc. if one has some to make, and light greens. Then the pistachio and apple greens are passed through; these must be boiled with only half the quantities of ingredients as for the other greens and within one hour; they are only given a few turns in the welding, and in the end they are passed again through it to unify the colour.

Then are welded the olives, if one has the opportunity to make some, and immediately afterwards greens are passed again, starting with the weakest in weld, giving them a few turns more; always observing to add water, as I have already said in the section on yellows, and keeping a brisk fire. At the end of the welding, if one has cabbage greens, sea greens, bronze greens, and woaded celadons to make, they are passed through, giving them 8 to 10 turns as a start, and as much when passing them again. These colours must not be boiled; the cloths must even be kept in a place where there are no boiled ones.

Verd noir[39] Black green	Woad ground of a King's blue Black greens and dark greens are saddened with logwood or copperas, by putting 2 lb logwood per end, and a bundle of flax-leaved daphne into the furnace, and as soon as the bath gets warm so that one can keep one's hand in it, greens are passed through it until the desired shade is obtained. This saddening is called "music" (*musique*), it must be done with caution, for fear of not getting a uniform colour	[p. 58]
Verd brun Dark green		
Verd obscur Obscure green	Woad ground of a Turkish blue	
Verd de canard Duck green	Woad ground of a high King's blue	
Verd d'herbe remply Full-bodied grass green	Woad ground of a King's blue	
Vert d'herbe Grass green	Woad ground of a dark azure blue	
Emeraude Emerald	Woad ground of an azure blue	[p. 59]
Perroquet Parrot	Woad ground of a sky blue	

Vert clair Light green	Woad ground of a dainty blue
Vert gay Gay green	Woad ground of a milky blue
Vert naissant Nascent green	Woad ground of an off-white blue

I have said that pistachio greens must be boiled only with half he quantities for a boiling, washed in the fulling mill, and passed in the welding bath after the greens here above.

[p. 60]

Verd de pistache Pistachio green	Woad ground of a sky blue
Verd de pomme Apple green	Woad ground of an off-white blue

I have said somewhere else that cabbage greens, sea green, bronze green, celadons etc. must not be boiled, and must be made at the end of the welding.

Verd de chou Cabbage green	Woad ground of a dark azure blue
Vert de mer Sea green	Woad ground of an azure blue
Verd bronzé Bronze green	Woad ground of a King's blue And saddened with logwood
Verd céladon Celadon green	Woad ground of a milky blue

Shades of greens are the same as those of blue, which vary infinitely according to the woad ground and the welding they are given.

[p. 61] **Olives**

Olives are made in a fresh bath but more commonly in a reused bath of cinnamons, or any other dye bath with little madder, as follows, the cloth having received the woad ground and the welding necessary for the shade that is desired.

Put 2 or 3 [lb] soot, and one pound common madder into the vessel.

Let it boil well for one hour, heave in the cloth, and turn it through for another hour, heave out, then sadden with copperas, increasing the dose until the shade you want is reached. These colours must be washed in the river and scoured in the fulling mill.

Olive brun Dark olive	Woad ground as for a dark parrot green Saddened with copperas
Olive pourrie Rotten olive	Woad ground as for a light green Saddened with less copperas

Olive	Woad ground as for parrot or sky blue, saddened etc.
Olive verte Green olive	Woad ground as for a light green
Olive claire Light olive	Woad ground as for a gay green [p. 62]
Olive sèche Dry olive	Woad ground as for a gay green

There undoubtedly are many other shades of this colour, from the strongest I give until the last. The woad ground makes the difference, with the saddening.

On Prussian Blue (*bleu de Prusse*) and Saxon Green (*vert de Saxe*)

The composition for Prussian blue and for Saxon green being the same, I announce the two colours together. Saxon green will however be dealt upon in a separate section.

The necessary quantity of oil of vitriol is put into a well fired faïence pot, a glass bottle, or even a stoneware pot well glazed inside, observing that it should be of a good quality. To make sure of this, see what I say about it at folio 24.[40] Put this pot near a moderate fire, and as soon as it is warm strew in 2 ounces of powdered indigo *flor* or of any other superior quality per pound of oil of vitriol, mark weight. Stir this liquor well in the pot, from time to time, with a little stick, airing it – to prevent the pot from bursting. Take care to do this for three quarters of an hour, or just under one hour, after which time the composition should be perfected. Stopper it now, and use it after two hours, the next day, or within the next 15 days: it will be equally good.

It must be observed that the indigo is reduced by half in the solvent, and that one must calculate to make neither too much nor too little of this chemick[41] than needed, because these colours are not often requested, particularly today.

For instance, if it is 5 pieces of Londrins Seconds that you have to dye, 3 ½ lb oil of vitriol and 9 ounces indigo will give you 3 ¾ lb of chemick, enough to dye your 5 ends blue. But since it is better to waste ¼ of chemick than not have enough of it when needed, my advice is to make a little more of it.

The boiling for Prussian Blue[42]

[p. 63]

4 lb Rome alum
2 lb white tartar — per end for 2 hours. Wash in the stocks

Throw out one quarter of this boiling bath, cool it down to the point where your hand can stay in it, put in ¾ lb of this chemick per end, turn the goods rapidly and heave them out within a quarter of an hour. If you leave longer in this colour, it gets dull. If during this time you see that the colour does not come up, heave out onto the reel and strew a little “flour” into the bath (this is calcined powdered alum) and pass the cloth in it again for two or three turns.

Prussian blues are often made without any boiling, but it must only be done in reused baths for such blues (this procedure will never be successful with a fresh bath). Wet the cloth in the same bath, put one pound of chemick at once, and 1 ½ lb flour, and heave out within a quarter of an hour. This colour made in this way looks fresher.

Bleu de Prusse Prussian blue	Boiled and ¾ lb chemick Alum helps to the success of this colour and gives it a slightly green hue
Bleu de Prusse Prussian Blue	Without boiling. 1 lb chemick in the warm dye bath Only 3 pieces can be done at a time, in order to get a uniform colour

Sometimes, darker or lighter Prussian blues are required, one just needs to increase or diminish the dose of chemick.

[p. 64] **Saxon Greens**

Prussian blue passed through a concentrated bath of old fustic results in Saxon green. This is how this beautiful colour used to be made at first. Nowadays it is done in a shorter, less expensive and more secure way.

The boiling for Saxon Green

4 ^{lb} Rome alum
2 ^{lb} white tartar — per end during 2 hours, wash in the stocks

Put from 17 to 18 ^{lb} chipped old fustic in a bag into the same bath, let it boil for 1½ hour. Take it out, cool the bath down to the point where your hand can stay in it. Put in 1 ^{lb} ¼ chemick per end, the same whose recipe I have just given in connection with Prussian blues, let the goods be turned rapidly, then slowly, and heave out when it is about to boil. Mr. Albert puts the dose of chemick in two times, that is to say, he puts 2/3 of it at first and after two or three turns, he lets the goods be heaved out and he puts the remaining 1/3. I have always practised it in the same manner, and I consider this way to be the best because it gives a more uniform colour. But however you make it, if the colour does not come up, resort to the use of "flour", following what I have said about Prussian blues.

A lazy and hurried dyer will make Saxon greens without any boiling, but he will always make a bad colour that will have even less fastness.

Vert de Saxe Saxon green	1 ¼ ^{lb} chemick in 2 dips. When one has several lots of cloth to do, one can re-use the same bath, reducing the quantity of old fustic by 4 to 5 ^{lb}, depending on its quality
Vert de Saxe Saxon green	¾ ^{lb} chemick, ditto
Verd de pomme[43] Apple green	
[p. 65] *Vert de pomme Saxe* Saxon apple green	Boiled with the greens here above Throw out 1/3 or ½ of the bath in which you have greened, let it cool down, heave in, turning rapidly until the bath starts boiling, let it boil for a ¼ of an hour
Céladon Saxe Saxon celadon	Boil in a reused dye bath for greens without any addition; in the same bath, ¾ ^{lb} old fustic simmered, a little chemick. Heave in when the bath is quite cool
Bleu de blanchi Saxe Saxon off-white blue	Boil for ¾ hour in a bath of Prussian blues without adding anything, and if the bath is not concentrated enough, add just a little chemick. The bath quite cool

Celadons with vitriol of Cyprus and verdigris

Although this colour is not in demand any longer today, and has no fastness whatsoever, I did not think I should proscribe it from this memoir.

Before undertaking the celadons with vitriol, provide yourself with two new nets made with thin strings, that you will scour in a boiling for scarlet before using them.

The vessel being well scrubbed, let it be filled up to one and a half span from the top[44], and as soon as it is warm to the point where your hand can stay in it, put in 2 ½ ^{lb} of first quality white soap per end, that will have been cut into small pieces, and put into a cauldron with river water, that is heated a little to get it well dissolved, during eight to nine hours.

Then put it into the vessel, let it be well stirred, and as soon as it starts making a lot of foam and it is about to
[p. 66] boil, put in one of the nets I have mentioned above, and which must only be used for this operation, which is called soaping. This net is used in the vessel to prevent the cloth from getting in contact with it, which would produce big reddish stains on it.

As soon as this soapy bath starts boiling, your cloth being scoured in the stocks, and washed in hot water to allow it to be evenly wet throughout, heave it in and leave it for one hour. While you are doing this soaping, have another vessel prepared to do the greening in it; you will have it filled up only to one *palm* and a half from the top. The soaping being finished, let the cloth be carried to the vessel prepared for the greening, which must be composed of 2 ½ [lb] vitriol of Cyprus, pounded and dissolved in a separate cauldron, and then poured into the vessel. Stir well, put in the second net that must only be used for this operation, heave in the cloth, turning rapidly during half or three quarters of an hour, observing to keep the bath just hot.

Then heave out, and let it be streamed.

Céladon
Celadon — With vitriol of Cyprus and soap

Celadons with Vitriol without soap

Fill the vessel with a fresh bath; as soon as it is lukewarm, put in two bucketfuls of a weld bath with 3 [lb] of vitriol of Cyprus that will have been well dissolved separately in a cauldron and move the goods in for one hour, in other respects observing all that I have just said here above.

Céladon
Celadon — With vitriol without soap

Celadons with verdigris [p. 67]

Dissolve 4 [lb] verdigris in hot water in a separate cauldron and pour it into the vessel as soon as the bath is warm. Stir well, and move the cloth in it during one hour, do not have it washed in the stocks at any other stage than before the operation, to get them thoroughly wet.

Céladon
Celadon — With verdigris

Céladon
Celadon — With blue vitriol

On tawny colours (*fauves*) or colours with sanderswood, and other browns (*marrons*)

Put sanders and *redoul*[45] and gallnuts into the furnace, all according to the shade you want to make.

Let it all boil during one hour, heave in the cloths that you must then leave to boil for another hour.

Then sadden them with copperas. The bath is to be let to cool down before heaving in.

Marron
Brown
3 [lb] sanders
5 [lb] *redoul*
½ [lb] gallnut — per end
Saddened with copperas

Marron plus clair
Lighter brown
3 [lb] sanders
3 [lb] *redoul*
½ [lb] gallnut — per end
Saddened with copperas

Marron rougeâtre [p. 68]
Reddish brown
1 [lb] fernambuco wood
4 [lb] sanders
1 [lb] *redoul*
½ [lb] gallnut — per end
Saddened with very little copperas

Gerofle[46] *rempli* Full-bodied clove	4 lb sanders 6 lb *redoul* ½ lb gallnut Saddened with copperas in several dippings	per end. Some clove colours are made with madder and weld, saddened with logwood
Musq[47] Musk	2 lb sanders 8 lb *redoul* ½ lb gallnut	per end. This colour needs some logwood in the galling bath; put 2 lb of it per end

Plums (*prunes*)

Plum colours are commonly made in a fresh bath, but you will always get best results by reusing a boiling bath for scarlet, such opportunity never fails to arise, and one should wait for it. Consequently, put the red sanders and gallnut into this bath, to make the shade of plum colour you want; let the cloth boil for one hour. If the shade requires some *redoul*, add it into this galling bath, and first pass the cloth in this bath for another hour before saddening with copperas.

The dark plum colours, the crimsoned plums and the plums tending towards brown, are made in the same way, by putting as much common madder as red sanders in the galling liquor.

The plum colours of a reddish tinge or with cochineal are made in a totally different way because they must be boiled and then dyed with cochineal. I shall deal with them afterwards.

Prune Plum	2 lb sanders 1 lb *redoul* ½ lb gallnut Saddened with copperas	per end in a boiling bath for scarlet
[p. 69] *Prune clair* Light plum	1 lb sanders 1 lb *redoul* ½ lb gallnut Saddened with copperas	per end in a boiling bath for scarlet
Prune grisâtre Greyish plum	2 lb sanders 2 lb *redoul* 1 lb tartar Saddened in the same way as above	per end in a fresh bath
Prune cramoisillé Crimsoned plum	1 ½ lb sanders 1 ½ lb common madder 1 lb *redoul* Saddened ditto	per end in a boiling for scarlet

Plum colours with cochineal

This colour is boiled with greens, yellows, cinnamons etc. according to opportunities, and failing these, in a fresh bath with

4 lb Rome alum 2 lb red tartar	per end during 2 hours. Wash in the stocks

Plum with cochineal (*prune à la cochenille*) in a reused madder bath

¼ lb cochineal
½ lb gallnut
¼ lb brazilwood — per end. Boiled for a ½ hour in the bath

Take part of this bath, put ½ lb madder to infuse in it per end and when it is dissolved put it into the bath with the cochineal

And use it once cooled down, like for madder reds. Let them boil for one hour, and sadden them with copperas, three at a time to get the colour uniform.

By only increasing or diminishing the quantity of cochineal, you will get a darker or lighter shade.

Hazelnut colours (*noisettes*)

[p. 70]

This is a colour that has no fixed shade, each dyer has his own taste to make it, this is why I shall give several manners to make a hazelnut colour, either dark, or light, greenish, or reddish, etc.

Noizette Hazelnut	Wet the cloth in the stocks 1 ¼ lb gallnut ¼ lb old fustic ¼ lb logwood 1 lb common madder 1 lb alum Sadden with copperas	 per end. Let the whole boil for ¼ of an hour per end. Pass the cloth in it for ½ an hour
Noizette Hazelnut too brown[48]	This colour is usually made in a reused bath of other hazelnuts, adding ¼ old fustic, and saddened with soot	
Noisette rougeâtre Reddish hazelnut too red[49]	In a reused madder bath, simmer a little Fernambuco wood 2 lb soot, the cloths are moved in it for half an hour and saddened slightly	
Noisette doux Soft hazelnut	In a reused bath of other hazelnuts, even those above Simmer ¼ lb old fustic, a little soot, and sadden very slightly	
Noisette clair Light hazelnut	Wet the cloth ½ lb *redoul* ¼ lb logwood ¼ olf fustic	 per end. Simmered and boiled in a bag for ½ an hour. Let the bath cool down, heave in and let boil for a ½ hour. Sadden with copperas in several dips
Noisette biche Doe hazelnut	In a reused finishing bath of scarlet Wet the cloths in the undyed state ¼ lb madder 1 ½ lb soot	 per end
Noisette foncé Dark hazelnut	In a reused madder bath, 2 or 3 lb soot and ¼ lb old fustic per end and saddened with copperas	

[p. 71]

(*Noisette verdâtre*)
Greenish hazelnut — In the same reused bath, 2 lb or 3 lb soot and 1 lb old fustic per end
Saddened ditto

Tobaccos and Coffees (*tabacs* and *caffés*)

Tobacco and coffee colours must be boiled and welded like cinnammons, reddened with
4 lb madder
½ lb gallnut
2 lb old fustic — per end and saddened until the desired colour is obtained.
Greenish tobacco requires a light woad ground

Tabac
Tobacco — 3 lb madder. Saddened with soot and copperas

Tabac brun
Dark tobacco — 4 lb madder. Saddened with copperas only

[p. 72] *Tabac doré*
Golden tobacco — 4 lb madder. Saddened with less copperas

Tabac d'Espagne
Spanish tobacco — Dyed with weld and madder like above
Slighly darkened with soot

Tabac verdâtre
Greenish tobacco — Woad ground of a dainty blue (*bleu mignon*)
Same process etc.
Saddened with soot

Coffee colours (*caffés*)

Boil the coffee colours like the greens. Wash in the stocks after the boiling.

4 lb madder per end in the finishing, heave out, and put 5 lb *redoul* in a bag into the same bath, and 1 lb gallnut per end. Let boil for one hour, let the bath cool down, heave in the cloths and move them in the boiling bath for another hour; heave them out, air them, and sadden until the colour is to your taste.

Observe to use the bath for the finishing as for madder reds, that is quite cool; greenish coffee requires a light woad ground and some soot.

Caffé
Coffee

Café verdâtre
Greenish coffee

[p. 73] ### Blacks (*noirs*)

Blacks in the manner of Sedan[50]
Let the cloth be given a woad ground of a strong King's blue. Fill the vessel with a volume of water proportionate to the number of cloth ends you have to dye and put in
7 lb chipped logwood in a bag
1 lb powdered gallnut
3 lb old fustic and the content of 2 sieves of *redoul* — per end
in a bag

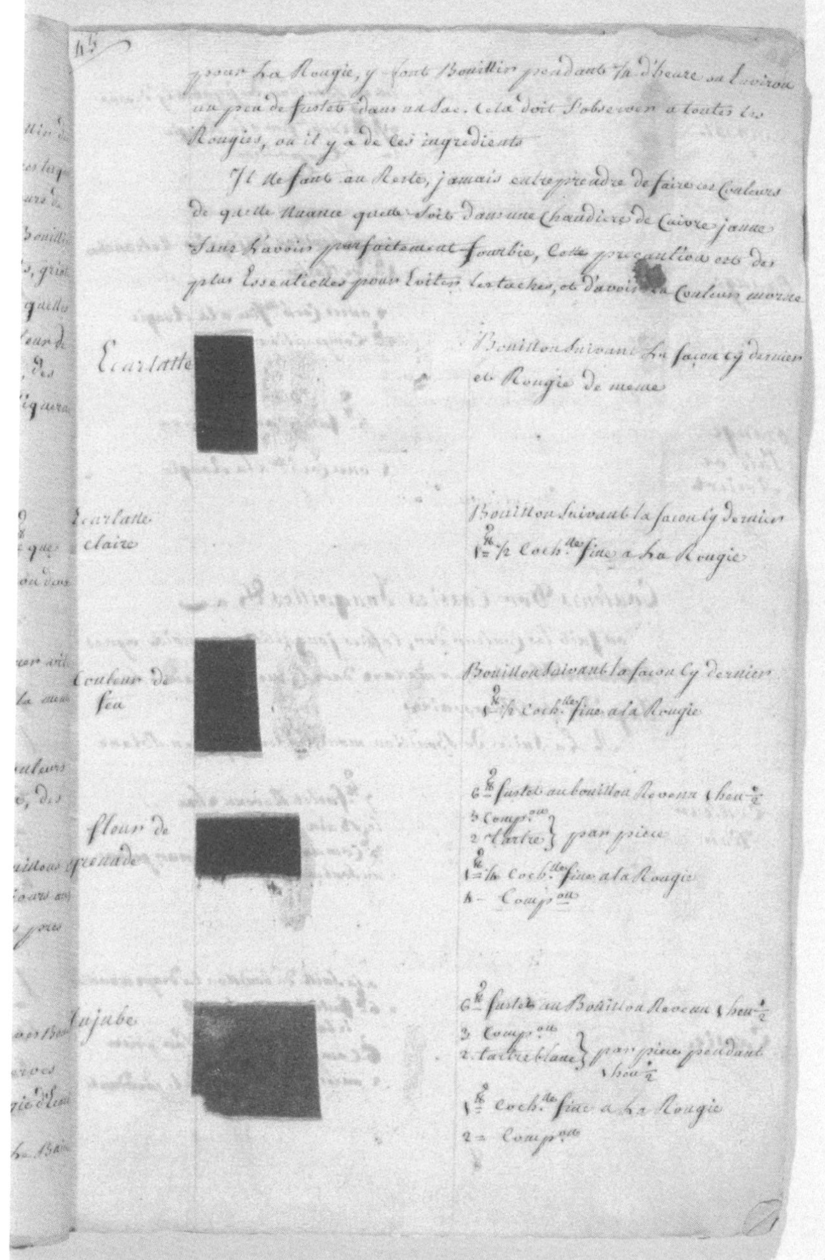

45

pour la Rougie, il faut Bouillir pendant 1/2 d'heure ou environ un peu de fustet dans un sac. Cela doit s'observer a toutes les Rougies, ou il y a de ces ingredients

Il ne faut au Reste, jamais entreprendre de faire ces Couleurs de quelle nuance quelle soit dans une Chaudiere de Cuivre jaune sans l'avoir parfaitement fourbie, Cette precaution est des plus Essentielles pour Eviter les taches, et d'avoir la Couleur unie

Ecarlatte	Bouillon suivant la façon cy dernier et Rougie de meme
Ecarlatte claire	Bouillon suivant la façon cy dernier 1 1/2 lb Coch.lle fine a la Rougie
Couleur de feu	Bouillon suivant la façon cy dernier 1 1/2 lb Coch.lle fine a la Rougie
fleur de grenade	6 lb fustet au bouillon Revenu 1 heu 1/2 3 Comp.tion 2 tartre } par piece 1 1/2 lb Coch.lle fine a la Rougie 4 Comp.tion
Jujube	6 lb fustet au Bouillon Revenu 1 heu 1/2 3 Comp.tion 2 tartre blanc } par piece pendant 1 heu 1/2 1 1/2 lb Coch.lle fine a la Rougie 2 Comp.tion

44

Bleu Pers en faux	fonds de pastel d'un Bleu de Roy foncé 2 lb bois de Campeche } par piece dans un 2 bois de Bresil } Bleu frais Bouilli une heure, avant de mettre dedans l'etoffe pour tromper les etoffes dedans jusques a la nuance que vous souhaites
Bleu Pers	Passé sur un Cochenillage de Violets
Bleu turq	Lavé au foulon seulement et passé sur l'eau chaude
Bleu de Roy	Lavé au foulon seulement et passé sur l'eau chaude
Bleu Pané	Passé sur la suite d'un Bouillon de Couleur de feu ou de jujube, on y ajoute un peu de fustet quand le bain est trop tiré
Bleu d'azur	Lavé au foulon et passé sur l'eau chaude
Bleu Celeste	Lavé au foulon et passé sur l'eau chaude
Bleu mignon	Idem
Bleu de Lait	Idem
Bleu de Blanchy	Idem

46

Langouste — meme Bouillon des jujubes cy dessus
2 1/2 ℔ coch.lle fine a la Rougie
2 — Composition

Orange — meme bouillon des jujubes Retranchés
1 ℔ de tartre
2 onces Coch.lle fine a la Rougie
2 ℔ Composition

Orange Pale ou Abricot — 5 ℔ fustet au Bouillon
1 once Coch.lle a la Rougie

Couleurs Dor Cassies Jonquilles &c.

on fait les Couleur dor, Cassies jonquilles, chamoix apres le Bouillon D'écarlatte, en mettant dans le meme Bain les Ingrédients Necessaires

A La Suite du Bouillon mouillés les draps en Blanc

Couleur D'or — 7 ℔ fustet Revenu 2 heures
le Bain Rafraichi
2 ℔ Composition
un soubçon de garance } par piece

Cassie — a la Suite du bouillon les draps mouillés
6 ℔ fustet revenu 3 heures
le bain rafraichi
6 ℔ Composition } par piece
2 onces coch.lle sur tout le pendant

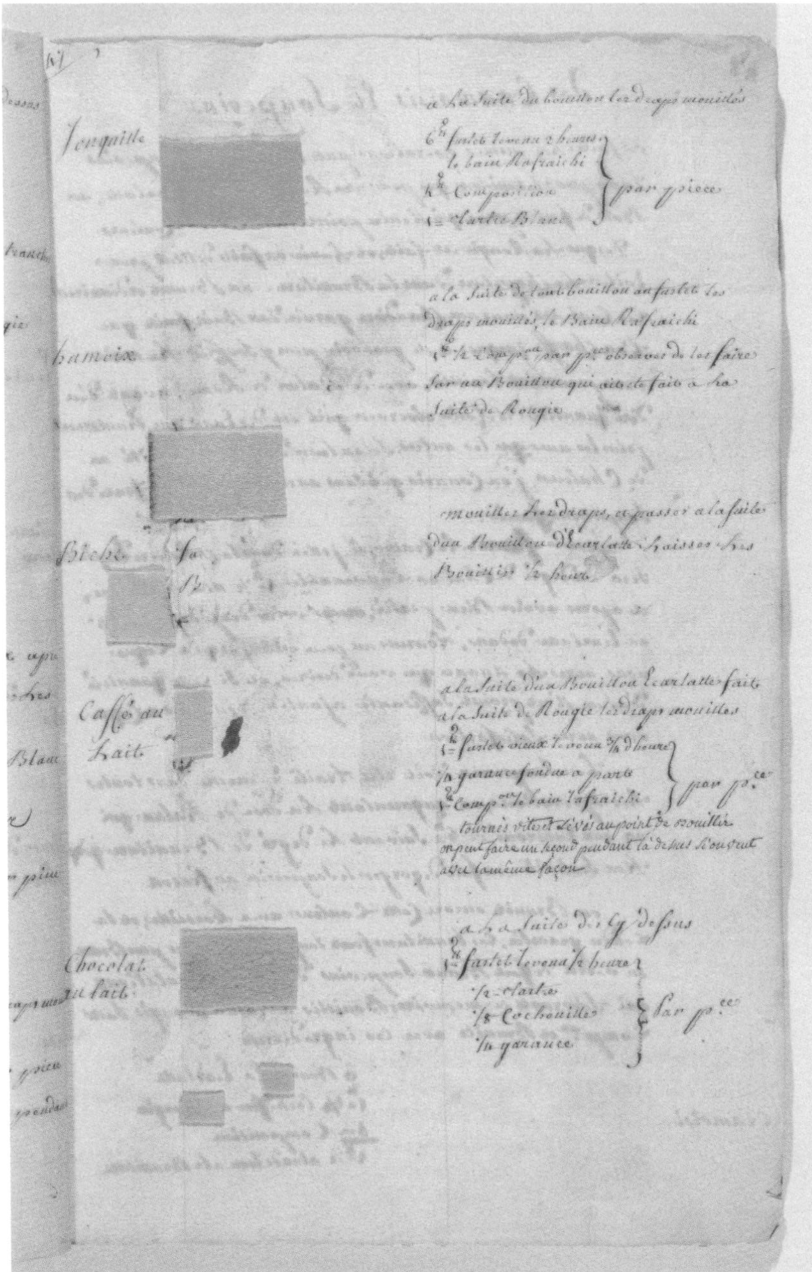

47

Jonquille — a La Suite du bouillon les draps mouillés
6 ℔ fustet revenu 2 heures
le bain Rafraichi
4 ℔ Composition
1 — tartre Blanc } par piece

Chamoix — a la Suite de tout bouillon on fustet les draps mouillés, le Bain Rafraichi
1 ℔ 1/2 Compon par p.ce observer de les faire sur un Bouillon qui ait ete fait a la suite de Rougie

Biche — mouillez les draps, et passer a la suite dun Bouillon D'écarlatte Laissez les Bouillir 1/2 heure

Caffé au Lait — a la suite dun Bouillon Ecarlatte fait a la suite de Rougie les draps mouillés
1 ℔ fustet vieux revenu 3/4 d'heure
1/4 garance fondue a parte
1 ℔ Compon le bain rafraichi } par p.ce
tournés vite et levés au point de bouillir
on peut faire un second pendant la dessus de suite avec la meme façon

Chocolat au lait — a la suite de cy dessus
1 ℔ fustet revenu 1/2 heure
1/2 — tartre
1/8 Cochenille
1/4 garance } par p.ce

118

des Cramoisis Et Soupevins

on fait les memes operations aux Cramoisis, et Soupevins tant pour le bouillon que pour la Rougie qu'aux Ecarlatte, au Bois de fustel pres, qu'il n'entre point dans ces deux Couleurs.

Desque la Rougie est faite, et Lavée on fait deliter pour Eviter la Chapeture dans la Bruniture. on Brunit ordinairement ces Couleurs dans une Chaudiere garnie d'un Bain fraix que l'on chauffe jusques a ne pouvoir plus y Souffrir la main sans pourtant bouillir, avec de l'alun de Rome; avant d'en dire la quantité il faut observer qu'il est des Eaux qui brunissent plus les unes que les autres, selon leur degré de Vivacité ou de Chaleur, j'en connois qui sans aucun ingredient font des Soupevins foncés.

pour faire le Cramoisi, jettés dans la Ch[audie]re lorsqu'elle sera au degré de Chaleur convenable 1 lt ½ a[lun de] Rome, et apres avoir Bien pallié, mettes votre draps de [Cramoi]si teint en Ecarlatte dedans, tournes un peu vite jusqu'a ce que vous aures la nuance que vous desires, et si cette quantité d'alun n'est point suffisante ajoutés en jusques a ce que vous aures satisfaction.

Le Soupevin doit etre traité de meme dans toutes ses operations, en augmentant la dose de l'alun qui va aux Environs de 6 lt suivant le degré de Bruniture qu'on souhaite. on fait degorger le Soupevin au foulon.

on Brunit encore cette Couleur avec l'orseille, et la cendre gravelée; ces bruniturres font toujours beaucoup plus Beau. on verra de fort Beaux Soupevins dans les annotations qui suivront ce memoire, Bouillis a l'alun Rougis sans Compo.on et Brunits avec les ingredients

Cramoisi

Bouilli en Ecarlatte
1 lt ¾ Coch.lle fine a la Rougie
6 lt Composition
1 lt ½ alun de Rome a la Bruniture

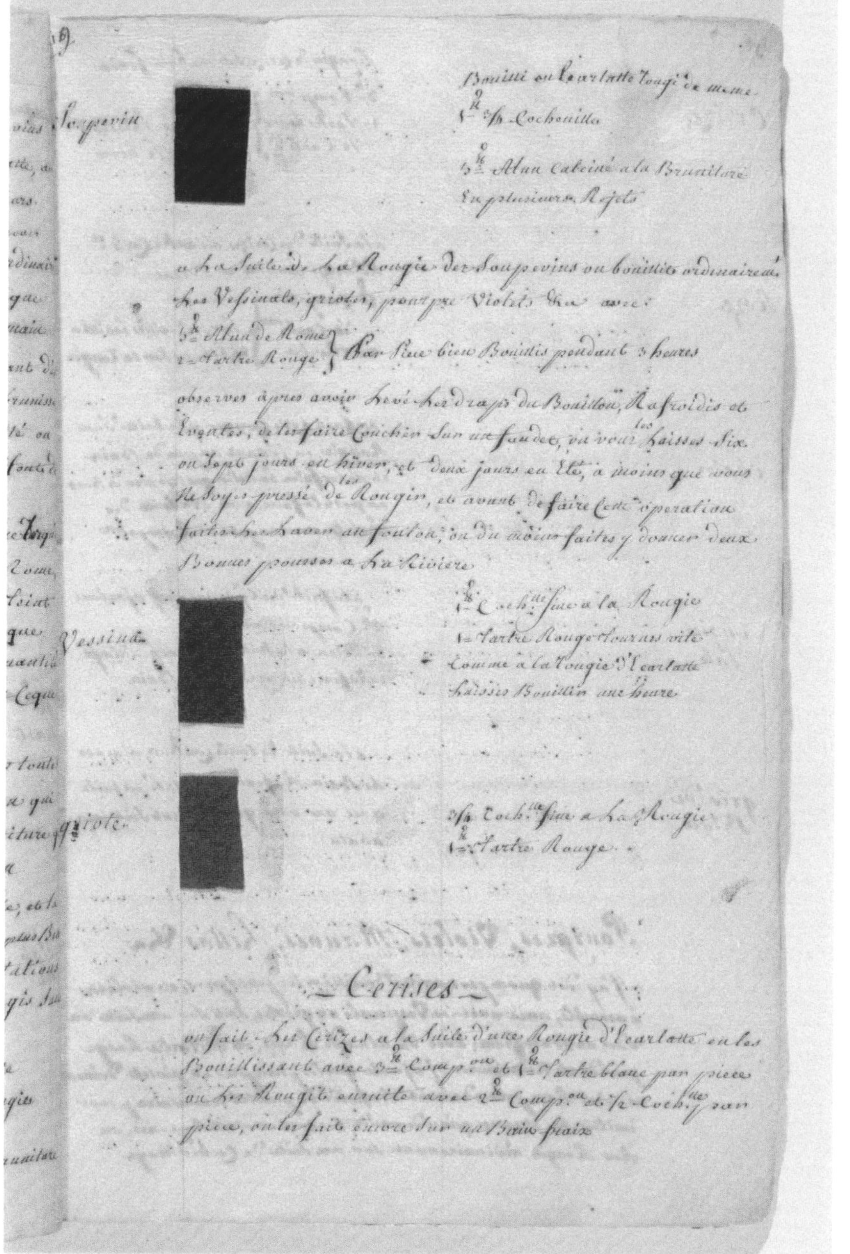

[1]19

Soupevin

Bouilli en Ecarlatte rougi de meme
1 lt ¾ Cochenille
6 lt Alun Calciné a la Bruniture
En plusieurs Rejets

a la suite de la Rougie des Soupevins ou bouillits ordinairement les Vesinals, griotes, pourpre Violets &a avec
6 lt Alun de Rome }
2 tartre Rouge } Par Piece bien Bouillis pendant 3 heures

observes apres avoir levé les draps du Bouillon, Rafroidis et Eventes, de les faire Coucher sur un fondet, ou vous les laisses six ou sept jours en hiver, et deux jours en Eté, a moins que vous ne soyes pressé de les Rougir, et avant de faire cette operation faites les Laver au foulon, ou du moins faites y donner deux Bonnes pousses a la Riviere

Vessinat

1 lt Coch.lle fine a la Rougie
1 tartre Rouge tournes vite
Comme a la Rougie d'Ecarlatte
Laisses Bouillir une heure

griote

¾ Coch.lle fine a la Rougie
1 lt tartre Rouge

Cerises

on fait les Cerizes a la suite d'une Rougie d'Ecarlatte en les Bouillissant avec 3 lt Compo.on et 1 lt tartre blanc par piece on les Rougit ensuite avec 2 lt Compo.on et ½ Coch.lle par piece, on les fait encore sur un Bain fraix

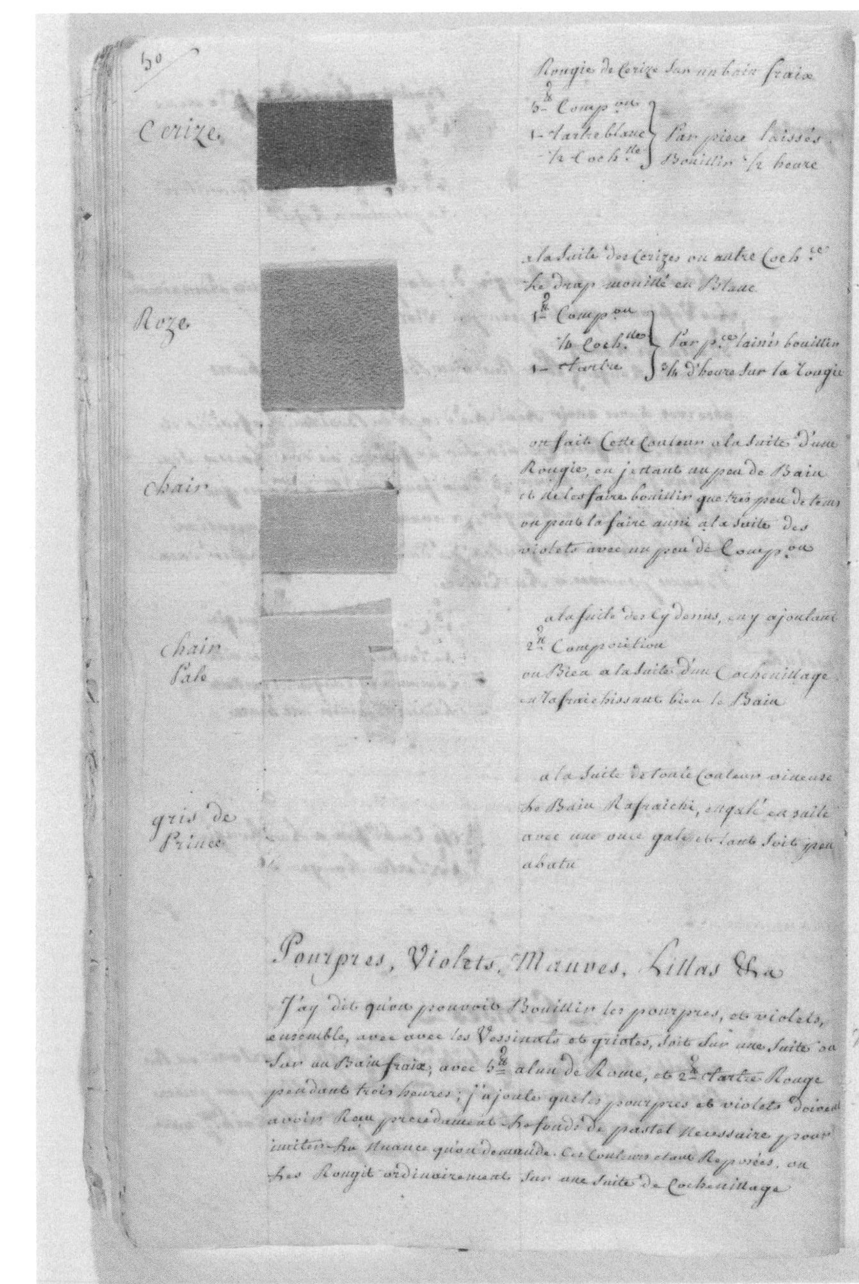

50

Cerize — Rougie de Cerize sur un bain fraix
6 [onces] Comp. ou
1 [livre] tartre blanc } Par piece laissés
½ Coch.[lle] } Bouillir ½ heure

Roze — a la suite des Cerizes ou autre Coch.[lle]
le drap mouille en blanc
1 [once] Comp. ou
⅛ Coch.[lle] } Par p[ie]ce laissés bouillir
1 [livre] tartre } ¾ d'heure sur la Rougie

chair — on fait cette Couleur a la suite d'une
Rougie, en jettant un peu de Bain
et de les faire bouillir que tres peu de tems
on peut la faire aussi a la suite des
violets avec un peu de Comp.[on]

chair pale — a la suite des Cy dessus, en y ajoutant
2 [onces] Composition
ou Bien a la suite d'un Cocheuillage
en rafraichissant bien le Bain

gris de Prince — a la suite de toute Couleur vineuse
le Bain Rafraichi, en y[.] en suite
avec une once gale et tant soit peu
abatu

Pourpres, Violets, Mauves, Lillas &[c]a

J'ay dit qu'on pouvoit Bouillir les pourpres, et violets,
ensemble, avec avec les Vessinats et griotes, soit sur une suite ou
sur un Bain fraix, avec 6 [onces] alun de Rome, et 2 [onces] tartre Rouge
pendant trois heures; j'ajoute que les pourpres et violets doivent
avoir Reçu precedament le fond de pastel necessaire pour
imiter la nuance qu'on demande. Ces Couleurs etant Rep[...]ées, on
les Rougit ordinairement sur une suite de Cocheuillage

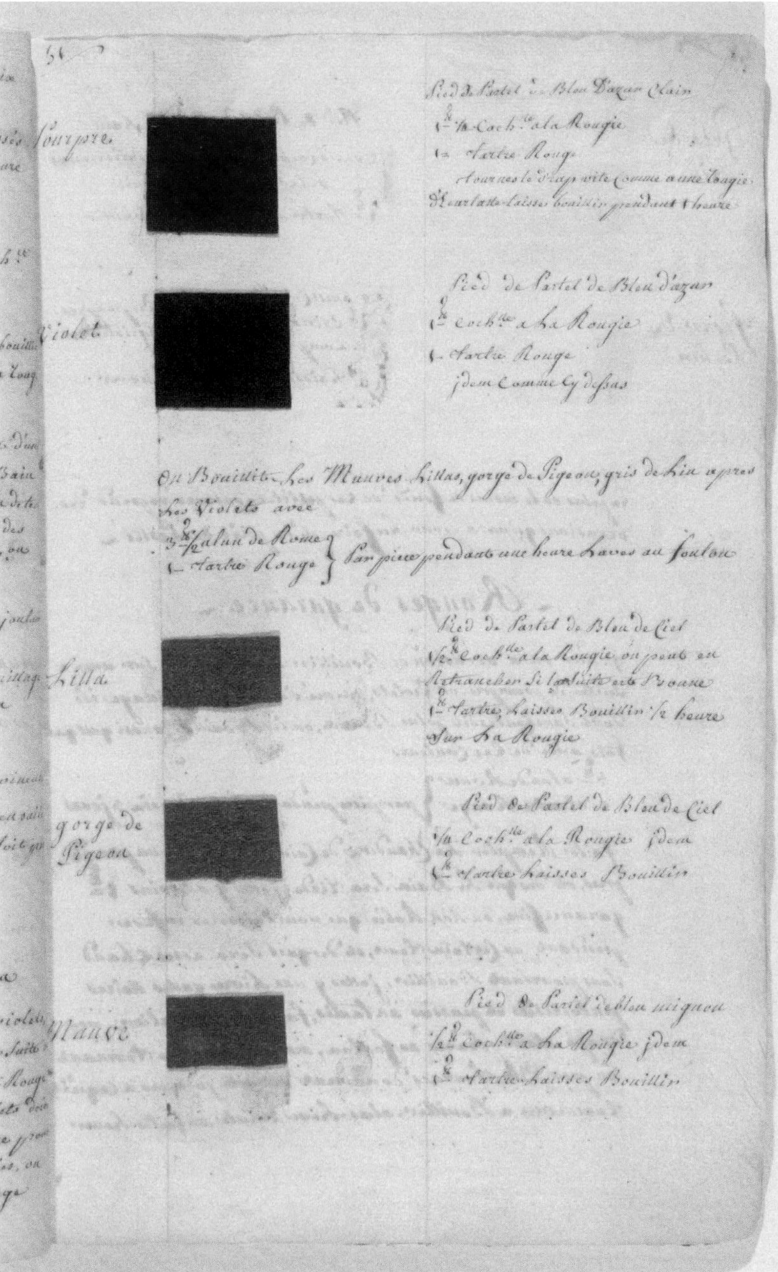

51

Pourpre — Pied de Pastel ... Bleu d'azur Clair
1 [once] ½ Coch.[lle] a la Rougie
1 [livre] tartre Rouge
tournes le drap vite Comme a une Rougie
d'Ecarlatte laissés bouillir pendant 1 heure

Violet — Pied de Pastel de Bleu d'azur
1 [once] Coch.[lle] a la Rougie
1 [livre] tartre Rouge
Idem Comme Cy dessus

On Bouillit les Mauves Lillas, gorge de Pigeon, gris de Lin apres
les Violets avec
3 [onces] ½ alun de Rome
1 [livre] tartre Rouge } Par piece pendant une heure Lavez au foulon

Lilla — Pied de Pastel de Bleu de Ciel
½ [once] Coch.[lle] a la Rougie on peut en
Retrancher si le Bain est Bonne
1 [livre] tartre, Laissés Bouillir ½ heure
sur la Rougie

gorge de Pigeon — Pied de Pastel de Bleu de Ciel
½ [once] Coch.[lle] a la Rougie Idem
1 [livre] tartre Laissés Bouillir

Mauve — Pied de Pastel de bleu mignon
½ [once] Coch.[lle] a la Rougie Idem
1 [livre] tartre Laissés Bouillir

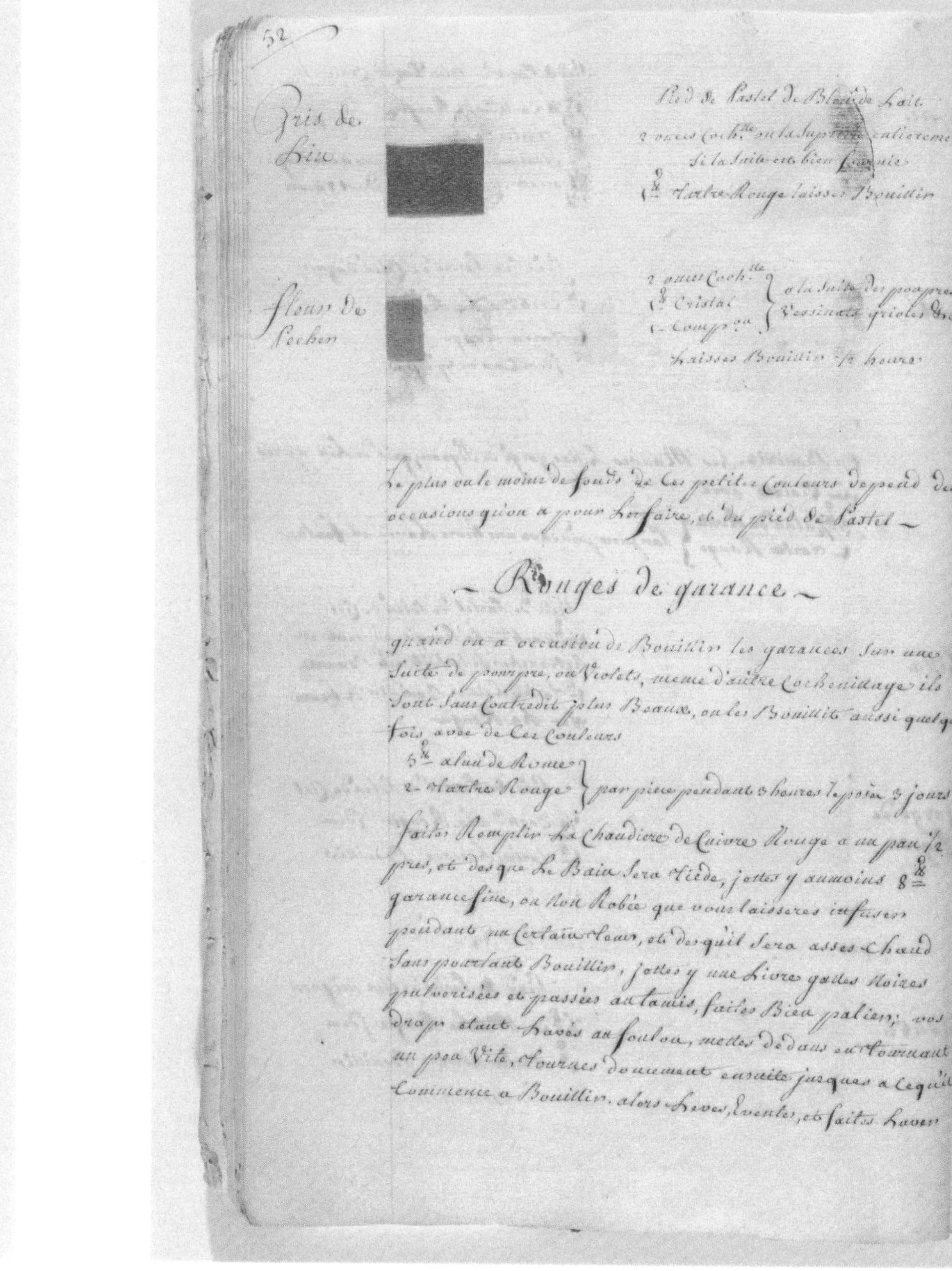

52

Gris de Lin

Pied de Pastel de Bleu de Lait
2 onces Coch.^lle ou la supprime entierement si la suite est bien fournie
1 ℔ Arbre Rouge laissés Bouillir

fleur de Pecher

2 onces Coch.^lle } a la suite des pourpres
1 ℔ Cristal } Vessinets griolés &c
1 Comp.^on }
Laisses Bouillir ½ heure

Le plus ou le moins de fonds de ces petites Couleurs depend des occasions qu'on a pour les faire, et du pied de Pastel –

– Rouges de garance –

quand on a occasion de Bouillir les garances sur une suite de pourpre, ou Violets, meme d'autre Cochenillage ils sont sans contredit plus Beaux, on les Bouillit aussi quelque fois avec de ces Couleurs

5 ℔ alun de Rome }
2 – d'arbre Rouge } par piece pendant 3 heures le poser 3 jours

faites Remplir la Chaudiere de Cuivre Rouge a un peu ½ pres, et des que le Bain sera tiede, jettes y au moins 8 ℔ garance fine, ou non Robée que vous laisserez infuser pendant un certain temps, et des qu'il sera asses Chaud sans pourtant Bouillir, jettes y une Livre galles Noires pulverisées et passées au tamis, faites Bien palier; vos draps etant Lavés au foulon, mettes dedans en tournant un peu vite, tournes doucement ensuite jusques a ce qu'il Commence a Bouillir. alors Leves, Eventes, et faites Laver

53

Garance

Mr Albert dans sa methode suit le meme procedé, avec la difference qu'il met 4 ℔ Vinaigre a la place des galles, d'autres mettent 1 ℔ Comp.^on Cet acide fait Beaucoup plus brillant, mais il est plus difficile a unir

Garance Cramoisillé

meme procedé qu'aux garances faites revenir dans un bain frais 1 ℔ fernambouc pendant 1 heure rafraichisses le Bain et passes y les garances jusqu'a la Nuance que vous souhaittes; faites cette operation sur son meme Bain, vous vous en trouverés mieux.

Rouge Bruni

les operations sont les memes avec la difference que vous pourques 2 ℔ garance jettes sur le meme Bain un peu de Couperose, bien palier et tourner le Drap; on augmente la dose jusqu'a la Nuance qu'on desire en observant de fermer la porte du fourneau; on doit le pratiquer de meme pour toutes les brunitures prises trop chaud ou mal unie

on brunit les Rouges brunes sur noir avec le Bois d'inde en plusieurs rejets

Couleur de Roy

Pied de Pastel de Bleu de Ciel
Bouilli Comme les Cy dessus et Reposé –
6 ℔ garance a la Rougie
1 ℔ galles
Bruni avec la Couperose

Couleur de Roy sans Bruniture

Pied de Pastel de Bleu d'azur
6 ℔ garance Commune }
1 ½ fernambouc } par piece faites
½ galles } Bouillir ½ heure sur la Rougie Laves
a la Riviere sans Bruniture

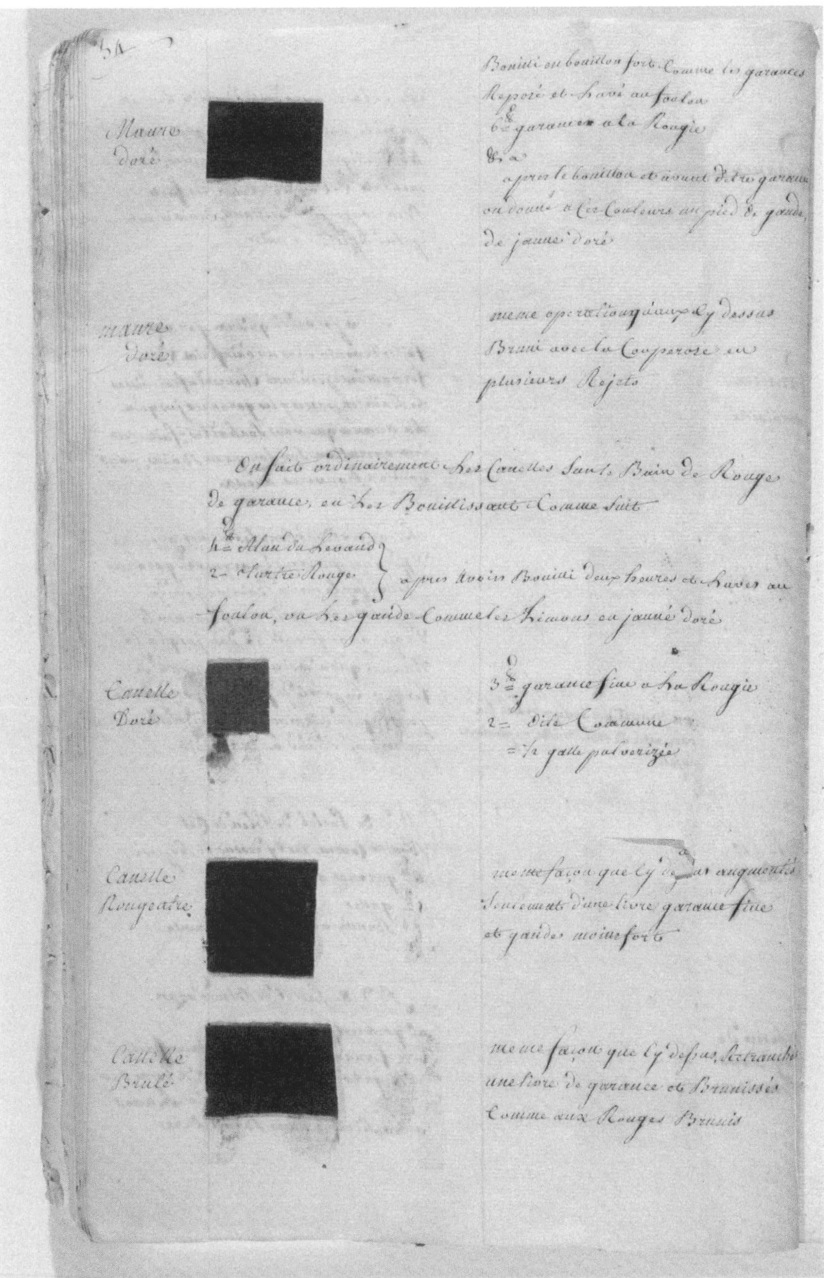

54

Maure doré — Bouilli ou bouillon fort Comme les garances Reposé et lavé au foulon
6 lb garance a la Rougie
&c a
apres le bouillon et avant d'etre garancé on donne a Ces Couleurs un pied de gaude, de jaune doré

Maure doré — meme operation que cy dessus Bruni avec la Couperose en plusieurs Rejets

On fait ordinairement les Canelles Sans le Bain de Rouge de garance, on les Bouillissent Comme suit
4 lb Alun du Levant }
2 — Tartre Rouge } apres avoir Bouilli deux heures et lavés au foulon, on les gaude Comme les Limons en jaune doré

Canelle Doré — 3 lb garance fine a la Rougie
2 — dite Commune
— 1/2 galle pulverizée

Canelle Rougeatre — meme façon que cy dessus augmentés seulement d'une livre garance fine et gaudés moins fort

Canelle Brulé — meme façon que cy dessus Retranché une livre de garance et Brunissés Comme aux Rouges Brunis

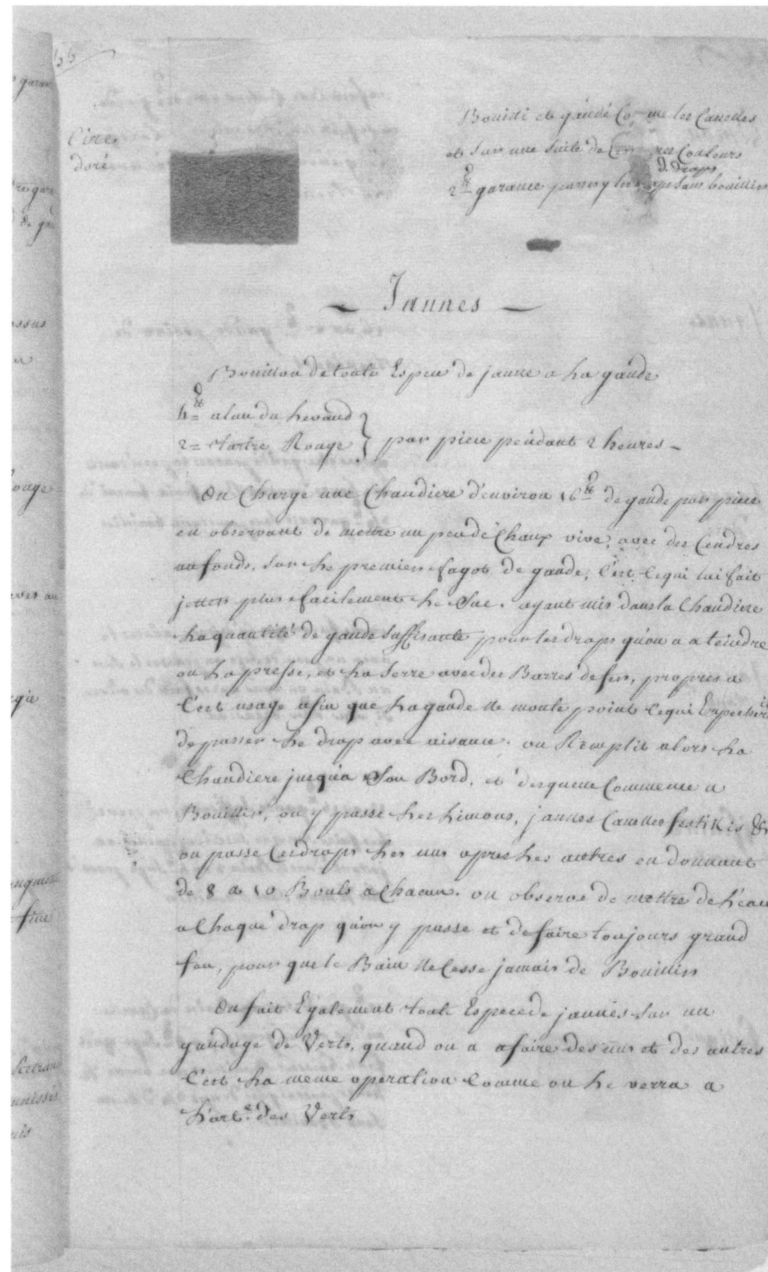

55

Cire doré — Bouilli et gaudé Comme les Canelles et Sur une suite de [illegible] Couleurs de draps 2 lb garance par [illegible] les draps dans bouillir

— Jaunes —

Bouillon de toute Espece de jaune a la gaude
4 lb alun du levant }
2 — Tartre Rouge } par piece pendant 2 heures —

On Charge une Chaudiere d'environ 16 lb de gaude par piece en observant de mettre un peu de Chaux vive, avec des Cendres au fonds, sur le premier fagot de gaude, C'est ce qui lui fait jetter plus facilement le Sac. ayant mis dans la Chaudiere la quantité de gaude suffisante pour les draps qu'on a a teindre on la presse, et la serre avec des Barres de fer, propres a Cet usage afin que la gaude ne monte point ce qui l'empecherait de passer le drap avec aisance. on Remplit alors la Chaudiere jusqu'a son Bord, et des qu'elle Commence a Bouillir, on y passe les Limons, jaunes Canelles [illegible] &c on passe Ces draps les uns apres les autres en donnant de 8 a 10 Bouts a Chacun. on observe de mettre de l'eau a chaque drap qu'on y passe et de faire toujours grand feu, pour que le Bain ne Cesse jamais de Bouillir

On fait Egalement toute Espece de jaunes sur un gaudage de Verts, quand on a a faire des uns et des autres C'est la meme operation Comme on le verra a l'art. des Verts

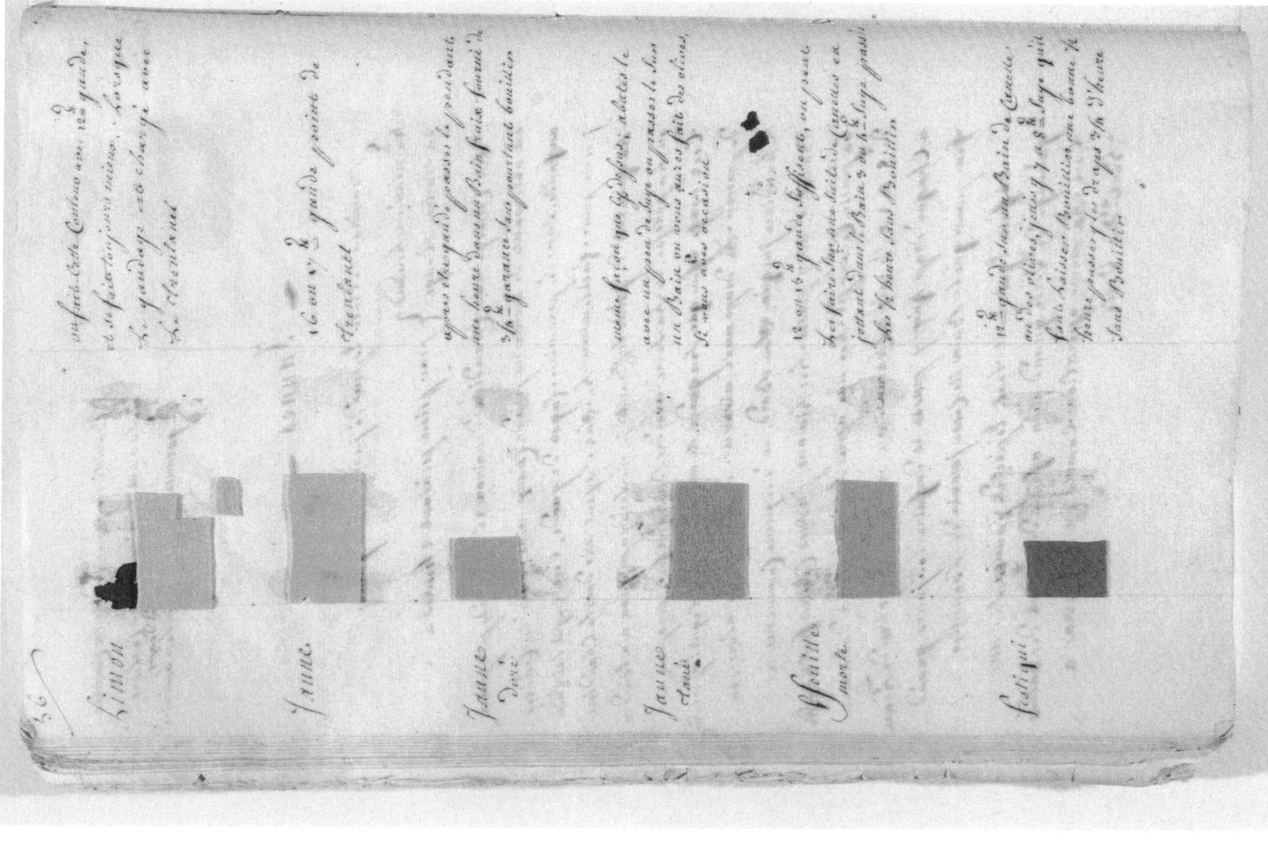

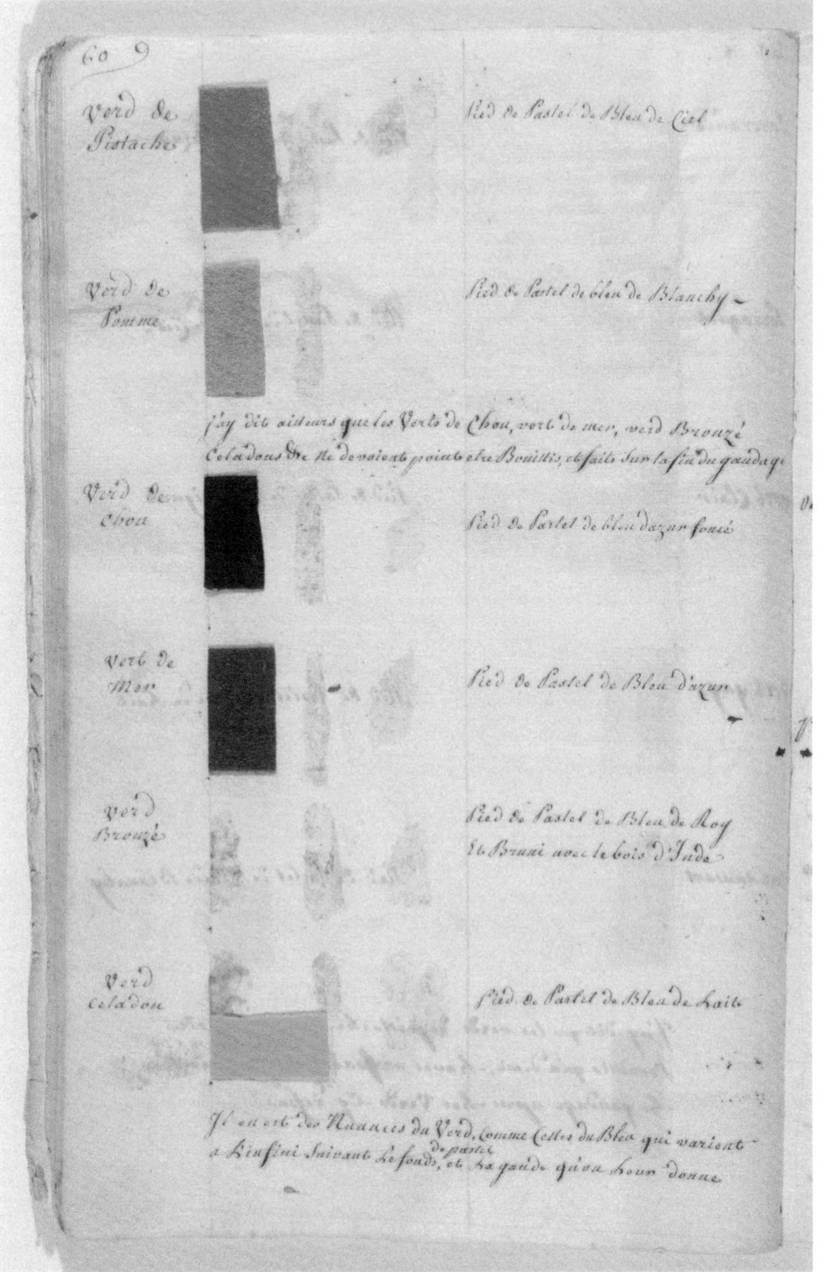

60

Verd de Pistache — Pied de Pastel de Bleu de Ciel

Verd de Pomme — Pied de Pastel de bleu de Blanchy

J'ay dit ailleurs que les Verds de Chou, vert de mer, verd Brouzé
Celadons &c ne devoient point etre Bouillis, et faits sur la fin du gaudage

Verd de Chou — Pied de Pastel de bleu dazur foncé

Vert de Mer — Pied de Pastel de Bleu d'azur

Verd Brouzé — Pied de Pastel de Bleu de Roy
Et Bruni avec le bois d'Inde

Verd celadon — Pied de Pastel de Bleu de Lait

Il en est des Nuances du Verd, comme celles du Bleu qui varient
a Linfini suivant le fonds de pastel, et la gaude qu'on leur donne

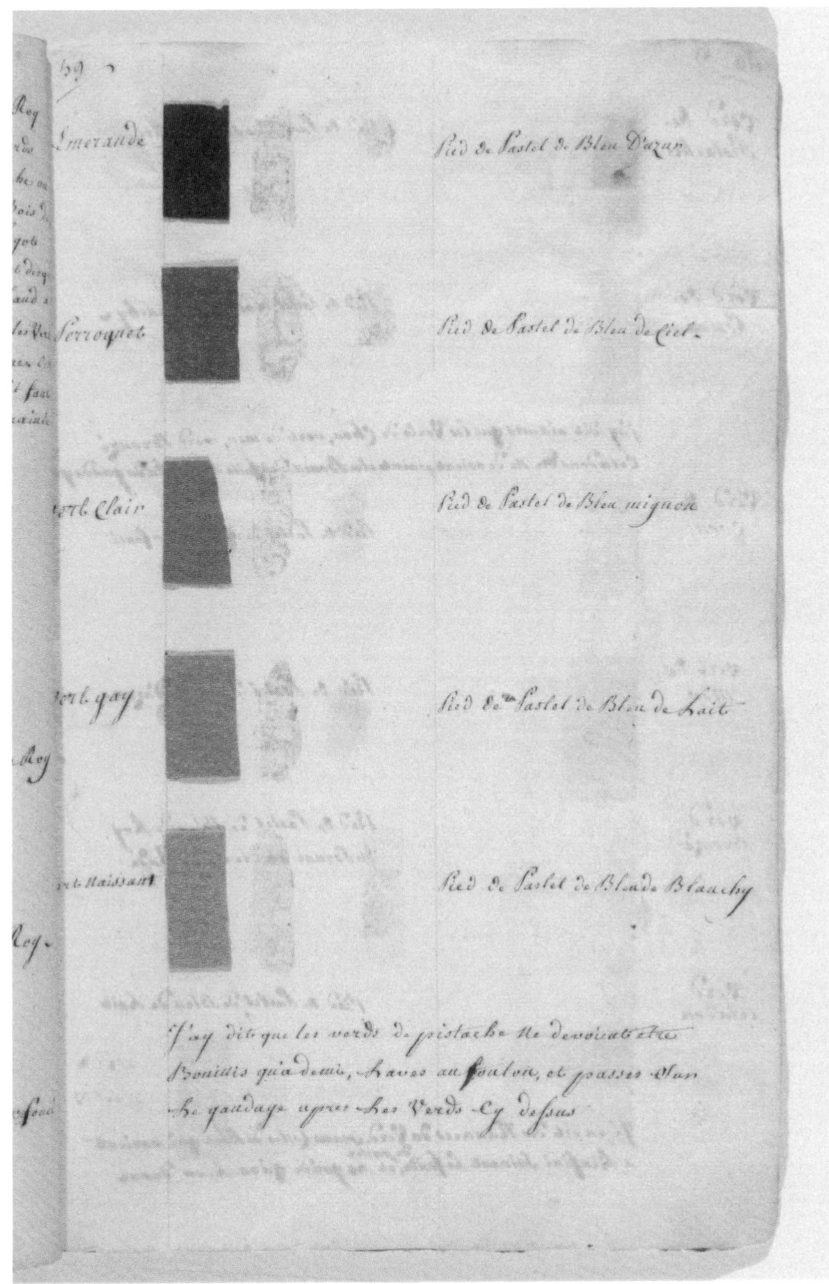

59

Emeraude — Pied de Pastel de Bleu D'azur

Perroquet — Pied de Pastel de Bleu de Ciel

Vert Clair — Pied de Pastel de Bleu mignon

Vert gay — Pied de Pastel de Bleu de Lait

Vert Naissant — Pied de Pastel de Bleu de Blanchy

J'ay dit que les verds de pistache ne devoient etre
Bouillis qu'a demi, laves au foulon, et passes sur
le gaudage apres les Verds cy dessus

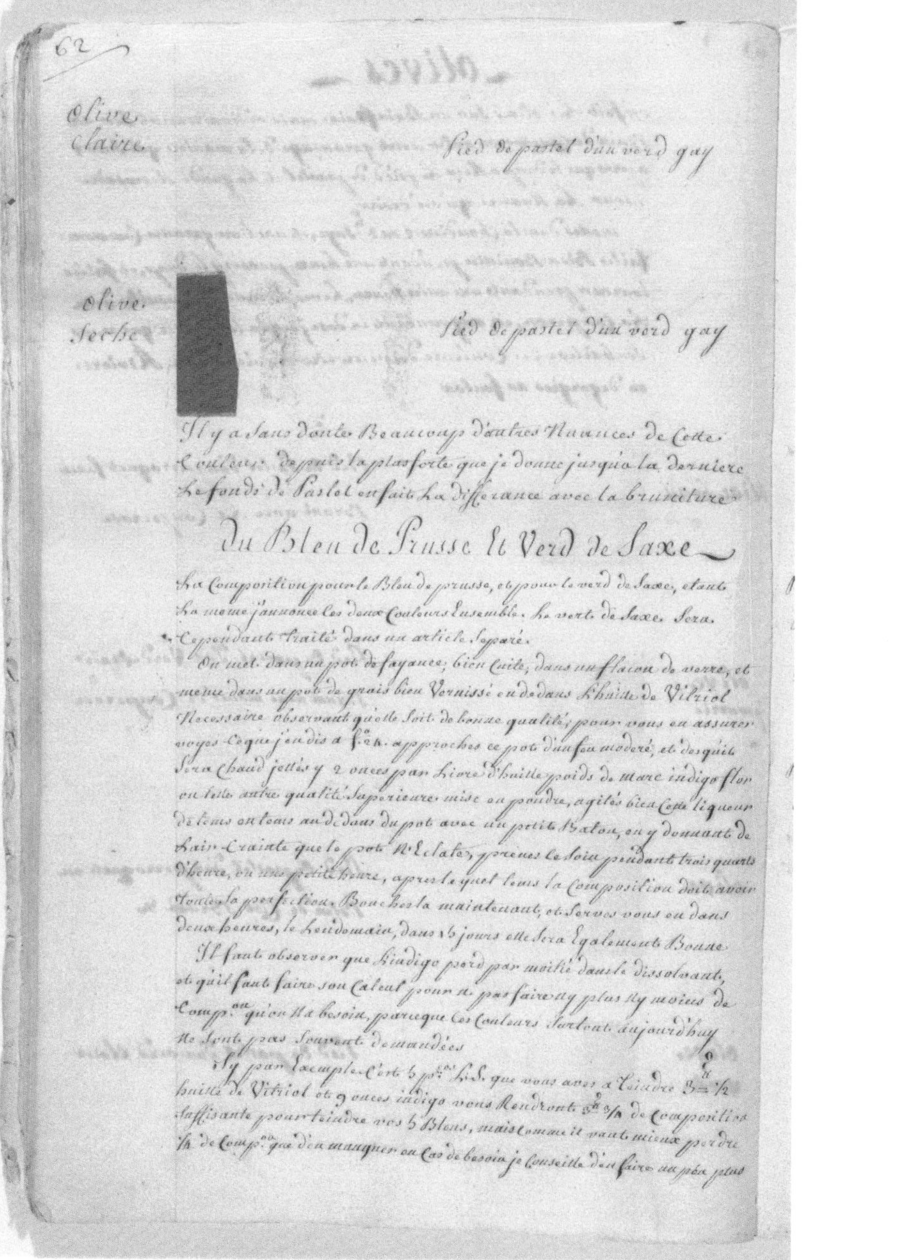

Olive Claire — Pied de pastel d'un verd gay

Olive Seche — Pied de pastel d'un verd gay

Il y a sans doute Beaucoup d'autres Nuances de Cette Couleur depuis la plus forte que je donne jusqu'a la derniere Le fond de Pastel en fait La difference avec la bruniture

du Bleu de Prusse et Verd de Saxe

La Composition pour le Bleu de prusse, et pour le verd de Saxe, etant La meme j'annonce les deux Couleurs ensemble. Le vert de Saxe Sera Cependant traité dans un article Separé.

On met dans un pot de fayance, bien Cuite, dans un flacon de verre, et meme dans un pot de grais bien Vernissé en dedans L'huile de Vitriol Necessaire observant qu'elle Soit de bonne qualité, pour vous en assurer voyes ce que j'en dis a f.o 14. approches ce pot d'un feu moderé, et des qu'il Sera Chaud jettés y 2 onces par Livre d'huille poids de marc indigo flor ou telle autre qualité Superieure mise en poudre, agités bien Cette liqueur de tems en tems au dedans du pot avec un petit baton, en y donnant de L'air crainte que le pot N'Eclate, prenes le Soin pendant trois quarts d'heure, ou une petite heure, apres Lequel tems la Composition doit avoir toute sa perfection. Bouches la maintenant, et Serves vous en dans deux heures, le Lendemain, dans 15 jours elle Sera Egalement Bonne

Il faut observer que L'indigo perd par moitié dans le dissolvant, et qu'il faut faire son Calcul pour ne pas faire ny plus ny moins de Comp.on qu'on en a besoin, parceque ces Couleurs Surtout aujourd'huy ne Sont pas Souvent demandées

Sy par Exemple C'est 6 p.ces L.d. que vous aves a teindre 3.tt ½ huille de Vitriol et 9 onces indigo vous Rendront 3.tt ½ de Composition suffisante pour teindre vos 6 Bleus, mais comme il vaut mieux perdre ½ de Comp.on que d'en manquer en Cas de besoin je Conseille d'en faire un peu plus

olives

on fait Les olives sur un Bain fraix mais ordinairement sur un Bain de Cuvettes, ou autre petit garançage de la maniere qui suit, apres que le drap a Recu Le pied de pastel et La gaude Necessaire pour La Nuance qu'on desire

mettes dans la Chaudiere 2 ou 3.tt Saye, et une livre garance Commune faites Bien Bouillir pendant une heure, passés y le drap, et faites tourner pendant une autre heure, Leves, Brunisses ensuite avec La Couperose, en augmentant la dose jusqu'a la Nuance que vous Souhaites. Ces Couleurs doivent etre Lavées a La Riviere et degorgées au foulon

Olive brun — Pied de Pastel d'un perroquet foncé Bruni avec La Couperose

Olive pourrie — Pied de pastel d'un Verd Clair Bruni avec moins de Couperose

Olive — Pied de pastel d'un perroquet ou Bleu de Ciel Bruni &c.

Olive verte — Pied de pastel d'un verd Clair

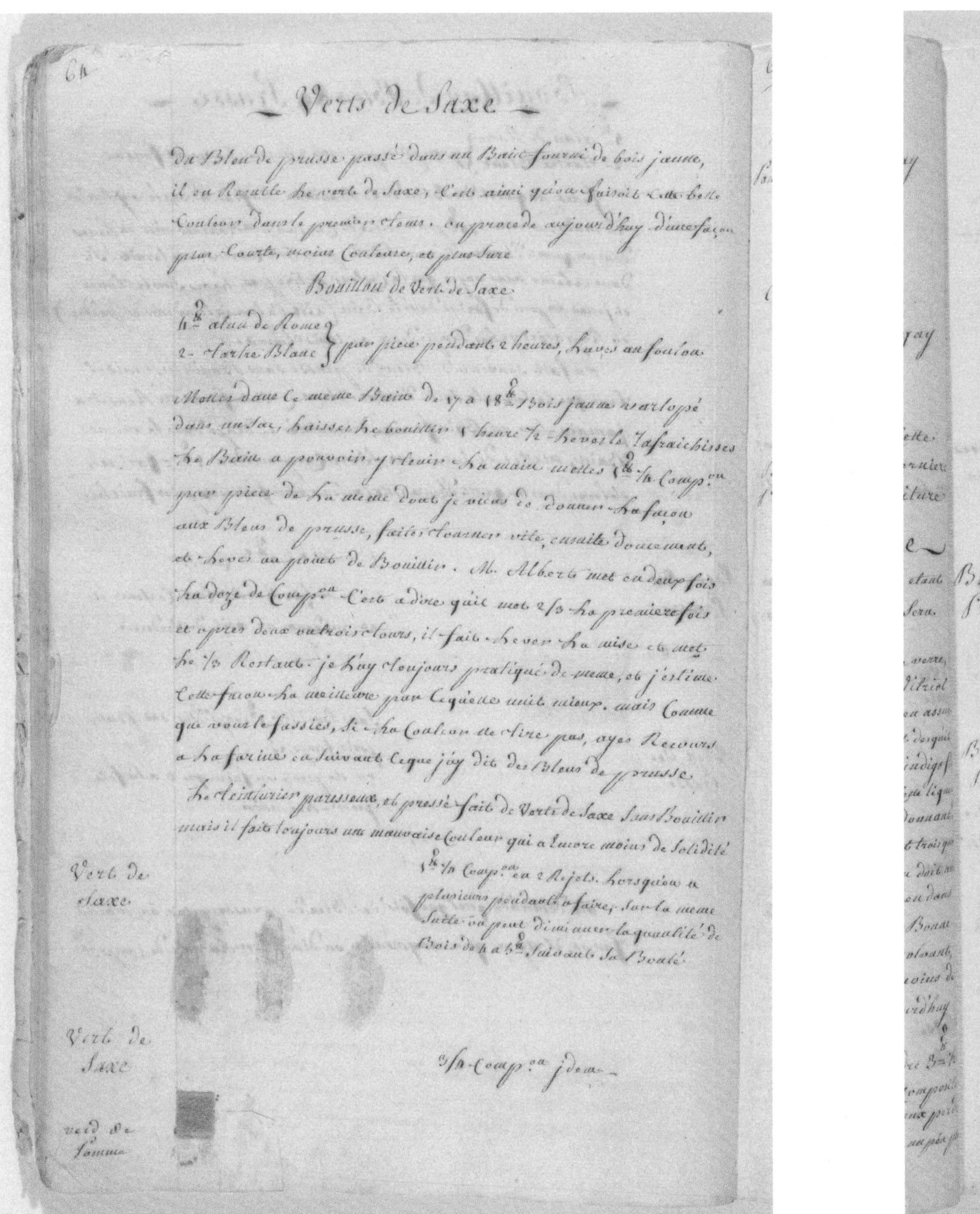

64

Verts de Saxe

Du Bleu de prusse passé dans un Bain fourni de bois jaune, il en Resulte le vert de Saxe, C'est ainsi qu'on faisoit cette belle Couleur dans le premier tems. on procede aujourd'huy d'une façon plus Courte, moins Couteuse, et plus Sure

Bouillon de Vert de Saxe

4 lb alun de Rome
2 — tartre Blanc } par piece pendant 2 heures, Lavés au foulon

Mettez dans le meme Bain de 17 a 18 lb bois jaune varlopé dans un Sac, laissez le bouillir 1 heure 1/2 levez le, rafraichissez le Bain a pouvoir y tenir la main mettez 1 lb 1/4 Compo.on par piece de la meme dont je viens de donner la façon aux Bleus de prusse, faites tourner vite, ensuite doucement, et levez au point de Bouillir. M. Albert met en deux fois la doze de Compo.on C'est a dire qu'il met 2/3 la premiere fois et apres deux ou trois tours, il fait lever la mise et met le 1/3 Restant. je l'ay toujours pratiqué de meme, et j'estime cette façon la meilleure par ce qu'elle unit mieux. mais comme que vous le fassiés, si la Couleur ne tire pas, ayez Recours a la farine en suivant ce que j'ay dit des Bleus de prusse.

Le teinturier paresseux, et pressé fait de Verts de Saxe sans Bouillir mais il fait toujours une mauvaise couleur qui a encore moins de Solidité

Vert de saxe — 1 lb 1/4 Compo.on en 2 Rejets. lorsqu'on a plusieurs pendant a faire, sur la meme Suitte on peut diminuer la quantité de Bois de 4 a 5 lb Suivant sa bonté.

Vert de Saxe — 3/4 Compo.on idem

verd de Pomme

Bouillon de Bleu de Prusse

4 lb alun de Rome
2 — tartre Blanc } par piece pendant 2 heures Lavés au foulon

Jettez un quart du Bain de ce Bouillon, Rafraichissez le a y tenir la main, mettez 3/4 lb par piece de cette Compo.on tournés vite et Levez dans un quart d'heure, si vous laissez plus cette Couleur, elle ternit. Si dans ce tems vous voyez que la couleur ne tire pas, Levez sur le tour et jettez un peu de farine dans le Bain (c'est l'alun calciné mis en poudre) et Repassez y les draps deux ou trois tours.

On fait souvent de Bleus de prusse sans Bouillir, mais il faut que ce soit a la suite d'autres Bleus (cette façon ne Reussira jamais sur un Bain frais) mouillés les draps dans le meme Bain, mettez de suite une Livre Composition 1 lb 1/2 farine, et levez dans un quart d'heure cette couleur ainsi faite est plus fraiche

Bleu de Prusse — Bouilli & 3/4 lb de Compo.on
L'alun fait Rosir cette Couleur et luy donne un oeil de verdeur

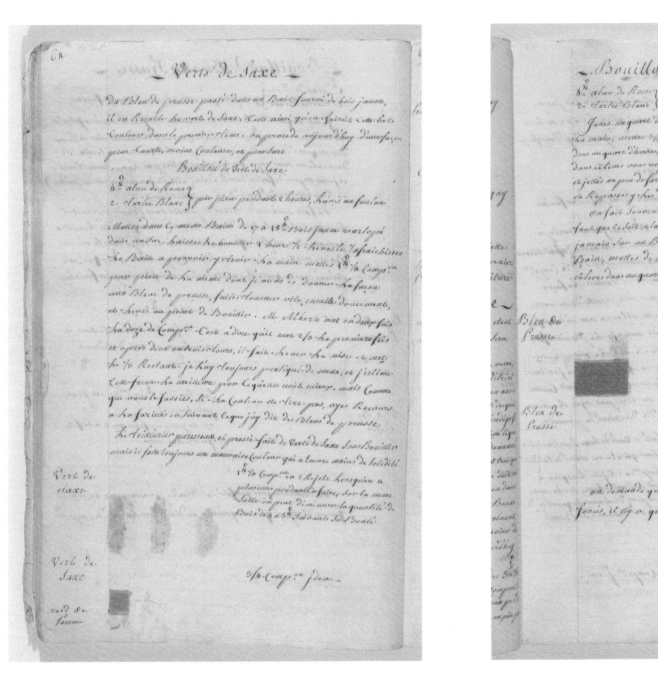

Bleu de Prusse — Sans Bouillir 1 lb Compo.on sur un Bain pris tempéré
on ne peut en faire que 2 a la fois pour les unir

on demande quelquefois de Bleu de prusse plus ou moins foncés, il n'y a qu'a augmenter ou diminuer la doze de Compo.on

67

Celadons au Vert de gris

faites fondre dans un Chauderon a part 4 onces vert de gris dans de l'eau chaude, que vous ferez jetter dans la Chaudiere, dès que le Bain sera tiede, Bien palier, et y tourner le drap pendant une heure, ne faites point laver au foulon, c'il n'est avant l'operation pour les Bien Mouiller

Celadon	au Vert de gris
Celadon	au Vitriol Bleu

Des Couleurs fauves ou au Sendal & autres Marrons

garnisés la Chaudiere, avec le Sendal, et le Redou et les galles, le tout suivant la Nuance que vous desirez faire laissez Bouillir le tout pendant une Bonne heure, passez y les draps que vous devez laisser Bouillir une autre heure Brunissez les ensuite avec la Couperose le Bain pris tres temperé

Marron — 3 livres Sendal, 6 — Redou, ½ galle } par piece. Bruni avec la Couperose

Marron plus clair — 3 livres Sendal, 3 — Redou, ½ galle } par piece. Bruni avec la Couperose

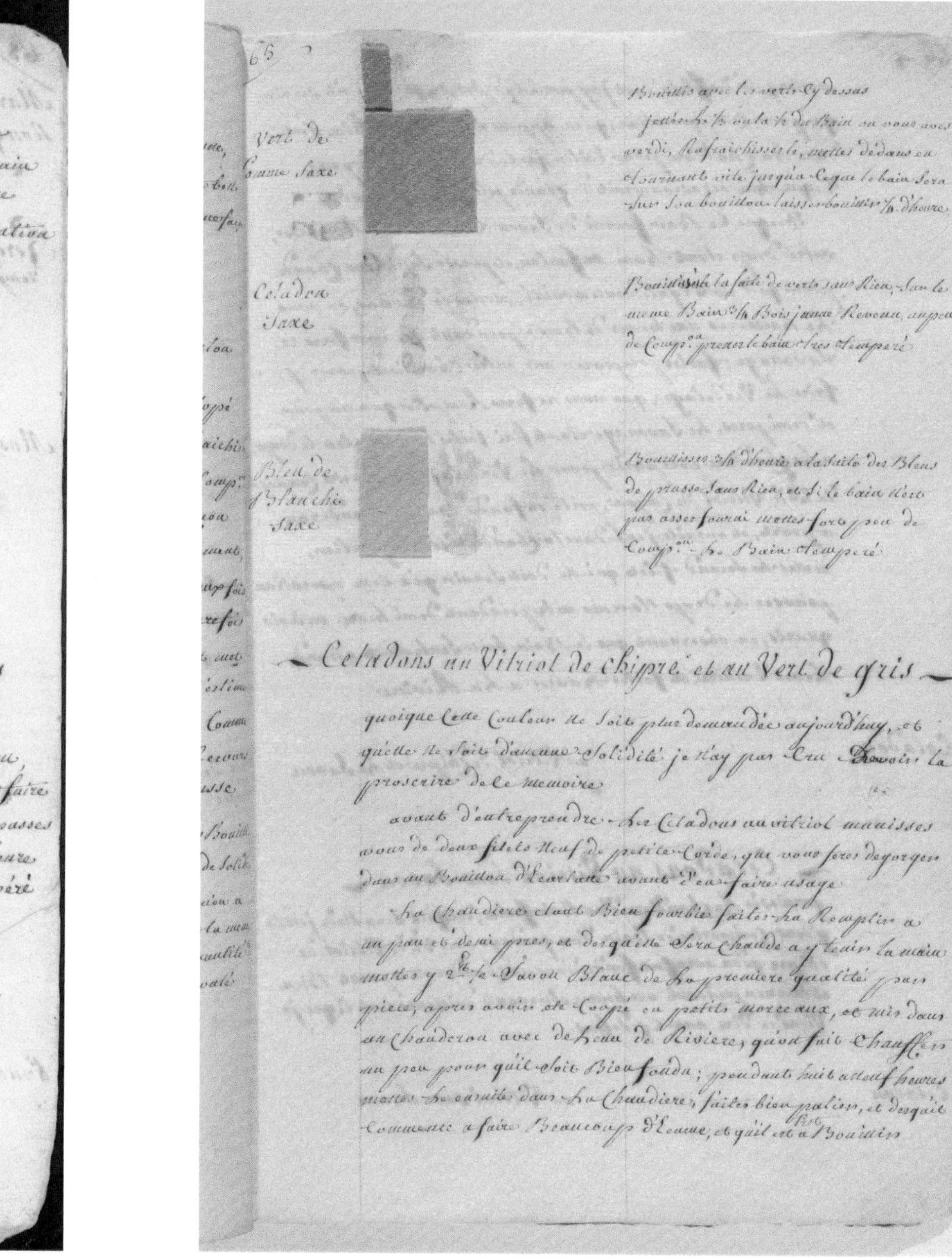

66

Vert de Comme Saxe — Bouillis avec les verts cy dessus jettés la ½ ou la ⅓ du Bain ou vous avez verdi, Rafraichissez le, mettez dedans en tournant vite jusqu'a ce que le bain sera sur son bouillon laissez bouillir ¾ d'heure

Celadon Saxe — Bouillissés la suite de verts sans Rien, dans le meme Bain ½ Bois jaune Revenu, un peu de Couperose prenez le bain tres temperé

Bleu de Blanchi Saxe — Bouillissez ¾ d'heure a la suite des Bleus de prusse sans Rien, et si le bain n'est pas assez fourni mettez fort peu de Couperose le Bain temperé

Celadons au Vitriol de chipre et au Vert de gris

quoique cette Couleur ne soit plus demandée aujourd'huy, et qu'elle ne soit d'aucune Solidité je n'ay pas cru devoir la proscrire de ce memoire

avant d'entreprendre les Celadons au vitriol munissez vous de deux filets neuf de petite Corde, que vous ferez degorger dans un Bouillon d'Ecarlatte avant d'en faire usage

La Chaudiere etant Bien fourbie faites la Remplir a un pan et demi pres, et dès qu'elle sera chaude a y tenir la main mettez y 2 livres ½ Savon Blanc de la premiere qualité par piece, apres avoir ete coupé en petits morceaux, et mis dans un Chauderon avec de l'eau de Riviere, qu'on fait chauffer un peu pour qu'il soit Bien fondu; pendant huit a neuf heures mettez le ensuite dans la Chaudiere, faites bien palier, et dès qu'il commence a faire Beaucoup d'Ecume, et qu'il est prêt a Bouillir

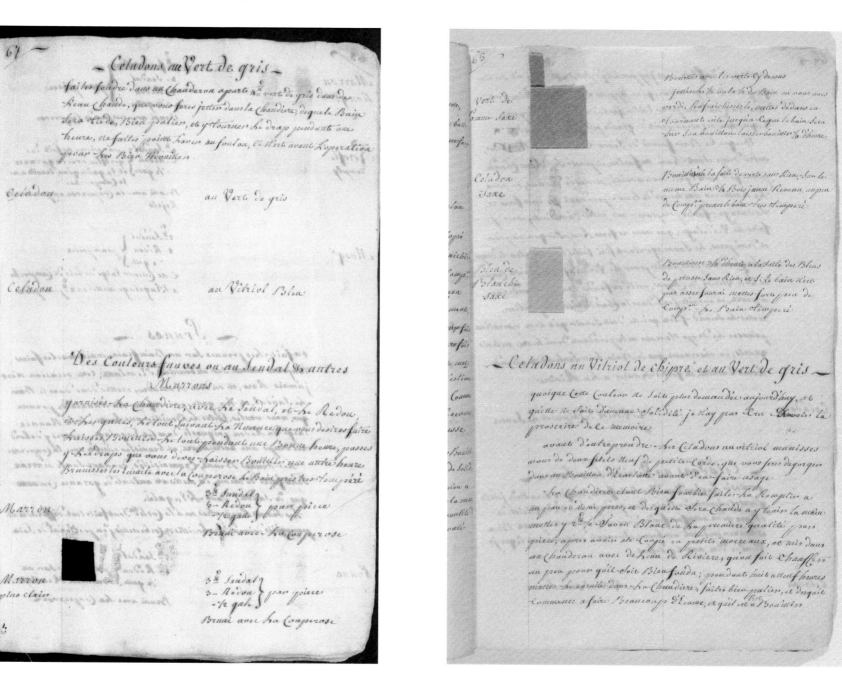

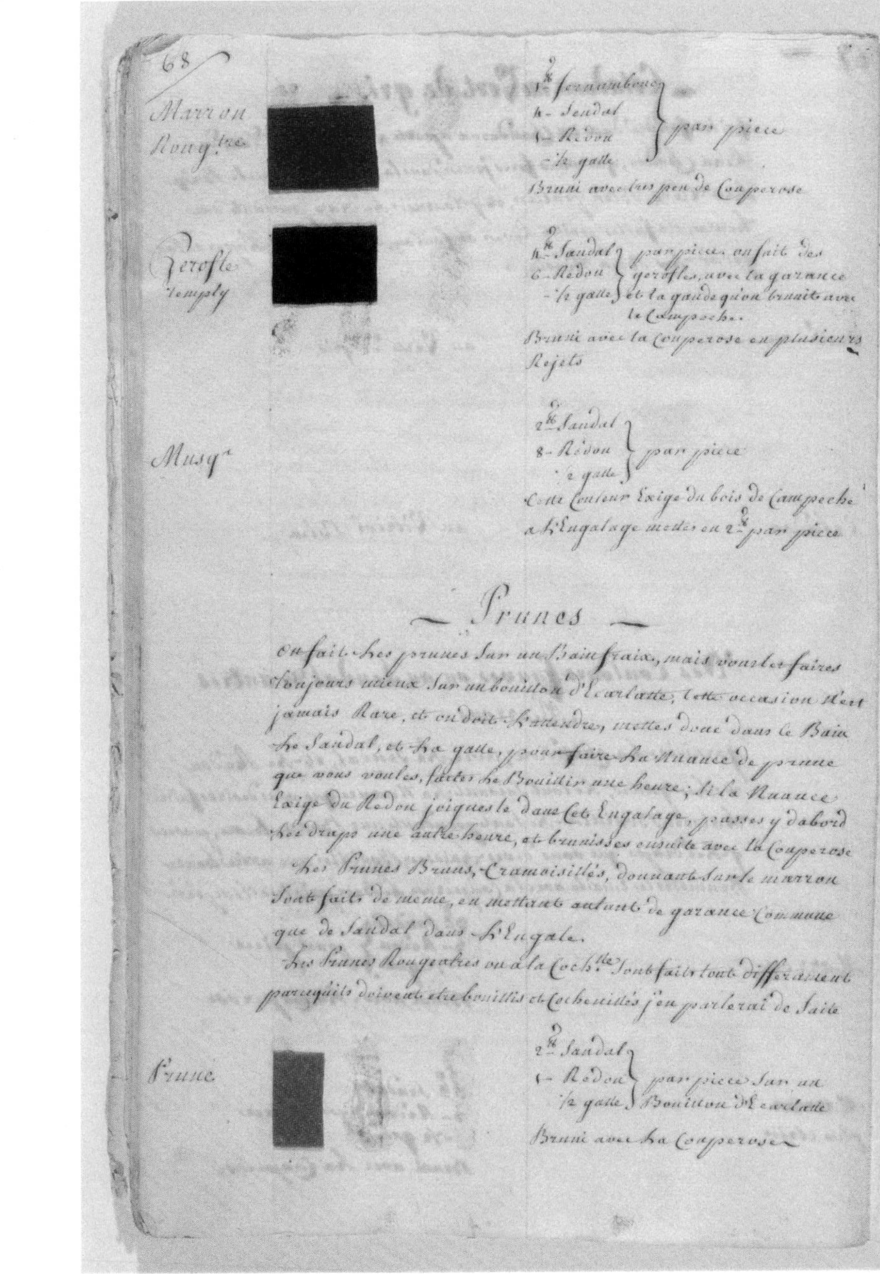

68

Marron Roug.tre — 1 lb fernambouc, 4 – Sandal, 1 – Redon, 1/2 gatte } par piece
Bruni avec tres peu de Couperose

Gerofle Tempty — 4 lb Sandal, 6 – Redon, – 1/2 galle } par piece on fait des gerofles, avec la garance et la gaude quon brunit avec le Campeche.
Brunie avec la Couperose en plusieurs Rejets

Musq.e — 2 lb Sandal, 8 – Redon, 1/2 galle } par piece
Cette Couleur Exige du bois de Campeche a L'Engalage mettes en 2 lb par piece

— Prunes —

On fait Les prunes sur un Bain fraix, mais vous les faires toujours mieux sur un bouillon d'Ecarlatte, cette occasion n'est jamais Rare, et on doit L'attendre, mettes donc dans le Bain Le Sandal, et La galle, pour faire La Nuance de prune que vous voules, faites Les Bouillir une heure; si la Nuance Exige du Redon joignes le dans cet Engalage, passes y dabord Les draps une autre heure, et brunisses ensuite avec la Couperose

Les Prunes Bruns, Cramoisillés, donnant sur le marron Sont faits de meme, en mettant autant de garance Commune que de Sandal dans L'Engale.

Les Prunes Rougeatres ou a la Coch.lle sont faits tout differament parcequ'ils doivent etre bouillis et Cochenillés j'en parlerai de Suite

Prune — 2 lb Sandal, 1 – Redon, 1/2 galle } par piece sur un Bouillon d'Ecarlatte
Bruni avec La Couperose

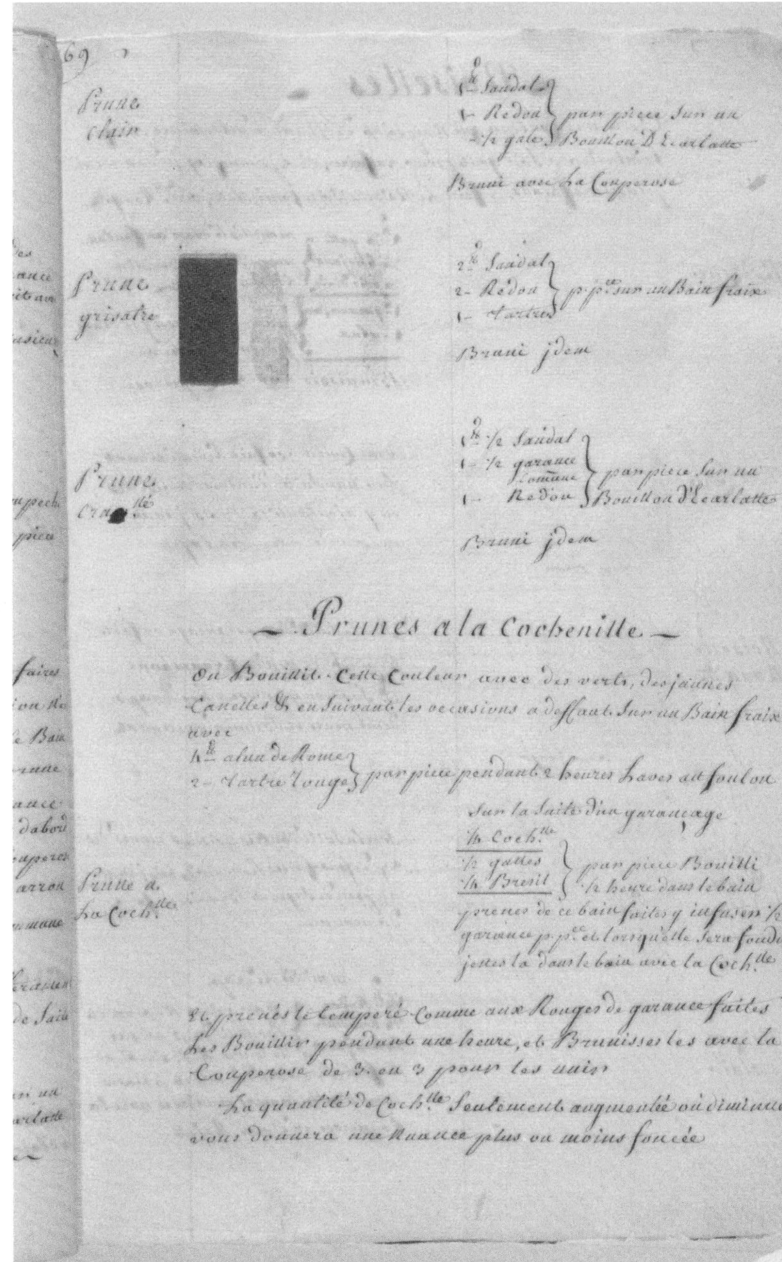

69

Prune clair — 1 lb Sandal, 1 – Redon, 1/2 gale } par piece sur un Bouillon D'Ecarlatte
Bruni avec La Couperose

Prune grisatre — 2 lb Sandal, 2 – Redon, 1 – tartre } p. p.ce sur un Bain fraix
Bruni idem

Prune Cra.ssé — 1 lb 1/2 Sandal, 1 – 1/2 garance comune, 1 – Redon } par piece sur un Bouillon d'Ecarlatte
Bruni idem

— Prunes a la Cochenille —

On Bouillit Cette Couleur avec des verts, des jaunes Canelles &c en Suivant les occasions a deffaut sur un Bain fraix avec
4 lb alun de Rome, 2 – tartre rouge } par piece pendant 2 heures Laves au foulon

Prune a La Coch.lle — Sur la Suite d'un garançage
1/4 Coch.lle, 1/2 galles, 1/4 Bresil } par piece Bouilli 1/2 heure dans le bain
prenes de ce bain faites y infuser 1/2 lb garance p. p.ce et lorsqu'elle sera fondue jettes la dans le bain avec la Coch.lle

Et prenés le tempé Comme aux Rouges de garance faites Les Bouillir pendant une heure, et Brunisses les avec la Couperose de 3 ou 4 pour les unis

La quantité de Coch.lle Seulement augmentée ou diminuée vous donnera une Nuance plus ou moins foncée

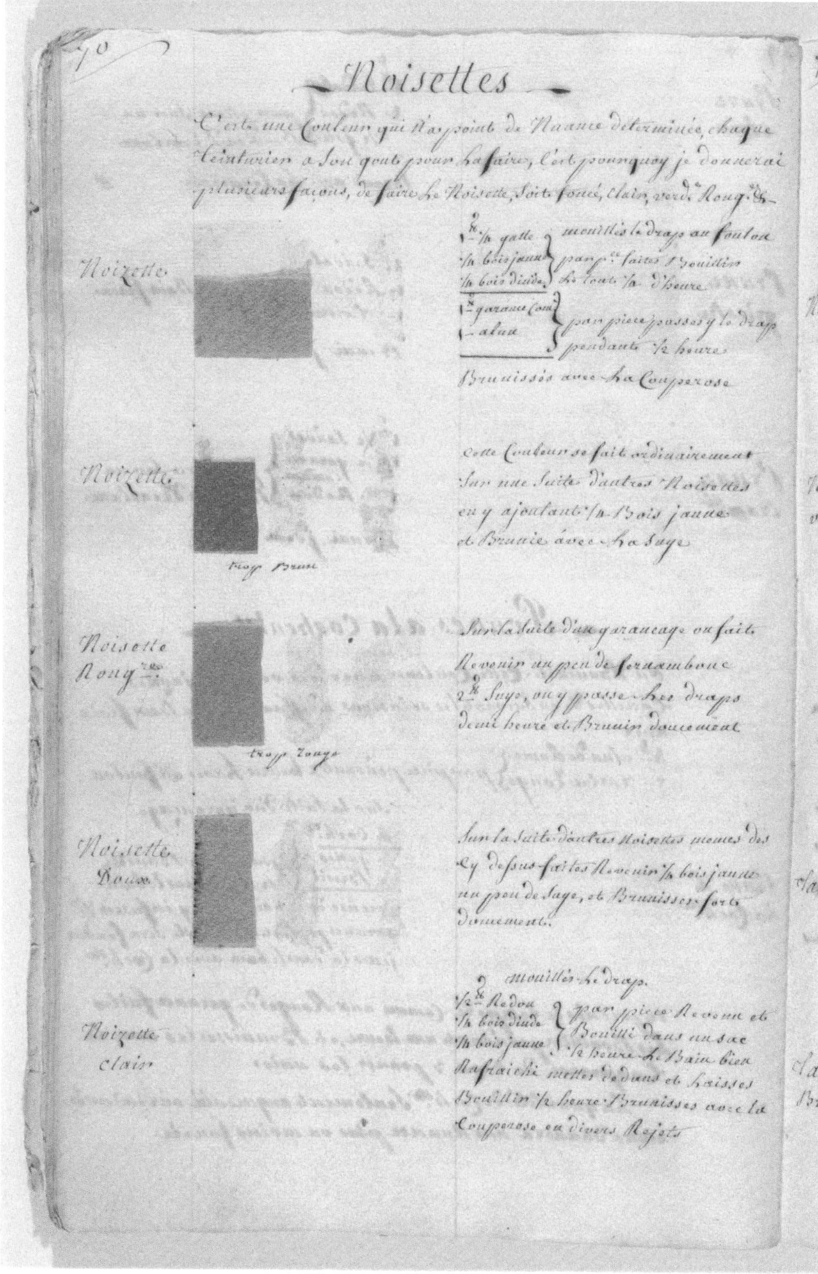

70

Noisettes

C'est une Couleur qui n'a point de Nuance déterminée, chaque Teinturier a son goût pour la faire, c'est pourquoy je donnerai plusieurs façons de faire le Noisette, soit foncé, clair, verd.re Roug.re &

Noisette

1℔ ½ galle
½ bois jaune
½ bois d'inde } mouillés le draps au foulon par pièce faites bouillir le tout ¾ d'heure

1℔ garance com.
1 alun } par pièce passés y le draps pendant ½ heure

Brunissés avec la Couperose

Noisette

Cette Couleur se fait ordinairement sur une suite d'autres Noisettes en y ajoutant ½ Bois jaune et Brunie avec la Suye

trop brun

Noisette Roug.re

Sur la suite d'un garançage on fait Revenir un peu de fernambouc 2℔ Suye, on y passe les draps demi heure et Brunir doucement

trop rouge

Noisette Brun

Sur la suite d'autres Noisettes memes des cy dessus faites Revenir ½ bois jaune un peu de Suye, et Brunisser fort doucement.

Noisette clair

mouillés le drap.
½℔ Redou
½ bois d'inde
½ bois jaune } par pièce Revenu et Bouilli dans un sac ½ heure le Bain bien Rafraichi mettes dedans et laisses Bouillir ½ heure Brunisses avec la Couperose en divers Rejets

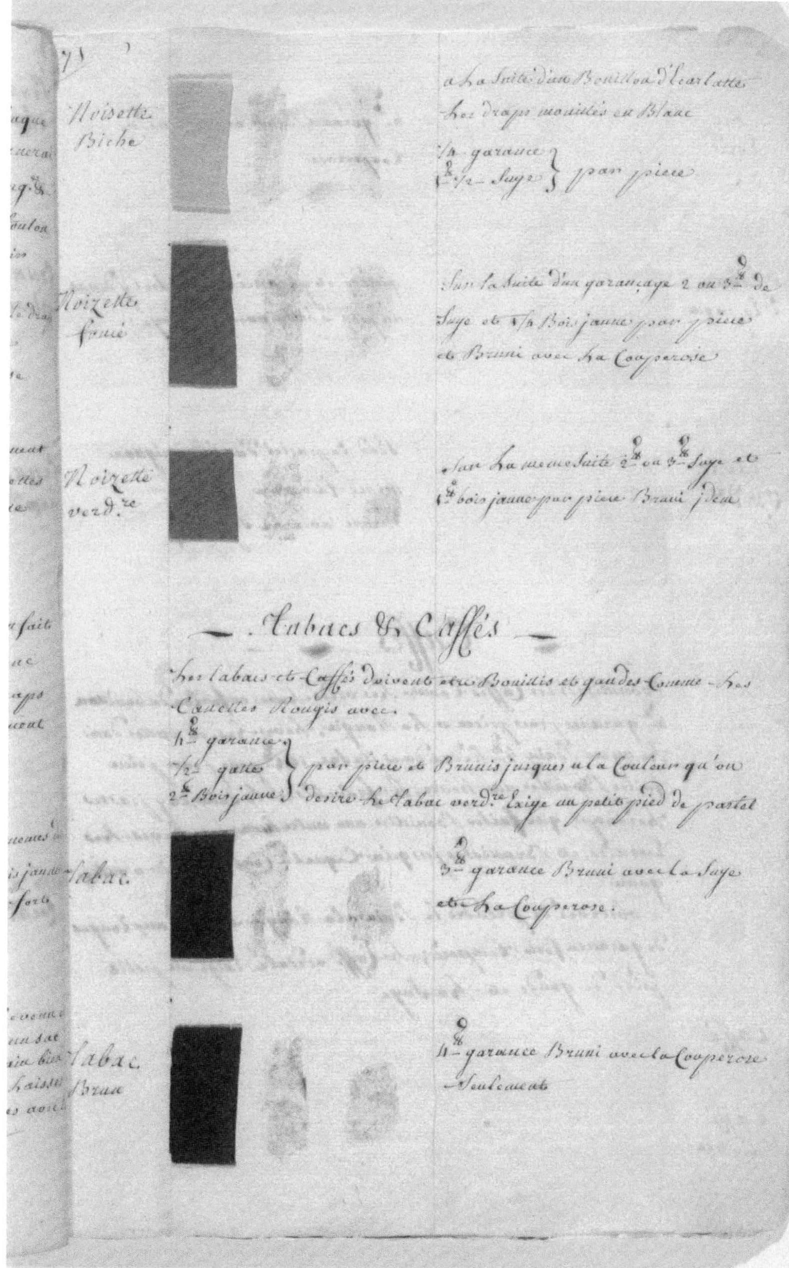

71

Noisette Biche

a la suite d'un Bouillon d'Ecarlatte les draps mouillés en Blanc
½ garance
1℔ ½ Suye } par pièce

Noizette foncé

Sur la suite d'un garançage 2 ou 3℔ de Suye et ½ Bois jaune par pièce et Bruni avec la Couperose

Noizette verd.re

Sur la meme suite 2℔ ou 3℔ Suye et 1℔ bois jaune par pièce Bruni idem

Tabacs & Caffés

Les tabacs et Caffés doivent etre Bouillis et gaudes comme les Canelles Rougis avec.
4℔ garance
½ galle
2℔ Bois jaune } par pièce et Brunis jusques a la Couleur qu'on desire le tabac verd.re Exige un petit pied de pastel

Tabac

3℔ garance Bruni avec la Suye et la Couperose.

Tabac Brun

4℔ garance Bruni avec la Couperose seulement

Noirs

Noirs a La facon de Sedan

faites donner au drap le pied de pastel d'un gros Bleu de Roy garnisses la Chaudiere d'eau de la Contenance proportionnée des pieces de Draps que vous aves a teindre et mettes y

7^{lb} Bois de Campeche varlopé dans un sac } par piece
1^{lb} galles pulverisées }

9^{lb} Bois Jaune et 2 Lamis Redou dans un sac faites Bouillir le tout pendant six bonnes heures, Rafraichisses ensuite le Bain, et passes y les draps pendant deux heures en tournant doucement et sans Relache (il faut que le Bain soit toujours sur le Bouillon) leves les draps, et Eventes les Bien, C'est le premier feu. donnes y encore deux feux meme facon en Rafraichissant le Bain; apres le troisieme feu Rafraichisses le encore, et mettes y 1^{lb} Couperose par piece fondue a part, passes les draps dans Cette Brunitture Chaleur moderée pendant une heure et demi Rafraichissant toujours le Bain, en les levant et Eventant chaque demi heure

Mettes encore 1^{lb} Couperose par piece dans la Chaudiere, et passes les draps dans Cette Seconde Brunitture Comme dans la premiere vous verres apres si vos draps sont d'un Beau Noir, et reviendres a une troisieme Brunitture, si vous ne les trouves pas tels. C'est au jugement de Celui qui dirige la Couleur d'augmenter ou diminuer la Couperose. Vos draps etant d'un Beau Noir faites les laver a la Riviere; a deux Bonnes prousses, et de suite au foulon a l'eau seule, apres quoy si vous Connoisses un fouloneur avec le savon blanc en Etat de les Epurer de ce savon, qui est tres difficile a bien tirer, sans quoy il vaut mieux les priver de ce dernier Lavage

finalement passes vos draps sur un gaudage de 3^{lb} par piece bien Chaud sans Bouillir, pourvû que le gaudage ait Bouilli auparavant.

Noir — a la facon de Sedan

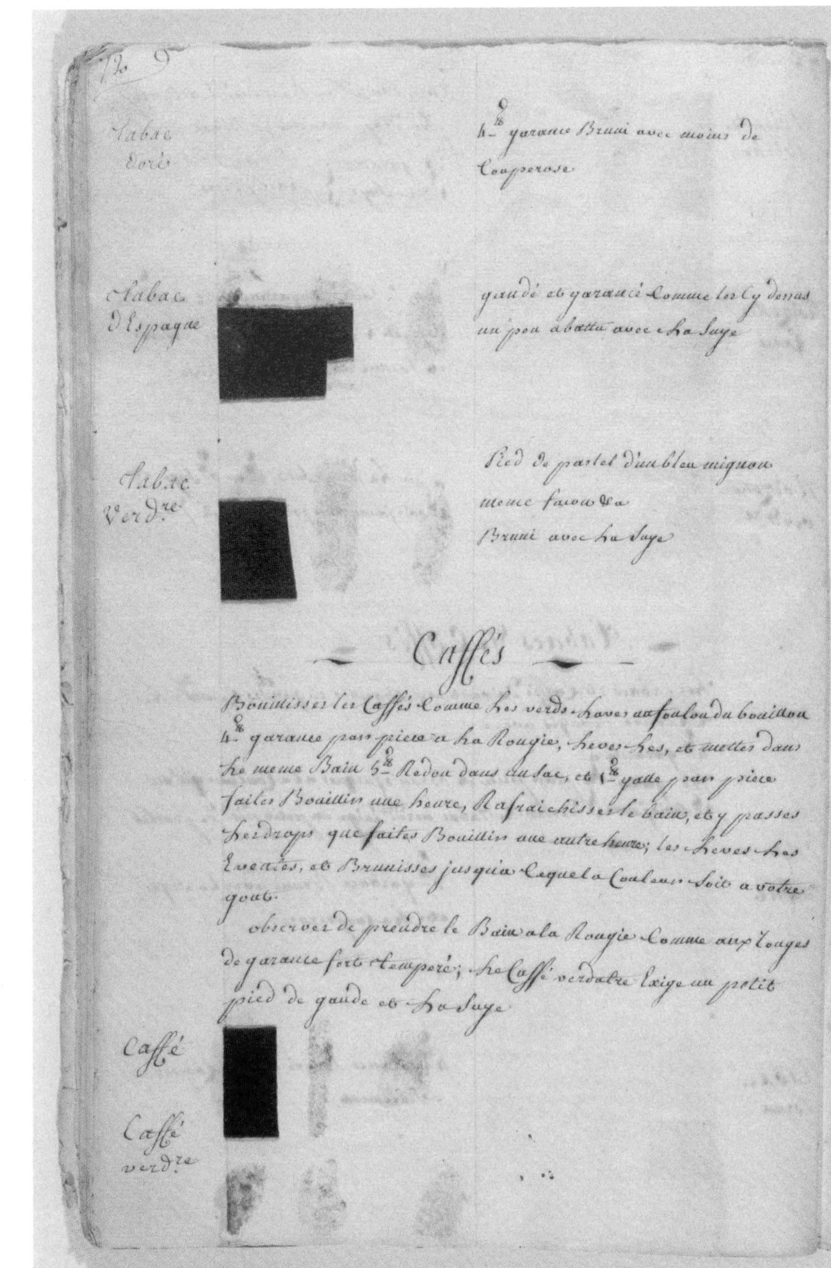

12

Tabac doré	4^{lb} garance Bruni avec moins de Couperose
Tabac d'Espagne	gaudé et garancé Comme les Cy dessus un peu abattu avec La Suye
Tabac verdâtre	Pied de pastel d'un bleu mignon meme facon et a Bruni avec La Suye

Caffés

Bouillisses les Caffés Comme les verds. laves au foulon du bouillon 4^{lb} garance par piece a la Rougie, leves les, et mettes dans le meme Bain 5^{lb} Redou dans un sac, et 1^{lb} galle par piece faites Bouillir une heure, Rafraichisses le bain, et y passes les draps que faites Bouillir une autre heure; les leves les Eventes, et Brunisses jusqu'a ce que la Couleur soit a votre gout.

observes de prendre le Bain a la Rougie Comme aux Rouges de garance fort tempéré; le Caffé verdatre Exige un petit pied de gaude et La Suye

Caffé

Caffé verdâtre

Gris de taupe Ardoises &ca

faites donner un petit pied de pastel aux taupes Ardoises &ca
Engalés Comme Suit

½ galles	par piece Revenu ½ heure
¼ Saudal	
3 on. garance	

tournés y les draps pendant une heure et Brunissés avec la Couperose jusqu'à la nuance que vous cherchés

taupe	Pied de pastel d'un Bleu d'azur 6.on Couperose pour la Bruniture
Ardoise	Pied de Pastel de Bleu Celeste ½ galles Bruni avec la Couperose
gris de Plomb	Pied de Pastel idem ¼ galles Bruni idem
gris de Rat	Pied de Pastel de Bleu mignon ¼ galles Bruni idem
gris de Rat plus f.cé	Pied de Pastel de Bleu de Blanchy ¼ galles ¼ Bois d'Inde Bruny idem

gris de rat Rouge	4l Sur la Suite d'un Bouillon de Coch.lle 1/2 galles, et un peu de bois d'inde 2 onces garances Bruni avec la Couperose
gris de rat clair	Sur un Bain fraix 1/2 Redou 1/4 Bois D'Inde 1/8 Sage Bruny idem
Agathe	Pied de Pastel d'un Bleu de Ciel passé sur la Suite d'un pendant de Violets. mieux sur une Suite de Bruniture de Soupevins
Agathe clair	Pied de pastel de Bleu de Lait Passé sur la Suite d'une bruniture des Soupevins a Lalun
gris de Prince	a la Suite de toute Couleur Vineuse avec un peu de galle et tant soit peu abattu
Rat sur Noisette	Mouillés le Drap en Blanc 1/2 Redou, 1/2 bois jaune, 1/4 Bois d'inde, 1/4 galle } Par pièce Revenu 1/2 heure Le bain Rafraichi, faites Bouillir le drap 1/2 heure Brunisses sur le meme Bain avec 1/4 Couperose par piece

Gris de Castor — a la Suite d'un Bouillon Ecarlatte ou Cramoisy ½ gatte par piece tournes Le drap ½ heure Relever le et Brunissés avec La Couperose

Gris de Perle — a la Suite d'un Bouillon Ecarlatte et jettés Le ½ du Bain 2 onces galle et brunissés avec tres peu de Couperose et avec precaution

Gris D'argent — a la Suite d'un bouillon Ecarlatte, jettés La ½ du Bain 1 once galle, et brunissés fort doucement avec La Couperose

Il en est de toutes ces petites Nuances de gris Comme de Celles du Noisette, elles varient toujours Selon Le gout d'un Chacun et Les occasions qu'on a pour Les faire. Cette Reussiroit Bien Sur La Suite d'un Bouillon ou autre Bain, qui N'aura aucune Vivacité Si on est obligé de les faire Sur un Bain fraix.

Instruction Sur les Debouillis des Couleurs en faux teint

Ce N'est point asses qu'un teinturier, ait acquis la Connoissance et La propriété des drogues qui lui Sont Necessaires, et qu'il est Reussi a les Employer avec Succes, il doit encore distinguer les bonnes Couleurs d'avec les fausses, ce qu'il fera par le moyen des debouillis Suivants

Pour mettre quelque ordre dans Cette petite instruction, je donnerai les Couleurs en 3 Classes, pour lesquelles je fixerai Les Ingrediens qui doivent etre employés dans leurs debouillis ce qui formera trois articles

Les Couleurs de la premiere Classe doivent etre debouillies
avec L'Alun de Rome Art^e. 1
Celles de la Seconde Classe avec le Savon art^e. 2
Celles de la troisieme Classe avec le tartre Rouge 3

Il Ne Suffit point d'employer dans les debouillis les ingrediants que je viens de Nommer, mais il faut a quelque chose près que la quantité et le tems en Soit determinés pour Se Convaincre avec Certitude de ce que l'on fait

Bouillon a la Suite d'une Rougie

Ecarlatte

6 ℔ Composition } Sur p.ce bouilli 2 heures
2 – tartre blanc } Laves a la pousse
Son

Rougie

1 – 3/4 Coch.lle garbelée
4 – Comp.on avec Son

Pied de pastel de bleu de Roy

Vert d'herbe

Bouillon
3 ℔ alun du Levaud } p.ce 2 heures
2 – tartre Rouge } Lavé au foulon
Le gaudage chargé par moitié de trentanel

Bouillon en Blanc

Cramoisi

6 ℔ Composition } par piece 2 heures
2 – tartre blanc } Laves a la pousse

Rougie

1 ℔ 1/2 Coch.lle avariée
3 – Comp.on
Bruni avec 1 ℔ alun de Rome sur 6 p.ces

Agathe

Pied de Pastel d'un bleu de Ciel passé sur la suite d'une Bruniture des Soupevins avec l'alun Calciné

Couleur de feu

Bouillon a la suite
6 ℔ Composit.on } par piece 2 heures
4 – Sel.. } Lavé au foulon
1 – 1/2 tartre blanc }

Rougie

1 – 1/2 Coch.lle avariée
5 – Composition bouilli 1/2 d'heure sur la rougie

Soupevin

Bouillon façon ordinaire
6 ℔ Composition } par p.ce 2 heures
2 – tartre Rouge } Laves a la pousse

Rougie

1 – 3/4 Coch.lle avariée
3 – Comp.on
Bruni avec 1 ℔ alun Calciné en divers Rejets

Soupevin Royal

Bouillon
4 ℔ alun de Rome
1 – Sel Commun
2 – tartre Rouge } par p.^ce 2 heures ½
1 – Composition } Lavés au foulon

Rougie
1 – 3/4 Coch.^lle avariée
1 – tartre Rouge } Bouilli une heure
½ – Composition } sur la Rougie
Bruni avec 4 ℔ orseille preparée dans un Chauderon

Soupevin Superieur

Bouillon a la suite de Rougie
6 ℔ alun despagne } par p.^ce 3 heures et reposé
2 – tartre Rouge } 24 heures si on le garde plus
Longtems en Eté il se chapelette defaut
irremediable Lavé au foulon

Rougie
1 ℔ 3/4 Coch.^lle fine
4 – Comp.^on } Bouilli 1 heure sur la Rougie
Bruni avec une livre Cendre gravelée en plusieurs
Rejets le bain pris temperé

Amarante Royal

Bouillon en tout point Comme les soupevins cy dessus
Reposé 6 jours en hiver, Lavé au foulon

Rougie a la suite d'autre
1 ℔ 3/4 Coch.^lle
1 – tartre blanc } Bouilli 1 heure sur la Rougie
1 – Comp.^on
Bruni avec L'eau Chaude seulement

Vessina

Bouillon
5 ℔ alun de Rome } Par p.^ce 3 heures Reposé
2 – tartre Rouge } 7 jours Lavé au foulon

Rougie a la suite d'autre
1 – ¼ Coch.^lle avariée
2 – Tartre Rouge
½ – Comp.^on Bouilli 1 heure sur la Rougie

Garance

Bouillon
5 ℔ alun de Rome } par piece 3 heures Reposé
2 – tartre Rouge } 8 jours, Lavé au foulon

Rougie
8 ℔ garance non Robée
¼ galle pulverisée
La garance pilée mise dans la Chaudiere venant d'etre
garnie d'eau a eté infusée jusques que le bain a eté
d'une chaleur temperée levée sur le point de bouillir

Vert Bronzé

Pied de pastel de Bleu de Roy Lavé au foulon, passé sur La fin d'un gaudage, et Ensuite Bruni avec le bois d'inde

Langoute

Bouillon sur la suite d'un bouillon Ecarlatte
4 ℔ furlet
1 – tartreblanc } par piece 2 heures
4 Composition }
Rougie sur un bain frais
5 onces Coch.^lle avariée
2 ℔ Comp.^on

Caffé verd.^re

Bouillon
3 ℔ Alun du Levant }
1 – tartre } p. p.^ce 2 heures
passé sur La fin d'un gaudage
Rougie
3 ℔ garance Commune }
1/2 galle } Bruni avec la sauge et La Couperoze

Orange.

Bouillon a la suite d'un bouillon Ecarlatte
Rougie a la suite des jujubes
3 onces Coch.^lle
2 ℔ Composition

Verd de Saxe

Bouillon
4 ℔ alun de Rome } par piece 2 heures
2 – tartreblanc } Lavez de suite au foulon
16 ℔ bois Bois jaune varlopé Levenu une heure et demi dans le meme bain; le Bois Levé, et le Bain Rafraichi a pouvoir a peine y supporter les doigts, 1 ℔ 1/4 Comp.^on Bleu palien tourner vite Levez sur le point de bouillir

La Bonne huille de vitriol et la qualité superieure d'Indigo font La bonne Comp.^on et le brillant de Cette Couleur, qui est toujours plus vive et s'unit mieux sur son bouillon, ou sur une suite, il faut etre attentif quand on la fait de Voir qu'elle tire Vite a deffaut Relevez la sur le tour, et jettez dans le bain 1 ℔ 1/2 ou 2 Alun Calciné Reduit en poudre fine, Remettez dedans Lavez au foulon

Celadon Saxe

Bouillis a la Suite des verts de Saxe sans autre ingredient, et sur le meme bain 3/4 bois jaune par piece Revenu une heure le bain Rafraichi a pouvoir a peine y supporter la main 3 onces Comp.^on Leves sur le point de bouillir meme observation pour cette Couleur comme aux Verts de Saxe, si elle ne tire pas Releves sur le tour et jettes 1^lb farine par piece cette Couleur est tres difficile a unir

Jonquille vif

a la Suite d'un bouillon Ecarlatte 3 p.^ces
30^lb fustet Revenu 2 heures 1/2
6 – Cristal de tartre
30 – Comp.^on
Les draps toujours tournés 1/2 heure sans bouillir

Poudre a Canon des Anglais

Pied de pastel de bleu d'azur
1/2 galle par piece a l'Engalage
Bruni avec la Couperose

Roze

Sur un bain fourni de Comp.^on ou on avoit Repassé des Ecarlattes gatés, les draps bien mouilles au foulon

1^lb Composition
1/4 Coch.^lle avariée
1^lb tartre
} par piece bouilli une heure sur la Rougie

Comme il a eté fait plusieurs pendants sur le meme Bain avant de Repasser les Ecarlattes on peut tabler sur 1 – 1/2 de Composition

Lilla

Pied de pastel de bleu de Ciel clair
Bouillon
3^lb Alun d'espagne
1/2 – tartre
} p.^r p.^ce 2 heures

Rougie a la Suite de violets

1/4 cochenille
1/2 tartre a la Rougie
} Bouilli 1/2 d'heure

Gorge de Pigeon clair

Lavés en Blanc au foulon et a la Suite des Violets
2 onces Coch.^lle par p.^ce Lavés a la pousse
Bruniture 1/4 Bois de Campeche
preparé dans un Chauderon

Verts de Pomme Saxe

Bouillis avec de verts de Saxe
jottes le 1/3 du bain, ou ont ete faits les
verts de Saxe; le Bain Rafraichi, mis
dedans en tournant vite, et Levés
apres avoir bouilli qques minutes

Maure doré

Bouillon
4 ℔ alun du Levant
2 – Tartre Rouge } par p.^ce 2 heures
Reposé et lavé au foulon
gaudé sur le pied d'un jaune doré
6 ℔ garance a la rougie
Bruni avec tant soit peu de Couperose

Pain Bis

Les draps Bien Lavés au foulon, et a la
Suite de toute Couleur Brunie avec la Suye
7 a 8 ℔ Suye Bouilli pendant une heure et
demi, le Bain Rafraichi, les draps mis
dedans et levez sur le point de bouillir

Marron

2 ℔ Sandal
3 – Redon
1/2 galle } par p.^ce Revenu une bonne heure au Bouillon
Le bain Rafraichi, les draps passez dedans
pendant une heure en Bouillissant
Bruni ensuite avec la Couperose
Le Bain pris temperé

Couleur dor

a la Suite du bouillon Ecarlatte
Le drap mouillé en Blanc
6 ℔ fustet Revenu 3 h. le bain Rafraichi
6 – Compo.^on
2 onces Coch.^lle Bouilli 1/4 d'heure

87

Cassie

5 Pieces a la Suite d'un bouillon Ecarlatte
les draps mouillés en Blanc
7½ lb Tartre Revenu 2 h. le bain Rafraichi
2 lb Composition
2 — garance fine
levés au point de bouillir

Vert d'herbe

Pistache manqué Repassé sur la Cuve
Bouilli avec d'autres verts, et gaudé avec
moitié trentanel

Jaune

Bouillon
3 lb alun de Suede } p. pce lavé au foulon
1 — tartre Rouge }
7 lb gaude vieille donnes huit Bouts
Bien lavé au foulon

Caffé au Lait

5 Pieces a la suite d'un Bouillon Ecarlatte fait
a la suite d'une Rougie
5 lb tartre qui avoit servi a le bouillon Revenu
¾ d'heures, une livre de garance fondue apart
5 lb Compon le bain bien Rafraichi, et les
draps tournes vite jusques au point de bouillir
on peut faire un Second pendant meme
façon et notament la Compon

Chocolat au Lait

5 pieces sur la suite des 5 cy dessus le
meme sur tartre et 2 lb neuf Revenu
½ heure
2 lb Tartre } Sur les dernieres 5 pieces
2 onces Cochle } faites ensuite Comme
1 lb garance } les cy dessus

Biche Anglais

a la suitte d'un bouillon Ecarlatte les draps mouillés en blanc et le bain rafraichi

½ ℔ garance } pour piece
1 ℔ Saye }

Cette Couleur donne sur le Caffé au lait et peut passer pour telle

Vert de capre

Pied de pastel de bleu d'azur foncé 6 ou 7 Bouts sur le gaudage et tiré lavé au foulon

Pourpre Anglais

Bouillon

5 ℔ Comp.on } par piece 2 heures
2 – tartre blanc } sans interruption
1 – alun de Rome }

Rougie

1 ℔ ½ Coch.lle avariée
2 – ½ Comp.on bouilli ¼ d'heure sur la Rougie

Bruni sur un bain frais avec le bois de Campeche bouilli dans un Chauderon 24 heures a l'avance afin qu'il aît le tems de Reposer ce qui est necessaire la quantité n'est point fixée, on leve a la nuance qu'on desire

Cette Couleur est tres difficile a unir parceque la Composition ne s'accorde point avec le bois d'inde, C'est pourquoy on ne scauroit assés prendre de precautions quand on la brunit il est tres Consequant de Laisser Reposer le bain chargé de la substance du bois qu'on fait Bouillir a part parcequ'il est chargé d'un marc qui occasionneroit des placards. Cette Bruniture se fait sur un bain frais pris tres temperé, partie de la Braise tirée s'il y en a trop, et la porte du fourneau fermée. Si malgré cela on n'unit Comme il arrive quelque fois, jettés dans le meme Bain une livre alun par piece et passer y les draps en guise de debouilly

Violet Anglais

Bouillon
6 ℔ alun d'espagne
2 — tartre Rouge } p. p.ce 3 heures Reposé et bien lavé au foulon

Rougi comme le pourpre cy devant et Bruni de meme

Rat sur Noisette

Engalagé
½ Rodou
¼ bois de Campeche
¼ bois jaune
2 onces galle } Par p.ce Reseau et Bouilli dans un sac ½ heure

Le Bain bien Rafraichi, les draps mis dedans laisses Bouillir ½ heure
Brunis sur le meme bain avec ¼ Couperose par p.ce divisé en 3 Rejets rafraichissant toujours le Bain

Au Levant Cette Couleur de son Engalage, elle a ordinairement une verdeur sur le poil qui se dissipe aisement si on les fait laver au foulon non a la brosse si vous avez une Couleur a imiter qui Exige plus de Rougeur augmentés la galle

Limon

Bouillon
3 ℔ alun de Suede
1 — tartre Rouge } Par piece 2 heures Lavé au foulon

8 Bouts sur un gaudage avec ⅔ de frontanel

Jaune doré

Bouilli Comme les cy dessus Lavés au foulon
8 Bouts sur un gaudage chargé a raison de 16 ℔ gaude Lavé au foulon
¼ garance sur un bain fraix passé pendant une heure Lavé avant de bouillir

Cire doré

Bouilli avec les cy dessus gaudé de meme
2 ℔ garance non robée a la suite des jaunes dorés passé pendant une heure Lavé avant de Bouillir

Canottes

Bouillon
3 ℔ alun de Rome } par piece 2 heures
1 — Tartre Rouge } Lavé au foulon
gaudé Comme les jaunes dorés lavé au foulon
4 ℔ garance Nourobée infusée dans un bain
fraix jusques a une chaleur moderée
1/4 galle mettes dedans, fait dans
3/4 d'heure sans Bouillir

Festiqui

Bouilli avec les limons de lautre part
gaudé de meme
5 a 6 ℔ Sage dans le bain des Canottes
Bouilli demi heure
fait dans 3/4 d'heure

Taupe

Pied de Pastel de bleu de Ciel f.ce
Engalage
1/2 galle
1/4 Sandal } p. p.ce bouilli 1/2 heure
3 onces garance }
Le drap passé dans Cet Engalage a
Bouilli une heure
Bruni avec 1 ℔ Couperose lavé au foulon

Ardoise

Pied de pastel de bleu de Ciel
Engalé a la suite des taupes avec 1/2 galle
sur 6 p.ces les draps passés dans Cet
Engalage ont bouilli une heure
Bruni avec la Couperose

Rouge Bruni

Bouillon
5 ℔ alun de Rome } p. p.ce 3 heures
2 — Tartre Rouge } Reposé 4 jours
Rougie a la suite de garances
6 ℔ garance Nourobée
1/2 galle. Bruni avec 1/4 Couperose
en deux Rejets le bain tres temperé
il faut bien faire tirer la garance
et Laisser Bouillir 2 heures a Compter
depuis qu'on mets dedans

Couleur de Roy

Pied de partel d'aubleu mignon
Bouillon
5 lt alun du levand } par p.ce 2 heures reposé
2 – tartre Rouge } lavé au foulon et Rougis
6 lt de garance Bruni avec la Couperose
Comme les Rouges Bruni

Couleur de Roy

Pied de Partel de bleu d'azur
Bouillon
5 lt alun de Suède } p. p.ce 3 heure Reposé
2 – tartre Rouge } 3 jours lavé au foulon
Rougie
6 lt garance
1 – ½ fernambouc
¼ – galle. Bouilli ½ heure sur La Rougie
Point de Bruniture

Noizette Roug.re

a la suite d'un garançage fourni de fernambouc
2 lt suye par piece Revenu demi heure
Les draps passés dedans pendant autre demi heure
Bruni sans ingrédients sur la bruniture des Rouges Brunis

gris Cendré

a la suite d'un bouillon de Cramoisi
Les draps mouillés au foulon
2 onces galle p. p.ce Revenu demi heure
Les draps passés autre demi heure
Bruni avec une once Couperose

Roze Pale

Bien Lavé en Blanc au foulon
passé sur La suite de trois pendants de pourpres, ou violets faits sur un Bain frais

Jujube

Bouillon a la suite

5 ℔ tartre
4 – Comp.^on
1 – tartre blanc } par piece 2 heures

Rougie

3/4 – Coch.^lle avariée
2 ℔ Composition
a Bouilli 1/4 d'heure sur la Rougie

Prune sur truffe

1 ℔ Redon
1 – 1/2 Sandal
1 – 1/2 garance commun } Par p.^ce sur un Bouillon d'Ecarlatte Revenu 1 heure passés y les draps une autre heure

Bruni avec la Couperose

Bleu de Prusse

Bouillon

3 ℔ alun
2 – tartre blanc } p.^ce p.^ce 2 h lavé au foulon

Sur le meme Bain apres en avoir jetté un quart rafraichi et pris temperé
4 ℔ Comp.^on faite avec l'indigo St Domingue sur 5 p.^ces, tournes vite, et Leves dans 1/4 d'heure si on la laisse plus cette Couleur ternit beaucoup

Bleu de Prusse

une seule p.^ce faite sans Bouillon sur le Bain des cy dessus, apres avoir ete passée sur le tour dans ce meme bain pour la Mouiller
1 ℔ Comp.^on faite comme cy dessus depuis 20 jours
1 – 1/2 farine levée dans un quart d'heure

Bleu de Blanchi Saxe

Bouilli une heure sur la suite du Bouillon des Bleus de prusse
apres les Bleus jettés le 1/3 du Bain
Bien Rafraichi passes y les draps
Il faut observer de ne point les mettre a la Rame avec la forte chaleur

93

Violet d'Evêque		Pied de partet de bleu de Roy Bouilli avec de Soupevins 1 ℔ 1/2 Coch.lle fine par pièce
herbe foncé		Pied de partet de Bleu de Roy Bouillon 3 ℔ alun de Suede } p. p.ce 2 h. Lavés au foulon 2 – tartre rouge } gaudage 1/6 gaude 5/6 trentanel
[illegible]ort brun		Pied de Partet de Bleu d'azur foncé Bouilli avec les herbes cy dessus gaudé avec le trentanel seul et Bruni avec La Couperose
Soupevin		Bouillon sur un Bain fraix 5 ℔ Alun de spaque } par p.ce 3 heures reposé 2 – tartre Rouge } et Lavé au foulon Rougie a la suite 1 ℔ 1/2 Coch.lle sur 5 pièces 2 – tartre par pièce Bruni avec 1 ℔ Cendre gravellée par pièce
Amarante.		Bouillon avec les cy dessus Rougie apres les memes 7 ℔ 1/4 Coch.lle sur 5 pièces Bruniture sur un Bain fraix 1 ℔ fernambouc p. p.ce Revenu une heure, Le Bain Rafraichi 1 ℔ Cendre gravelée p. p.ce faits dans un seul Rejet

N° 94

Pourpre Violet	Pied de Partel de bleu d'azur foncé Bouillon 4 lb alun d'espagne 2 — tartre Rouge } p. p.ce 3 heures Reposé Rougie a La Suite des Soupevins 4 lb garance 1 — Alarie } Sur 4 p.ces infusé 3 heures une Livre tartre a La Rougie
Pourpre foncé	Bouillis avec les cy dessus Rougie a la Suite des dits 4 lb 3/4 garance 1 — Alarie } Sur 4 p.ces infusé dit une Livre tartre a la Rougie
Soupevin Royal	Bouillon a La Suite 3 lb Alun d'espagne 3 — Sel de Cardonne } par p.ce 2 heures Reposé 2 — tartre Rouge } et Lavé au foulon Rougie a La Suite 10 lb garbeau 10 — mamout } Sur 5 p.ces infusé quatre 5 Alarie } heures Ensemble 10 — tartre a La Rougie
Lilla ou gorge de Pigeon f.cé	Pied de partel de Bleu de Ciel Passé Sur La Suite d'un bouillon chargé d'alun pour Les mouiller Seulement passé Ensuite Sur la Suite des Violets
Couleur de feu	Bouillon a la Suite de Bouillon 25 lb Composition 12 — tartre Blanc } Sur 5 p.ces 2 heures 20 — fustet choisi } Rougie Sur Rougie 5 lb Coch.lle } Sur 5 p.ces infusé ensemble 2 — Alarie } 24 h. C. C. au Colorant. 14 lb Composition

9b

Vert de Mer f.cé

Pied de Pastel de petit Bleu de Roy
Sans bouillon, sur un gaudage tiré chargé
au trentanel
7 Bouts
7 dit de Rejet pour unir
passé a la suite d'une musique des verts
D'herbe cy dessous

Crize

a la suite d'une Rougie ou il a eté Repassé
de Soupevins, les draps passés ½ heure sur
le meme Bain Rafraichi

Rougie

7 lb ½ granille
20 Comp.on } sur 6 pieces

herbe très foncé

Pastel de Bleu de Roy f.cé
Bien Bouilli, et gaudé avec le trentanel
seul, qui lui donne mieux le gout
que la gaude
Bruni avec 1 lb campeche par piece

Bouillon de vert ord.re

6 onces Comp.on de bleu de prusse par p.ce
Lavé au foulon

Vert Naissant sans pastel

7 Bouts sur un Beau gaudage tout a la
gaude
9 Bouts de Rejet

Soupevin qui a eté fort recherché au Caire

Bouilli sur un Bain fraix 6 p.ces

5 ℔ Comp.on
2 – tartre blanc
1/2 – garance } p. p.ce bouilli 3 heures

Rougie

1 ℔ 3/4 garbeau
3 – Comp.on } par piece

Bruniture sur un bain fraix

1 ℔ 1/2 orseille d'herbe par p.ce et passes ensuite sur 4 bains d'eau chaude tous neuf

L'orseille d'herbe est la plus belle des Brunitures, elle Epargne le Colorant parcequelle donne du fonds a la Couleur, elle a encore l'avantage sur celle de terre encequ'on peut la laisser Bouillir sans se degrader, mais elle est plus sujette a chappeller, pour l'Eviter, il faut jetter le drap, en levant dans la chaudiere a coté pleine d'eau fraiche

Si vous voulés Rendre le soupevin plus Rempli, et plus vineux suprimés 2 ℔ Comp.on au bouillon, et mettre 2 ℔ alun

Pourpre Anglais f.t Beau

Pied de Pastel de bleu de Ciel

Bouillon

4 ℔ Comp.on
2 – tartre } Bouilli 2 heures

Rougie

1 ℔ Coch.lle
3 – Comp.on } par piece

Bruniture

6 ℔ Bois de Campeche p. p.ce Revenu 2 heures, passes les draps 1/2 heure Rafraichisses le Bain, jettés y 2 ℔ alun, 1 ℔ tartre p. p.ce passés les draps faites Bouillir une heure

Chamoix

Sur un Bain fraix

2 ℔ fustel par p.ce Revenu 1 heure 1/2
Bien Rafraichi le Bain

3 ℔ Cristal de tartre
8 – Composition } par piece

tournes vite Comme a une Rougie d'Ecarlatte 2 leves dans 1/4 d'heure

Let it all boil during at least six hours, then let the bath cool down, and pass the cloths in it for two hours, turning them gently and continuously (the bath must be kept constantly boiling). Heave out the goods, and air them well; this is the first blacking. Give them two more blackings in the same manner after letting the bath cool down. After the third blacking let it cool down again, and put in 1 ^{lb} copperas per end that will have been dissolved separately. Pass the cloths in this saddening, at moderate temperature during one hour and a half, constantly cooling down the bath, heaving them out and airing them every half hour.

Put 1 ^{lb} copperas more per end into the vessel; and pass the cloth in this second saddening as in the first one.

You will then see if your cloths are of a beautiful black, and will proceed to a third saddening if you do not find it so. It is up to the dyer's judgement to increase or diminish the quantity of copperas. Your cloths being of a beautiful black, send them to be streamed, two times, and immediately afterwards to be scoured in the stocks, first with water only and then with white soap, if you know a fuller who is good at washing the cloth clean from this soap, which is very difficult to entirely eliminate. Otherwise it is better to spare them this last scouring.

Finally, pass your cloth in a welding with 3 ^{lb} weld per end, quite hot without boiling, provided that the weld liquor has been boiled previously.

Noir
Black In the manner of Sedan

Blacks with tartar or in Albert's manner[51] [p. 74]

Boiling of blacks

1 ^{lb} red tartar
1 ^{lb} copperas per end boiled for 2 hours; wash in the stocks

Empty the vessel, and in a fresh bath

15 ^{lb} logwood
¼ ^{lb} verdigris per end, simmered for one hour. When you have series of blacks to make, subtract 4 to 5 ^{lb} of wood, the bath being concentrated enough

The logwood and verdigris having boiled for one hour, pass the cloths in the bath for two hours, keeping it boiling all the time.

Heave out, air them well, and pass them again in the same bath for two more hours. If they are of a beautiful black, let them be washed in the stocks and passed in a weld bath, which is more necessary for these than for any other kind of blacks, to dull the violet tinge that this quantity of logwood gives them.

Noirs
Blacks With tartar in Albert's manner

Woaded Blacks
Woaded black in the ordinary manner
Let the cloth be given a woad ground of an azure blue
Gall as follows

2 ^{lb} powdered gallnut
3 ½ ^{lb} logwood
1 ½ ^{lb} *redoul* per end. Simmered for two hours

Pass the cloths in this galling bath during 1 ½ hour boiling constantly, heave out and air well. Then dye them in a madder bath, with 3 ^{lb} common madder, and into the same bath, as saddening, put 3 ^{lb} copperas previously dissolved before being poured into the vessel.

Turn the cloth in it, heave out, and air well; give a second dip with ½ ^{lb} copperas, increasing the dose if the colour is not to your liking. Lastly, heave out the cloth, and put it back into the vessel, where it must spend the night. Observing that there should not be any fire left burning in the furnace; even remove the embers before saddening.

The next day, heave out the cloth, start having it scoured of the copperas in the stocks for ½ an hour. Let it be washed another time with soap if the fuller is good at scouring it from it afterwards. After this operation, pass it in a strong weld bath.

[p. 75] **Mole greys (*gris de taupe*), slates (*ardoises*), etc.**

Let a light woad ground be given to mole greys, slates, etc.
Let them be galled as follows

½ lb galls
¼ lb sanders
3 oz madder — per end, simmered for ½ an hour

Turn the cloth in it for one hour and sadden with copperas until the shade you wish is obtained.

Gris de taupe Mole grey	Woad ground of an azure blue 1 lb copperas for its saddening
Ardoise Slate	Woad ground of a celestial blue ½ lb gallnut Saddened with copperas
Gris de plomb Lead grey	Woad ground ditto ¼ lb gallnut Saddened ditto
Gris de rat Rat grey	Woad ground of a dainty blue ¼ lb gallnut Saddened ditto
Gris de rat plus foncé Darker rat grey	Woad ground of an off-white blue ¼ lb gallnut ¼ lb logwood Saddened ditto
[p. 76] *Gris de rat rougeâtre* Reddish rat grey	In a reused boiling with cochineal ½ lb gallnut, and a little logwood 2 oz madder Saddened with copperas
Gris de rat clair Light rat grey	In a fresh bath ½ lb *redoul* ¼ lb logwood 1 lb soot Saddened ditto
Agathe Agate	Woad ground of a celestial blue Passed in a reused bath of violets or better still, in a reused saddening for winesoups
Agathe clair Light agate	Woad ground of a milky blue Passed in a reused saddening for winesoups with alum

Gris de Prince Prince's grey	In a reused bath of any wine colour With a little gallnut and just slightly saddened
Rat sur noisette Rat on hazelnut	Wet the cloth in the undyed state 1 lb ½ *redoul* ½ lb old fustic ¼ lb logwood ¼ lb gallnut per end. Simmered for ½ an hour The bath cooled down, let the cloth boil in it for ½ an hour Sadden in the same bath with ¼ lb copperas per end
Gris de castor Beaver grey	In a reused boiling for scarlet or crimson [p. 77] ¼ lb gallnut per end Turn the cloth for ½ an hour. Heave it out and sadden with copperas
Gris de perle Pearl grey	In a reused boiling for scarlet and throw out 1/3 of the liquor 2 oz gallnut and sadden with very little copperas and with caution
Gris d'argent Silver grey	In a reused boiling for scarlet, throw out ½ of the liquor 1 oz gallnut And sadden very slightly with copperas

The same can be said about all these little shades of greys as about the shades of hazelnuts, they always vary according to one's taste and the opportunities one has to make them. Some would come out nicely in a reused boiling for scarlet or other etc., that would not have any vividness if one was obliged to make them in a fresh bath.

(III)

Instruction on the testing procedures for false colours[52]

It is not enough for a dyer to have acquired knowledge on the drugs that are necessary to him and on their properties, and to have managed to employ them with success. He must also distinguish the fast colours from the false ones, which he will do by using the following testing liquors.

To put some order in this brief instruction, I shall deal with the colours in 3 classes, for which I shall give the ingredients that must be employed in their test liquors, which will make up three parts.

Colours of the first class must be boiled

with Rome alum	part	1
those of the second class with soap	part	2
those of the third class with red tartar	part	3

It is not enough to employ the ingredients I have just mentioned in the testing liquors, but the quantities and times must also be determined fairly precisely, to be quite certain of what one is doing.

1 [p. 78]

Test with Rome alum

For one pound of water[53] put one ounce[54] of alum, and when the water reaches a rolling boil, put in the sample, which must weigh about one dram (*gros*),[55] and let it boil for a full quarter of an hour.

Observe to increase everything in proportion (and never to boil different colours together) when there are several samples.

In this boiling with Rome alum, test scarlets, flame colour, pomegranate flower, jujube, spiny lobster colours and crimsons, wine soups, purples, violets, mauves, lilacs, linen greys, blues, slate greys, and winy greys.

Alum intensifies scarlets and flame colours to a purple colour and turns the lighter shades violet; while it nearly completely strips off the scarlets made with brazilwood and it is even more efficient on the lighter shades of this false colour.

The pomegranate flower, jujube, spiny lobster colours will be greatly damaged by this test if young fustic has been used, and will instead turn to a purple shade if it has not. The use of this wood for such colours being tolerated today, they are not considered as false.

Fine crimson is not damaged at all by this test which only makes it a little more purple, but which destroys the highest shades of false crimsons.

False purples, violets, etc. do not resist to this test, that has no effect on fine purples and violets, but if these false colours have received a woad ground, which may happen, this woad ground being a fast dye stands the test, but the red dye disappears.

This test will have no effect on fast blues, while it will strip off nearly all the colour of the false ones.

Slate greys, winy greys, etc. are also destroyed by this testing liquor when they are false.

[p. 79] **2**

Test with soap

For one pound of water, put two drams of grated white soap that must be dissolved in this water. As soon as it boils, put in the sample that you will let boil during 5 minutes, always observing the proportion if there are several samples.

In this boiling with soap, test the daffodils, yellows, lemon colours and all the shades tending towards yellow, all kinds of greens from nascent green to the darkest one, madders, cinnamons, tobaccoes etc.

This testing liquor perfectly reveals whether all these colours are fast or false; it takes most of the colour of the false ones away but it does not alter the yellows made with weld and other fast ingredients.

Fast greens are not damaged at all by it but false ones become blue, if they have received a woad ground.

Madder dyes become more beautiful when passed in soap, but they lose all the colour due to brazilwood if some has been mixed in them. Cinnamons, tobaccoes, etc. are not altered by it if they are made in the best mode of dyeing.

3

Test with tartar

Observe to use the same doses, and same proportion as for the testing liquor with alum; the tartar must be well reduced into powder before being used.

This testing liquor will serve for tawny colours, or colours made with *redoul*, sanderswood, and soot, ingredients which give a great number of different shades. When sanders and soot have been employed in too big amount, the tartar liquor will reveal it easily.

Test for blacks

Blacks will be tested with one ounce of Rome alum and as much tartar, that you will let boil for a full quarter of an hour, together with the sample. Wash it then in fresh water. If it has had the necessary woad ground, the colour will remain of a dark blue, otherwise it will turn much greyer.

[p. 80] **(IV)**

Annotations on the colours made for the Levant with their patterns, processes, and observations[56]

The colours hereunder accompanied by their processes, are made at the Royal Manufacture of [ ][57] where waters are very bad, that from the river, already warm by itself, being further infected by the spring of Las Fons which predominates in it during three quarters of the year; this water also is nearly tepid, and it is moreover laden with a lot of aluminous salts.

The spring of La Bouilhete which is used by the fullers, is softish and beside the fact that it could not produce a firm and nervous fulling, cloths are never thoroughly scoured with it. It is perceivable that a soapy cloth much injures the beauty of the colour, undoubtedly gives more trouble to the dyer, and occasions extra costs. In the course of the present annotations, it will be seen in how many ways it has been necessary to experiment, to correct the bad quality of these waters in order to manage to make a beautiful scarlet, which they keep saddening right through to the wash place. There are certain times in the year, when 25 to 30 pounds of good tin liquor are necessary to exhaust a bath, in spite of the other ingredients that need has occasioned to discover; lastly, by themselves and without any help, they sadden the dark wine soups. Nevertheless, I think that my colours are good, but that better ones can be made with good waters.

Observations on the tin liquor for scarlet

Our dyers are accustomed to make their tin liquor for scarlet with two parts of fresh water, one part of common, and more or less violent, *aqua fortis*, and one ounce of tin per part of fresh water, without sal ammoniac: this process truly gives them a tin liquor, after it has settled down and is limpid and citrine, but which has little effect on the colorant. The great quantity of sediment it leaves at the bottom of the pot, often even some undissolved tin, should be enough to convince them that their way of operating is imperfect. Anyway, the lime in tin which is only corroded cannot be well dissolved without sal ammoniac. As a result, most of them make very ordinary colours: which shows that the best way is not always to compose the *aqua fortis* per third part; even though it would dissolve well with this dose, nevertheless the colour would not be any brighter, since experiments have proved that it is the most subtle part of the tin lime, that gives vividness to the colouring parts of cochineal, which by itself only gives a dull colour. I shall add that it should not be overlooked, that it is a very different matter to prepare tin liquor in summer, or in winter; in summer, the heat greatly helps, and in winter the dose of *aqua fortis* must be increased.

There is still another observation to be made about the different qualities of *aqua fortis*, which some people will use indifferently. The chemist who provides us does not always take care to provide it to us at the same degree of acidity. It is therefore necessary to know how to differentiate, in order to distinguish between an *aqua fortis* that is too violent and one which is too weak, and to work with them with caution not to expose oneself to unnecessary labour and expenses. I have employed some, which worked very slowly and nevertheless caused me to fail in the making of my tin liquors, by deciding to increase the dose because of the weakness of the reaction; and although I did manage to make strong tin liquors with it, that I used (not to my satisfaction), I advise not to use this kind, and to discard it because it is not of a quality appropriate for scarlet (it is the kind used by etchers, which, although slow, remains active for a long time). The liquor it gives after settling down remains whitish and clear like water, with a lot of yellowish sediment at the bottom of the pot. Eventually, I recognised it had to be proscribed because it never brings out the colour well, and because, by nature of its violence, it produces whitish stains on the cloths, which can only be remedied with the scissors. [p. 81]

Here is the way I have adopted to make tin liquor after many experiments, because it is the one with which I have been most successful.

For example, I take 6 ^{lb} good quality *aqua fortis* which I dilute with 12 ^{lb} river water in winter, or from the well in summer. Little by little, I dissolve 3 ounces sal ammoniac previously reduced to powder (it can be replaced by common salt). Then I put in one pound of very pure granulated tin: it dissolves more or less slowly because the *aqua fortis* does not always possess the same degree of acidity. This liquor then turns a lead grey verging to black, which is the certain sign of success. I leave it in this state to react quietly during three quarters of an hour, after which I stir the tin that has not finished dissolving with a stick, I add one pound of *aqua fortis* to it, which revives the reaction. I come back to this last operation again twice, every three quarters of an hour. After that I let it rest entirely for eight or ten hours. It becomes limpid, of a beautiful colour of gold solution, and with nearly no sediment at the bottom of the pot. I have experienced that this way of prolonging the work of the *aqua fortis* had more action on the tin, and that its most subtle parts did not evaporate as easily as in too precipitate a dissolution. Moreover, this liquor keeps for one month and more in summer, and two in winter, while the other gets milky and opaque and gets spoilt within three or four days.

Tin liquor made with saltpeter would be even preferable, but, its making taking a long time, it cannot be used when one has a lot of work to dispatch.

Ecarlatte [p. 82]

Scarlet

Boiling in a reused finishing bath for scarlet

6 ^{lb} tin liquor
2 ^{lb} white tartar
Bran — per end. Boiled for 2 hours
Wash in the stream

Finishing bath

1 ¾ ^{lb} cochineal garbled
4 ^{lb} tin liquor with bran

Vert d'herbe
Grass green

Woad ground of a King's blue
Boiling
3 lb alum from the Levant
2 lb red tartar per end 2 hours
Washed in the stocks
The welding bath loaded half and half with flax-leaved daphne

Cramoisi
Crimson

Boiling on undyed cloth
6 lb tin liquor
2 lb white tartar per end 2 hours
Wash in the stream
Finishing bath
1 ½ lb spoilt cochineal
3 lb tin liquor
Saddened with 1 lb Rome alum for 6 ends

Agathe
Agate

Woad ground of a celestial blue.
Passed in a reused saddening bath of wine soups made with calcined alum.

Couleur de feu
Flame colour

Boiling in a reused finishing bath
6 lb tin liquor
4 lb young fustic
1 ½ lb white tartar per end 2 hours
Washed in the stocks
Finishing bath
1 ½ lb spoilt cochineal
5 lb tin liquor, boiled for ¼ of an hour in the finishing bath

Soupevin
Wine soup

Boiling in the ordinary way
5 lb tin liquor
2 lb red tartar per end 2 hours
Wash in the stream
Finishing bath
1 ¾ lb spoiled cochineal
3 lb tin liquor
Saddened with 4 lb calcined alum in several dips

[p. 83] *Soupevin royal*
Royal wine soup

Boiling
4 lb Rome alum
1 lb common salt
2 lb red tartar
1 lb tin liquor per end 2 ½ hours
Wash in the stocks
Finishing bath
1 ¾ lb spoiled cochineal
1 lb red tartar
½ lb tin liquor, boiled for one hour in the finishing bath
Saddened with 4 lb orchil[58] prepared in a separate cauldron

Soupevin supérieur
Superior wine soup

Boiling in a reused finishing bath
6 lb alum from Spain
2 lb red tartar — per end for 3 hours and let to cool down for 24 hours. If kept longer in summer, it gets stained, which is an irreparable damage
Washed in the stocks
Finishing bath
1 ¾ lb fine cochineal
4 lb tin liquor, boiled for 1 hour in the finishing bath
Saddened with one pound of lees ash in several dips, in a lukewarm bath

Amarante royal
Royal amaranth

Boiling in all ways as for the above described wine soups
Let to rest for 6 days in winter, washed in the stocks
Finishing in a reused finishing bath
1 ¾ lb cochineal
1 lb white tartar
1 lb tin liquor, boiled for 1 hour in the finishing bath
Saddened with hot water only

Vessina
Vetch flower

Boiling
5 lb Rome alum
2 lb red tartar — per end for 3 hours. Left to rest for 7 days
Washed in the stocks.
Finishing in a reused finishing bath
1 ¼ lb spoiled cochineal
2 lb red tartar
½ lb tin liquor, boiled for 1 hour in the finishing bath

Garance
Madder

Boiling
5 lb Rome alum
2 lb red tartar — per end for 3 hours. Left to rest for 8 days
Washed in the stocks
Finishing
8 lb unstripped madder
¼ lb powdered gallnut
The madder, pounded and put into the vessel freshly filled with water, was let to infuse until the bath reached a moderate heat. Heave out when about to boil

Vert bronze [p. 84]
Bronze green

Woad ground of a King's blue
Washed in the stocks, passed at the end of a welding, and then saddened with logwood

Langouste
Spiny lobster

Boiling in a reused boiling for scarlet
4 lb young fustic
1 lb white tartar
4 lb tin liquor — per end for 2 hours
Finishing in a fresh bath
5 oz spoiled cochineal
2 lb tin liquor

Caffé verdâtre
Greenish coffee — Boiling
3 lb alum from the Levant
1 lb tartar per end for 2 hours
Passed at the end of a welding
Finishing
3 lb common madder
½ lb gallnut
Saddened with soot and copperas

Orange
Orange — Boiling in a reused boiling for scarlet
Finishing in a reused finishing bath for jujubes
3 oz cochineal
2 lb tin liquor

Vert de Saxe
Saxon Green — Boiling
4 lb Rome alum
2 lb white tartar — per end for 2 hours
Wash immediately in the stocks
6 lb chipped old fustic, boiled for one hour and a half in the same bath; the chips removed, and the bath cooled down to hardly be able to put in one's fingers.
1 ¼ lb tin liquor
Stir well, turn fast. Heave out when on the verge of boiling

The goodness of the oil of vitriol and superior quality of the indigo make the goodness of chemick and vividness of this colour, which is always brighter and more even using its boiling bath, or a reused dye bath of this colour. One must pay attention when making it that it should come up quickly. If it does not, heave the cloth out onto the reel, and strew in 1 lb ½ or 2 of calcined alum reduced into thin powder. Heave back in, wash in the stocks.

[p. 85] *Céladon Saxe*
Saxon celadon — Boiled as the Saxon greens without more ingredients, and in the same bath
¾ lb old fustic — per end. Boiled for one hour, the bath cooled down to hardly be able to put in the hand
3 oz tin liquor
Heave out when on the verge of boiling

Same observation for this colour as for Saxon greens: if it does not come up, heave the cloth out onto the reel and strew in 1 lb calcined alum powdered (*farine*) per end. This colour is very difficult to make even.

Jonquille vif
Bright daffodil — In a reused boiling for scarlet, for 3 ends:
30 lb young fustic, boiled for 2 ½ hours
6 lb crystals of tartar
3 oz tin liquor
The cloths constantly turned for ½ an hour without boiling

Poudre à canon des Anglais
English gunpowder — Woad ground of an azure blue
½ lb gallnut — per end in the galling bath
Saddened with copperas

Roze Pink	In a bath rich in tin liquor, in which spoilt scarlets had been passed a second time; the cloth thoroughly wetted in the stocks 1 lb tin liquor ¼ lb spoiled cochineal 1 lb tartar per end, boiled for one hour in the finishing bath Since several ends have been dyed in the same bath before putting the scarlets back in, one can reckon 1 ½ lb tin liquor
Lilas Lilac	Woad ground of a light sky blue Boiling 3 lb alum from Spain ½ lb tartar per end for 2 hours Finishing in a reused finishing bath for purples ¼ lb cochineal ½ lb tartar in the dye bath, boiled for ¼ of an hour
Gorge de pigeon clair Light dove breast	Washed in the stocks in the undyed state and dyed in a reused bath for purples [p. 86] 2 oz cochineal, washed in the stream Saddening ¼ lb logwood prepared in a separate cauldron
Verts de pomme Saxe Saxon apple greens	Boiled with Saxon greens Throw out 1/3 of the bath in which Saxon greens have been made; the bath cooled down; heave in turning fast, and heave out after boiling for a few minutes
Maure doré Golden Moor	Boiling 4 lb alum from the Levant 2 lb red tartar per end for 2 hours. Let to rest and washed in the stocks Weld ground as for golden yellow 6 lb madder in the finishing bath Saddened with just a little copperas
Pain bis Brown bread	The cloths well washed in the stocks, and in a reused bath of any colour saddened with soot 7 to 8 lb soot, boiled for one hour and a half The bath cooled down, the cloths heaved in, and heaved out when on the verge of boiling
Marron Chestnut brown	2 lb sanders 5 lb *redoul* ½ lb gallnut per end. Boiled for a good hour in the boiling bath The bath cooled down, the cloths passed in it during one hour, boiling Saddened then with copperas. The bath at moderate temperature
Couleur d'or Gold colour	In a reused boiling for scarlet The cloth wetted in the undyed state 6 lb young fustic, boiled for 3 h. The bath let to cool down 6 lb tin liquor 2 oz cochineal, boiled for ¼ of an hour

[p. 87] *Cassie*
Mimosa

5 ends in a reused boiling for scarlet
The cloth wetted in the undyed state
35 ^{lb} young fustic, boiled for 2 h. The bath let to cool down
25 ^{lb} tin liquor
2 ^{lb} fine madder
Heave out when on the verge of boiling

Vert d'herbe
Grass green

Failed pistachio green passed in the vat again
Boiled with other greens and welded with half and half weld and flax-leaved daphne

Jaune
Yellow

Boiling
3 ^{lb} alum from Sweden
1 ^{lb} red tartar per end. Washed in the stocks
17 ^{lb} old weld
Give eight turns. Well washed in the stocks

Caffé au lait
Coffee with milk

5 ends in a reused boiling for scarlet made in a reused finishing bath
5 ^{lb} young fustic that had been used for this boiling
Boiled for ¾ of an hour, one pound of madder extracted separately
5 ^{lb} tin liquor
The bath well cooled down, and the cloth turned fast until on the verge of boiling
One can dye a second lot in the same manner, inclusively concerning the tin liquor

Chocolat au lait
Chocolate with milk

5 ends, after the 5 above
Same bag of young fustic plus 2 ^{lb} new one, boiled for ½ an hour
2 ^{lb} tartar
2 ^{oz} cochineal
1 ^{lb} madder for the last 5 ends. Then proceed as above

[p. 88] *Biche Anglais*
English doe

In a reused boiling for scarlet
The cloths wetted undyed and the bath cooled down
¼ ^{lb} madder
1 ^{lb} soot per end
This colour tends towards coffee with milk and can pass for it

Vert de câpre
Caper green

Woad ground of a dark azure blue
6 or 7 turns at the end of a welding
Washed in the stocks

Pourpre anglais
English purple

Boiling
5 ^{lb} tin liquor
2 ^{lb} white tartar
1 ^{lb} Rome alum per end. 2 hours without interruption
Finishing
1 ¼ ^{lb} spoilt cochineal
2 ½ ^{lb} tin liquor, boiled for ¼ of an hour in the finishing

Saddened in a fresh bath, with logwood previously boiled in a caldron 24 hours beforehand to give it time to rest, which is necessary
The quantity is not fixed, one heaves out when the desired shade is reached

This colour is very difficult to dye evenly, because the tin liquor does not go well with logwood, this is why one cannot take too many precautions for the saddening.

It is of great consequence to allow time to rest for the bath that is loaded with the substance of the wood that has been boiled separately, because it contains some slur which could produce large stains. This saddening is made in a fresh bath at a very moderate heat, part of the embers being removed if they are too abundant, and the door of the furnace closed.

If in spite of this it does not dye evenly, as it sometimes happens, strew one pound of alum per end into the same bath and pass the cloth in it as for a testing bath.

Violet anglais [p. 89]
English violet

Boiling
5 lb alum from Spain
2 lb red tartar — per end for 3 hours. Let to rest and well washed in the stocks
Reddened as the purple above and saddened the same

Rat sur noisette
Rat on hazelnut

Galling
½ lb *redoul*
¼ lb logwood
¼ lb old fustic
2 oz gallnut — per end. Simmered and boiled in a bag for ½ an hour
The bath well cooled down, the cloths heaved in
Let boil for ½ an hour
Saddened in the same bath with ¼ lb copperas per end, divided into 3 dips, always letting the bath cool down

When this colour is heaved out of the galling bath, it usually shows a green tinge on the nap which is easily removed if the cloth is washed in the stocks and not in the stream. If you have to match a colour which requires more redness, increase the quantity of gall.

Limon
Lemon

Boiling
3 lb alum from Sweden
1 lb red tartar — per end for 2 hours. Wash in the stocks
8 turns in a welding bath with 1/3 flax-leaved daphne

Jaune doré
Golden yellow

Boiled as above
Washed in the stocks
8 turns in a welding bath loaded with 16 lb of weld. Washed in the stocks
¾ lb madder in a fresh bath, moved in it for one hour, heaved out before boiling

Cire doré
Golden wax

Boiled together with the above, welded in the same way
2 lb unstripped madder in the reused bath of golden yellows
Moved in it for one hour. Heaved out before boiling

[p. 90] *Canelles*
Cinnamons

Boiling
3 lb Rome alum
1 lb red tartar per end for 2 hours. Washed in the stocks
Welded as for golden yellows, washed in the stocks
4 lb unstripped madder, infused in a fresh bath up to a moderate heat.
¼ lb gall
Heave in, done in ¾ of an hour without letting boil

Festiqui[59]

Boiled with the lemons already described, welded in the same way
5 to 6 lb soot in the bath for cinnamons, boiled for half an hour. Done in ¾ of an hour

Taupe
Mole

Woad ground of a dark sky blue
Galling
½ lb gallnut
¼ lb sanders
3 oz madder per end boiled for ½ an hour
The cloth passed in this galling bath has boiled for one hour
Saddened with 1 lb copperas, washed in the stocks

Ardoise
Slate

Woad ground of a sky blue
Galled after the moles adding ½ lb gall for 6 ends. The cloths passed in this galling bath have boiled for one hour
Saddened with copperas

Rouge bruni
Saddened red

Boiling
5 lb Rome alum
2 lb red tartar per end for 3 hours. Let to rest for 4 days
Red finishing bath in a reused madder bath
6 lb unstripped madder
½ lb gallnut
Saddened with ¼ lb copperas in two dips, the bath at very moderate temperature
One must let the madder get well extracted and let boil for 2 hours from the time of heaving in

[p. 91] *Couleur de Roy*
King's colour

Woad ground of a dainty blue
Boiling
5 lb alum from the Levant
2 lb red tartar per end for 2 hours
Let to rest, washed in the stocks and reddened in the finishing bath
6 lb madder
Saddened with copperas as for saddened reds

Couleur de Roy
King's colour

Woad ground of an azure blue
Boiling
5 lb alum from Sweden
2 lb red tartar per end for 3 hours. Let to rest for 3 days, washed in the stocks
Red finishing bath
6 lb madder
1 ½ lb Fernambuco wood

1/4 lb gallnut. Boiled for a ½ hour in the finishing
No saddening

Noisette rougeâtre
Reddish hazelnut

In an old madder bath with Fernambuco wood added
2 lb soot per end
Boiled for half an hour
Boiled for another ½ an hour, stirring the cloth
Saddened without ingredients added, in a reused saddening bath for saddened reds

Gris cendré
Ash grey

In an old boiling for crimson
The cloths wetted in the stocks
2 oz gallnut per end
Boiled for half an hour
Boiled for another ½ an hour, stirring the cloth
Saddened with one ounce copperas

Roze pâle
Pale pink

Well washed in the stocks in the undyed state
Passed in a bath already used for three lots of purples or violets done in a fresh bath

Jujube

Boiling reusing the previous bath [p. 92]
5 lb young fustic
4 lb tin liquor
1 lb white tartar per end for 2 hours
Finishing
¾ lb spoilt cochineal
2 lb tin liquor
Boiled for ¼ of an hour in the finishing

Prune sur truffe
Plum on truffle

1 lb *redoul*
1½ lb sanders
1½ lb common madder per end in a reused boiling for scarlet
Simmered for 1 hour. Pass the cloths in it for another hour
Saddened with copperas

Bleu de Prusse
Prussian blue[60]

Boiling
3 lb alum
2 lb white tartar per end for 2 h. Washed in the stocks
In the same bath, after throwing out a quarter of it, cooled down and used at moderate temperature
4 lb chemick made with indigo from Saint-Domingue for 5 ends
Turn fast, and heave out within ¼ of an hour. If left longer this colour gets very dull

Bleu de Saxe
Saxon blue

Only one end, done without boiling, in the bath above, after turning it in that same bath to wet it
1 lb chemick made as above 20 days before
1 ½ lb calcined alum
Heaved out within one quarter of an hour

Bleu de blanchi Saxe
Saxon off-white blue

Boiled for one hour in the reused boiling for the Saxon blues

		After the blues, throw out 1/3 of the bath. Well cooled down, pass the cloths in it	
		One must observe not to stretch them in the tenters in strong heat	
[p. 93]	*Violet d'évêque*		
	Bishop's violet	Woad ground of a King's blue	
		Boiled with wine soups	
		1 ½ ^{lb} fine cochineal	per end
	Herbe foncé		
	Dark grass	Woad ground of a King's blue	
		Boiling	
		3 ^{lb} alum from Sweden	
		2 ^{lb} red tartar	per end for 2 h. Washed in the stocks
		Welding	
		1/6 weld	
		5/6 flax-leaved daphne	
	Vert brun		
	Dark green	Woad ground of a dark azure blue	
		Boiled with the grass greens above	
		Welded with flax-leaved daphne only and saddened with copperas	
	Soupevin		
	Wine soup	Boiling in a fresh bath	
		5 ^{lb} alum from Spain	
		2 ^{lb} red tartar	per end for 3 hours. Let to rest and washed in the stocks
		Finishing in a reused finishing	
		1 ½ ^{lb} cochineal for 5 ends	
		2 ^{lb} tartar per end	
		Saddened with 1 ^{lb} lees ash per end	
	Amarante		
	Amaranth	Boiling with the above	
		Finishing bath reusing same	
		7 ¼ ^{lb} cochineal for 5 ends	
		Saddened in a fresh bath	
		1 ^{lb} Fernambuco wood	per end. Boiled for one hour, the bath cooled down
		1 ^{lb} lees ash	per end. Done in one dip only
[p. 94]	*Pourpre violet*		
	Violet purple	Woad ground of a dark azure blue	
		Boiling	
		4 ^{lb} alum from Spain	
		2 ^{lb} red tartar	per end for 3 hours. Let to rest
		Finishing bath reusing the finishing of wine soups	
		4 ^{lb} *granille*	
		1 ^{lb} larch agaric	for 4 ends. Infused for 3 hours
		One pound tartar in the finishing	
	Pourpre foncé		
	Dark purple	Boiled with the above	
		Finishing reusing the finishing above	

4 ¾ lb *granille*
1 lb larch agaric for 4 ends, infused ditto
1 lb tartar in the finishing

Soupevin Royal
Royal wine soup

Boiling in the bath above
3 lb alum from Spain
3 lb salt from Cardona
2 lb red tartar per end. 2 hours
Let to rest and washed in the stocks.
Finishing reusing the finishing above
10 lb *garbeau*
10 lb Lamont[61]
5 lb larch agaric for 5 ends. Infused for four hours all together
10 lb tartar in the finishing bath

Lilas ou gorge de pigeon foncé
Lilac or dark dove breast

Woad ground of a sky blue
Passed in a reused boiling with alum, just to wet them
Then passed in the reused bath of purples

Couleur de feu
Flame colour

Boiling in a reused boiling for scarlet
25 lb tin liquor
12 lb white tartar
20 lb choice young fustic for 5 ends. 2 hours
Finishing in reused finishing
5 lb cochineal
2 lb larch agaric
14 lb tin liquor for 5 ends. Infused together for 24 hours
6 turns in the dye bath

Vert de mer foncé [p. 95]
Dark sea green

Woad ground of a light King's blue
Without boiling, at the end of a welding with flax-leaved daphne
7 turns
7 ditto backward, to get an even dyeing
Passed in the reused bath of *musique* for the grass greens below

Cerise
Cherry

In a reused finishing bath already re-used for wine soups
The cloths passed for ½ hour in the same bath cooled down
Finishing
7 ½ lb *granille*
2 oz tin liquor for 6 ends

Herbe très foncé
Very dark grass

Woad ground of a dark King's blue
Well boiled, and welded with only flax-leaved daphne, which better gives it the taste than weld
Saddened with 1 lb logwood per end

Vert naissant sans pastel
Nascent green without woad

Ordinary boiling for greens
6 ^{oz} chemick for Saxon blue — per end. Washed in the stocks
7 turns in a beautiful welding bath entirely with weld
9 more turns backward

[p. 96] *Soupevin qui a été fort recherché au Caire*
Wine soup which has been much in demand in Cairo

Boiled in a fresh bath, 6 ends
5 ^{lb} tin liquor
2 ^{lb} white tartar
½ ^{lb} madder — per end. Boiled for 3 hours
Finishing
1 ¾ ^{lb} *garbeau*
3 ^{lb} tin liquor — per end
Saddening in a fresh bath
1 ½ ^{lb} orchil — per end and then pass the cloth in successive fresh baths of hot water

Orchil is the most beautiful of saddening ingredients, it saves on the dyestuff because it gives body to the colour, moreover it has the advantage on cudbear that it can be let to boil without getting altered, but it is more susceptible to bleed. To avoid this, when heaving out, the cloth must be thrown into the vessel beside, filled with cool water.

If you want to make the wine soup fuller and more winey, subtract 2 ^{lb} tin liquor in the boiling, and put 2 ^{lb} alum.

Pourpre Anglais foncé beau
Beautiful dark English purple

Woad ground of a sky blue
Boiling
4 ^{lb} tin liquor
2 ^{lb} tartar, boiled for 2 hours
Finishing
1 lb cochineal
3 ^{lb} tin liquor — per end
Saddening
6 ^{lb} logwood — per end. Boiled for 2 hours, pass the cloths in it for ½ an hour
Let the bath cool down, strew in 2 ^{lb} alum, 1 ^{lb} tartar per end. Pass the cloths, letting boil for one hour

Chamois

In a fresh bath
2 ^{lb} young fustic — per end. Boiled for 1 ½ hour
The bath well cooled down
3 ^{lb} crystals of tartar
8 ^{lb} tin liquor — per end
Turn fast, as for a finishing bath for scarlet. Heave out within ¼ of an hour.

Notes

1 About currency and measure units in this translation of the *Memoirs on Dyeing* and their abbreviations: in Ancien Régime France, some units are used in the whole kingdom, some are used at provincial or local level. The identification of the units used in the *Memoirs* is based on research published in Cardon 2013. The metric equivalences to the ancient units of measurement mentioned in the text and the references to metrological sources are given in appendix 1. In this part, only the metric results of the conversions of the data expressed in ancient units in the text are given. The author of the manuscript extensively uses abbreviations, only some of which have been kept in this translation, not to make reading too cumbersome. The currency units respectively abbreviated "£" and "s" in this translation are the French *livre* of Ancien Régime (formerly designated as *livre tournois*) and the *sol* or *sou*. The *livre* was subdivided into 20 *sols*, each of 12 *deniers*, abbreviated "d". The weight unit abbreviated "lb" is the *livre poids de table* (tableweight pound) of Montpellier = 0.414 kg, unless stated otherwise in a note. It was subdivided into 16 ounces, here abbreviated "oz". The weight unit translated here by "hundredweight" is the *quintal*, which was composed of 100 pounds mark weight.

2 Tivoli, an ancient Italian town in Lazio, about 30 kilometres east-north-east of Rome.

3 Civitavecchia, a sea port on the Tyrrhenian Sea, located 80 kilometres west-north-west of Rome.

4 Tolfa, a town in the Mounts of La Tolfa, an extinct volcanic group, to the north-east of Civitavecchia.

5 Pozzuoli (correct name in French Pouzzoles). The misspelling of the place names of alum sites in Italy, "Trivoly, Pourrouzole", in the first *Memoir* have a simple explanation: the author does not know these places, he has never had time to travel to Italy.

6 This is the Roman scales' pound of Marseilles of 0.403 kg; hence a bale of Rome alum weighs from 121 to 141 kg.

7 Ancien Régime France had a system of custom duties between the different parts of the kingdom. Languedoc was one of the "*provinces reputées étrangères*" (provinces considered as foreign from the taxation point of view) where high duties were fixed both for import into and export to neighbouring provinces, such as Provence. Marseilles was in still another category "*provinces à l'instar de l'étranger effectif*", free of taxes on trade with foreign countries. But high custom duties had to be paid to import products from there into Provence and other regions of France.

8 The present spelling of the name of this Languedocian harbour is Sète.

9 The author means that, alum not being difficult to obtain in France, it is not worth importing this ordinary kind from so far.

10 82.800 kg.

11 Of course, the correct spelling would be York and Lancaster.

12 The Pyrenees.

13 In Southern France before the Revolution, the *viguerie* was the lowest administrative court dealing only with day-to-day affairs, and the territory under this jurisdiction.

14 The *toise* was a length unit composed of 6 *pieds* (feet), corresponding to the English fathom but unlike it, it was used both in land and sea contexts; 1 to 4 *toises* = 1.949 m to 7.706 m. The *lieue* (league) = 3.898 km, hence 4 lieues = 15.592 km.

15 Here again, geographical data are approximative, being borrowed from the entry on *bois d'inde* (one of the names of logwood in French at the time) in the *Dictionnaire universel françois et latin,* 1721, t.3, p. 941. As opposed to Jamaica and St Croix, Campeche, which has given its name to this dyewood in French, *bois de campêche*, is the continental, western province of Mexican Yucatan peninsula, the coast of which faces the vast Bay of Campeachy.

16 The monetary unit mentioned as pound in the *Memoirs* and abbreviated here as £, is the *livre-tournois*, subdivided into 20 *sols tournois* or 240 *deniers tournois*. For its value and for the metric value of the *quintal* and other weight measures, see Appendix 1.

17 Brazil from Lamon, mentioned by Savary des Bruslons as coming from All Saints' Bay.

18 This sentence is a faulty copy from Jean Hellot's treatise on wool dyeing. The correct original text is "they kill and dry what they intend to sell, and keep the rest to propagate it", Hellot 1750, p. 278.

19 The *arroba*, ancient Spanish weight measure = 11.512 kg. The equivalence 1 *arroba* = 27 to 28 pounds of Languedoc confirms that the pound used by the author is the ancient *livre grosse* of Montpellier of 414.6 g.

20 The arrest – actually dated 3 December 1712 – granted a duty free for a total of 210 *quintaux* markweight (= 10,279 kg) of mesteck cochineal per year, An. 1730, vol. 3, p. 198.

21 The mark was used in Ancien Régime France for precious goods and precious metals. It was composed of 8 ounces and weighed 244.75 gr = 3777.5 English grains, Doursther 1965, p. 230.

22 This section is, in its first part a copy, and then a summary of chapter 16 in Jean Hellot's treatise on *The Art of dyeing wool,* Hellot 1750, pp. 364–369. The author has never seen or used the insect, Polish cochineal, *Porphyrophora polonica* (Linné, 1758); about the biology, chemical contents and history of uses of Polish cochineal, Cardon 2014, pp. 616–624, 694–696.

23 What remains of the section on sal ammoniac in the manuscript is borrowed from the entry dedicated to this commodity in the *Dictionnaire universel de Commerce* (Universal Dictionary of Commerce) by Savary des Bruslons, but it is a faulty copy, the end of the original sentence actually being: "sal ammoniac pays twenty per cent of its value, when it is not imported directly into the Kingdom but has transited through foreign countries » (Savary des Bruslons and Savary 1748, pp. 131–135).

24 Kermes, the Coccid insect *Kermes vermilio* (Planchon,1864), source of the true scarlet dye, is actually a parasite of the kermes oak, *Quercus coccifera* L.

25 This part, including this oddball story, is borrowed from the entry on *vermillon* in the popular *Dictionnaire universel françois et latin* (Universal French and Latin dictionary) also called *Dictionnaire de Trévoux*, ed.1743, vol. 6, p. 718.

26 Two blank spaces have been left in the manuscript where the lower and upper price should have been indicated.

27 The setier was a measurement unit for volumes in Ancien Régime France. Its value varied according to places. The setier used in the text is the *setier* of Narbonne, equivalent to 71 litres, Abbe 1994, p. 86.

28 The "boiling" is the technical name used by William Partridge in his *Practical Treatise on dying of woollen* for mordant baths for all "furnace colours", Partridge re-ed. 1973; it is the exact counterpart of the French technical term "bouillon" used by the author of the Memoirs ad all other contemporaneous dyers.

29 1 *palm* at Bize = 24.6 cm, Abbe 1994, p. 90–91.

30 This again is the technical name used by Partridge for the dye bath for scarlet. The French term used in the *Memoirs* is *rougie*, literally "reddening bath".

31 This operation is described and its rationale well explained by William Partridge: "On covering the lists of cloth with webbing to prevent its taking colour – Cloth, intended for scarlet, or any other cochineal colour, is always girt-webbed, to prevent the lists from taking the dye, as it would, being heavy and coarse, absorb much of the cochineal. This operation is performed with thick cotton, or linen webbing, which, being doubled to half its breadth, is then wide enough to enclose the list when rolled up. The webbing is put round the list, so as to enclose it all, and is sewn on with small twine, passing through the cloth close to the list, and drawn tight over both. The stitches are about one-fifth of an inch apart, when the list is covered merely to save cochineal [...] Soon as a scarlet cloth is finished colouring, and has been partly cleaned by the streamers, it is put on a slatted scrave, that has been covered with a clean white cloth, and the girt-webbing is taken off. This is performed by women, who draw the threads out with hooks. After it is taken off, both the thread and webbing are well washed and hung up to dry for further use." Partridge re-ed. 1973, pp. 123–124. A scarlet standard from Languedoc, showing the undyed list, is illustrated in Part 3 of this book, chapter 3, fig. 6.1. In 18th century Languedoc, covering the lists with webbing and taking it off cost 8 £ for a whole bale of ten pieces of cloth, while the cost of cochineal, for only five pieces in the same bale, was 144 £ 10 s (AD34 C5552). The woman who sewed the webbing on was paid 4 *sols* per Londrin Second and the one who took it off, 2 or 3 *sols* (Marquié 1993, p. 139).

32 Yellow of the mimosa flower of *Acacia farnesiana* (L.) Willd.

33 These three lines have been added by another hand.

34 Span is the equivalent in English of *pan*, *empan* or *palm*, an ancient length unit commonly used in Languedoc. In the *Memoirs*, it is the *pan* of Narbonne which is used = 24.6 cm, Abbe 1994, pp. 90–91.

35 This character appears in several places in the *Memoirs on dyeing*. He is a local scientist and busybody who is further introduced in Part 3 of this book.

36 These comments have been added by the "other hand".

37 This colour name already figures in a regulation for dyers of tapestry wools of August 1667, *Statuts et Reglemens pour les Teinturiers en Soye, Laine et Fil* (Statutes and regulations for dyers in silk, wool and linen) (an. 1730, p. 380, art. 45). It is meant to imitate the colour of the Moors' complexion. In later texts the name appears as *mordoré*.

38 This colour name, derived from the name of the pistachio fruit in Greek (*fistiki*), is discussed in part 3, chapter 4 of this book.

39 The correct spelling in modern French is "*vert*". Although the author of the *Memoirs on Dyeing* writes in very good 18th century French, the spelling of some words is influenced by the local Occitan language, closer to Latin. Hence *verd*, derived from Latin *viridis*, instead of *vert* in some places.

40 Unfortunately, this is one of the pages that have been cut off from the manuscript.

41 "Chemick" is the name employed by Partridge for the solution of indigo in sulphuric acid, further discussed in this book in part 3, chapter 2, section on Saxon blue.

42 Prussian blue was the name often given to Saxon blue by French dyers until the end of the 18th century, see part 3, chapter 2.

43 This colour name corresponding to a tiny sample has been written by the "second hand".

44 "Span" used here as the equivalent of the Languedocian measure unit *palm*; 1½ *palm* = 34.9 cm.

45 *Redoul* is *Coriaria myrtifolia* L., a tannin-rich bush very common in damp places in Languedoc; see part 3, chapter 5.

46 Correct spelling in modern French *girofle*.

47 Correct spelling in modern French *musc*.

48 Comment written by the "second hand".

49 Comment written by the "second hand".

50 Sedan was a prestigious Royal Manufacture of broadcloth, created in this city close to the northern border of France. The quality of the black dye of Sedan cloths was famous.

51 About this Albert, see note 35 above and part 3.

52 This third Memoir, the shortest, is a version of Jean Hellot's *Instruction sur le déboüilli des laines, et etoffes de laine* (Instruction on the testing of wools, and wool textiles), the final part of his *Art de la Teinture des Laines* (Hellot 1750, pp. 617–631), improved, and reorganised and condensed by the author of the *Memoirs on Dyeing* to make it more user-friendly for clothiers and their dyers. Their purpose is opposite to Hellot's, however: to know how the quality of their dyes is going to be controlled, in order to avoid being caught using forbidden ingredients, or too little of the prescribed ones. This Memoir, it will be seen, again reveals the author's proficiency in practical chemistry.

53 This amounts to 0.41 L.

54 This amounts to 25.9 g of alum.

55 *Gros* was another name used in French for the *drachme* or dram, the 1/8th part of the ounce. In the Languedocian system of weight used by the author, 1 *gros* = 3.24 g. This testing liquor is highly concentrated, alum being used in proportion of 800 % of the dry cloth weight. This is double the proportion recommended by Charles-François de Cisternay Dufay, who had been entrusted to develop these testing procedures, later published by Hellot, his successor (Hellot 1750, pp. 33–37 and AN F/12/740). For the chemical rationale behind these testing procedures, see Cardon 2013, pp. 297–303.

56 This is the last of the four Memoirs grouped in the manuscript, and the most original part of this exceptional document.

57 The name of the manufacture is missing.

58 Also called lichen purple, see Cardon 2014, pp. 467–496 for an update of orchil-giving lichens and their chemistry.

59 As already mentioned, this colour name is derived from the name of the pistachio fruit in Greek (*fistiki*). It is further discussed in part 3, chapter 4 of this book.

60 As explained in Part 3, chapter 2 of this book, *bleu de Prusse* (Prussian blue) was the name commonly given by 18th century French dyers to Saxon blue, a new blue colour obtained from indigo carmine.

61 "Brazil from Lamon" is mentioned by Savary des Bruslons as coming from All Saints' Bay in Brazil, Savary des Bruslons 1723, p. 478. It therefore is only another provenance of brazilwood, *Caesalpinia echinata* Lamarck, as discussed in part 3, chapter 3.

Part III

Polyphony on Colours

In the critical edition of the French text of the *Memoirs on Dyeing*, I had chosen to present the explanations, complements of information and comments I thought could be useful, as a continuous, flowing counterpoint of notes, in the same manner as Kenneth Ponting had commented upon William Partridge's *Practical Treatise on Dying.*[1]

In this new volume, comments and analyses have been organised into chapters where all the information on each type of mordant or dye scattered across the four memoirs is discussed in a cohesive way. Some points of view from experts in the art of dyeing, contemporaneous with the *Memoirs on Dyeing* or from other periods, are also brought into the discussion where relevant. This will allow a wide reflection on the art of dyeing with natural dyes, at a time when it was reaching summits of skill.

4

Transparent, Crystalline: Water, Mordants, Minerals

Water

"At the Royal Manufacture of [...] waters are very bad, that from the river, already warm by itself, being further infected by the spring of Las Fons which predominates in it during three quarters of the year; this water also is nearly tepid, and it is moreover laden with a lot of aluminous salts."

It is significant that water should be the first element the author discusses, at the beginning of the last of his four memoirs. In *Annotations*, which is a kind of diary, we find the transcription of his observations on the results of his continual experiments to improve his dyes while, at the same time, managing the smooth running of different simultaneous dyeing processes permanently performed in his dye-house.

Water is the medium, and predominant ingredient, every dyer has as his composition base. In the author's case, his water resource is a mixture of stream and spring water in proportions that vary with the season and year which makes it a perpetual challenge to manage to reach the exact results expected, evidenced by the omnipresent and constantly changing pattern sheets, transmitted by the factors in Marseilles and the Levant. One can sense the stress it generated as he records "...in how many ways it has been necessary to experiment, to correct the bad quality of these waters in order to manage to make a beautiful scarlet, which they keep saddening right through to the wash place." The Anglo-American dyer William Partridge similarly describes his disarray when arriving in America and finding that dyeing processes he had always perfectly managed "would not produce the same colour in any two" places, which he attributes to "the effects of water; I had no conception when I left England, that water could have had so material an effect in the production of colour, as I have since found it to possess."[2] As highlighted by the author of the *Memoirs*, the issue was of particular consequence in scarlet dyeing, due to the high cost of the ingredients – cochineal, tin – and to the evolution of fashion in the course of the 18th century, when scarlets are wanted of an ever brighter red, to the point "that the eye can hardly bear its radiance".[3] The problem was that cochineal is sensitive to the pH of the dye baths: a mildly acidic water would help obtain an orangey, *vif* ("live") scarlet, in dyers' terms, while alkaline water more easily produces crimson to purplish reds, and similarly, water containing iron or copper salts could "sadden" the reds, giving purplish colours.

As already related in the first chapter of this book, the identification of the place where the author of the *Memoirs* was working ironically came from his mentioning the names of some springs he was using. Another consequence was that it made it possible to try and understand why he was so unhappy with "the bad quality of these waters". After a rainy period at the end of winter, when the springs which nowadays are dried-up most of the year were flowing again, I went to Bize, measured the temperature and pH of the river Cesse and of the two springs mentioned in the *Memoirs*, and sampled their waters to have them analyzed. The water of the Cesse did not prove particularly "warm", its temperature barely reaching 16 °C, measured right below the walls of the ancient manufacture. Upstream, the spring of Las Fons gushes forth at 20 °C from crevices at the bottom of a mossy basin, in magic clouds of bubbles; it could indeed be described as "tepid", especially compared with that of the Bouillette spring downstream, definitely cooler with its temperature of 15 °C. About pH measurements, the variation ranged from just below 7, for the water of the Bouillette (neutral pH corresponds to 7), to 7.4 and 7.5 respectively for the Cesse and the spring at Las Fons, slightly more alkaline. The results of the analyses performed by the laboratory "Développement Méditerranéen" at Saint-Hilaire de Brethmas (Gard, France), showed that there was no significant amount of either aluminium or iron in the water of the river, and none either in the two springs (quantities were all below the quantitative threshold of 10 mg/L). There was hardly any copper either: less than 0.001 mg/kg in the springs, just a little more (0.023 mg/kg) in the Cesse. Nothing, therefore, that could explain the author's negative judgment. Could it then be from the hardness of these waters that difficulties were coming? With its content of 119 mg/L of calcium, the water of the Cesse can indeed be classified as hard – albeit moderately so – and that of the spring of Las Fons is still harder (153 mg/L of calcium), although not to the point of being considered as "very hard". This is certainly why the dyed cloths were carried to the fulling mill of La Bouillette for washing off, because the spring there is less hard, containing only 81mg/L of calcium and, as mentioned by Kenneth Ponting, who "managed a dye house for some years", "the real answer to good washing off is a plentiful supply of soft water."[4]

The hardness and moderate alkalinity of the waters available at his manufacture may therefore explain the author's observations. The fact that "by themselves and without any addition they saddened the dark wine colours" was an advantage, and he did use this property to give the right shade to an *amarante royale* (royal amaranth) in a final bath of pure hot water, and to a *soupevin* (wine soup or wine colour) "much in demand in Cairo", saddened with four successive baths of pure hot water.[5] But this advantage hardly compensated the obstacle such waters presented to obtain really brilliant scarlets, until the last stage of washing off. This was mentioned by Jean Hellot as one of the main hazards in scarlet dyeing: "The major part of common waters do make it approach towards a pink colour, since they nearly always contain a gypsiferous or calcareous earth, and sometimes some sulfurous or vitriolic acid. It is such kind of waters that is commonly called *eau crue* (raw water); by this term is meant a water that does not dissolve soap, and in which it is very difficult to cook vegetables."

Hellot proposes some solutions to correct these waters, his favourite one being to add some *eaux sûres* (sour broth), obtained by leaving a bran decoction to develop an acidic fermentation.[6] Like him, the author of the *Memoirs on Dyeing* has tried diverse ingredients to correct the undesirable properties of the waters he has access to, and he includes information to that effect in several of the recipes of the last memoir. He has also learned how to finely adjust the proportions in his mordanting and dyeing processes to the seasonal variations inherent to the use of river water: "there are certain times in the year, when 25 to 30 pounds of good tin liquor are necessary to exhaust a bath." Such a perseverant approach allows him to conclude his comments on his water supply on a note of moderate – and modest – satisfaction: "nevertheless I think that my colours are good, but that better ones can be made with good waters."[7]

These comments, along with the limited insight gained from the analyses of the waters available at the Royal Manufacture of Bize, find a close, very interesting parallel in the recurrent discussions which can be found in English texts about the famous scarlets of the Stroud region, the beauty of which was often attributed to some properties of the waters. At the beginning of the 18th century, Daniel Defoe, in his *Tour thro' the whole Island of Great Britain*, expresses no doubts about this: "the manufacture of white cloths was planted in Stroud Water in Gloucestershire for the sake of the excellent water there for the dyeing scarlets, and all colours that are dyed in grain, which are better dyed there, than in any other part of England, some towns near London excepted."[8] In her seminal book on *The Cloth Industry in the West of England*, Julia de Lacy Mann, using unpublished geographical surveys, made an attempt to correlate the undisputable reputation for quality of this scarlet, and of other dyes made in this part of Gloucestershire, with the geology of the region and the composition and properties of its water resources. "The harder the water" – she sums up – "the more difficult it is to produce an even colour when dyeing in the piece. The streams which make up the Stroudwater system derive from a substratum of limestone with an admixture of Cotswold sands, and water from the latter is softer and more suitable for dyeing than that from the limestone. In the streams the two are mixed but well water differs in different places as to the amount of solid matter it contains. No Gloucestershire water is really soft, but it is appreciably softer than what comes from the chalk, as much Wiltshire water does". All this evidence, however, does not really clarify the question and for her, the fact remains that "the property which was believed to make the water of the Stroud region so suitable for dyeing red has never been identified."[9]

Practical, experienced dyers proposed another approach to this issue of water supply. Partridge, from a family of clothiers and dyers at Bowbridge, by the River Frome (once also known as the Stroudwater), records that "it was the opinion of my father, and his predecessors in the same business, who have been eminent dyers for more than a century, that none but soft water could be used for dying; and this in direct opposition to their own daily practice; for they had all this time been making use of spring water, that was very hard, in preference to water from a fine mill stream, that ran between the dye-houses, and was remarkably soft. And I am convinced they have owed their celebrity, purely to this circumstance. My practice in America has convinced me of this important fact, that any water with the exceptions before mentioned [spring waters laden with "metallic oxids and marine salts"] may be used successfully by the dyer, with one proviso – that it is always in the same state: it is on this account that springs are better calculated for the purpose than mill-streams."[10] However, "the most important inference to be drawn from these facts" – he concludes – "is that, for dyers to become eminent, they must be stationary, they must continue to practice in one situation, and with one kind of water, that by these means alone, can they be able to obtain perfection in the art."[11]

In the same line of thought, Samuel Rudder, in his *New History of Gloucestershire*, published in 1779, finds the nature of the water of the Stroud basin less important than the local dyers' expertise: "from Stroud, superfine broadcloths are sent away either white (that is undyed) or dyed in the cloth and in particular great quantities are dyed scarlet, for which branch of trade the place is noted. The beauty of their colours is very great, to the perfection of which the Froom water has been erroneously supposed to contribute, for it is most assuredly owing to the skill of the artist."[12]

This comment would apply as aptly to the author of the *Memoirs on dyeing* and to his perseverant struggling with his fluctuating water supply.

Alum

It is a fortunate circumstance that the alphabetical order adopted by the author in his first *Memoir*, dealing with the dyestuffs and mordants he currently uses, makes him start his description with alum.

Alum is an essential ingredient for dyers "in the best mode of dyeing".[13] At the time when the *Memoirs on dyeing* were written, it was still one of the two most important mordants (Table 7.1). Since the dawn of the art of dyeing, empirical intuition had inspired dyers to use mordants when dyeing with the majority of natural dyestuffs whose colorants, due to their chemical structure, would not normally bind strongly enough with textile fibres to give intense, fast dyes. The mordants mostly used in such cases were metallic salts that facilitated the formation of a complexion between colorant, metal and the fibre to be dyed.[14]

Alum was one of the most anciently and most massively used mordants, both because the diverse aluminium salts that were commonly referred to as "alum" were quite easily available – aluminium being the most common metal found in the earth's crust – and because mordanting with aluminium did not alter the colour of the dyes.[15] However, alum was not necessary and was never used in a number of dyeing processes. For instance, it was not of any use for blue dyeing in the woad vat and later, in the woad and indigo vat. It was not needed either for the vast number of beige and brown dyes obtained from plant sources rich in tannins, that functioned as efficient mordants, especially since they were often post-mordanted – "saddened" – with iron salts to produce grey and black colours. When the *Memoirs* are written, a further significant limitation in the use of alum has recently been adopted in common dyeing practices, since a newly discovered tin mordant – discussed in more details below – is now replacing it for the whole range of bright reds and scarlets dyed with American cochineal, and all the lighter "French colours" created in an aim to exhaust the dye-baths made with this costly ingredient. The author of the *Memoirs* even uses this tin mordant instead of alum for dyeing the more purplish crimsons and wine colours with cochineal, in contradiction with the regulations stipulating that these colours should be obtained by pre-mordanting the cloths with alum and tartar.[16] However, Hellot, after describing this orthodox process of crimson dyeing, mentions two other recipes for crimsons in which tin and alum mordants are combined, either by adding the two in the same mordant bath, or by mordanting wool successively in two mordant baths, the first one with tin liquor and the second with alum, both with tartar added. He describes the former process as being common practice in Languedoc for the cloths exported to the Levant and indeed, this type of mixed mordant is described in the last part of the *Memoirs* for a *pourpre anglais* (English purple) without a woad ground, in which the blue component is logwood.[17] More generally, the author's use of tin mordant in cases for which other dyers would use alum is one of the ways he has found to counterbalance the dulling effects of his waters.

In spite of the above mentioned restrictions, not only is alum still employed at the time by all dyers – including the author – as the main mordant for a large range of colours, but it is also used – albeit generally in smaller proportions – in final baths called *brunitures* (saddening), to finely tune the shades of crimsons and wine colours. In his last *Memoir* recording his personal experiments, the author further shows that alum – in this case the purified, calcined sort – can also serve to give the greyish-purplish tinge of blue agate to a previously woaded cloth.[18]

His choice of alums of different provenances for particular dyes and saddening processes (Table 7.2) raises a number of questions, particularly since important centres of industrial alum production, such as the numerous alum shale mines of central Europe, of the Saar region in the Rhineland, and of the Meuse valley, near Liège, are not mentioned at all. An examination of the kinds of alums mentioned in the first *Memoir*, as compared to those recommended for actual recipes in the following *Memoirs*, does suggest some unexpected answers which will be proposed as concluding remarks.

Unsurprisingly, however, the first kind of alum mentioned, and the one used in the highest number of processes described in the *Memoirs*, is "Rome alum" from the alunite mines of the hills around La Tolfa, near Civitavecchia, north of Rome. This Rome alum is also the first described in the *Dictionnaire universel de Commerce (Encyclopedic Dictionary of Commerce)* by Jacques Savary des Bruslons, the bible of European merchants in the 18th century.[19] It had been the most important source of this commodity in international trade since the start of the exploitation of the mines towards the end of the 16th century, when the Papacy intended to finance a crusade against the Turks by the sale of this alum to all Christian countries.[20] By the 18th century, Rome alum was still considered the best, not only by the author of the *Memoirs*, but by all contemporary experts.[21] This was due, most of all, to its consistent quality, allowed by a production that had been organised on massive scale and standardised from the beginning. A large part of this section is borrowed – and may have been dictated to the author's secretary – from the entry on alum in the first volume of the *Dictionnaire domestique portatif, contenant toutes les connoissances relatives à l'oeconomie domestique et rurale* (Portable Domestic Dictionary, including all knowledge related to rural and domestic economy) published in Paris in 1762.[22]

It has been mentioned in the first chapter of this book that this case of plagiarism had helped situate the drafting of the manuscript after this date. The first-hand data that are added on the fluctuations of the price of Rome alum confirm such dating: the author complains about the recent considerable rise in its price in Marseilles (in the order of two-and-a-half times superior to the average price which had remained stable for the preceding years). This is evidently in relation to the disruption of trade across the Mediterranean occasioned by the Seven Years' War (1756–1763), which made the sea route between Civitavecchia and Marseille risky. The Languedocian clothiers are not the only ones to suffer from the situation. The Dutch, who also import massive quantities of Rome alum, record the same sudden rise in its price, starting in 1756 and going on during the following years, in contrast with the long period of stability that had prevailed previously, between 1701 and 1755.[23] For the clothiers of the west of Languedoc, this rise was made worse by the taxes and expenses they still had to pay to get their alum transported from Marseille to their manufacture. Rome alum was conditioned in bales weighing 300 to 350 "Marseille pounds" (121 to 141 kg).[24] The customs duty for import out of Marseille into the neighbouring regions of France, fixed per 100 Marseille pounds, added 9 % to the highest buying price (40 Marseille pounds) mentioned by the author.[25] This duty was heavier for Languedocian buyers than if it had been calculated per "Montpellier pound", the one they commonly used, because the Marseille pound was lighter (11 g less), therefore they paid proportionately more for a given weight of any commodity measured with this "foreign" weight. Adding the transport costs from Marseille to Languedoc, by sea as far as the harbours of Sète or Agde, and then by the canal of Languedoc or by wagons to the manufactures, the final price of the product had risen by 15 % when it arrived. The same applied, of course, to the other sorts of alum imported via Marseille.

What alums are these? The other alums from Italy mentioned in the first *Memoir* would probably not be worth the trouble and expense, that from the region of Tivoli not having ever been exploited, at least on a big enough scale to be exported, and the native alum from the Phlegraean Fields around Pozzuoli being too inconsistent in quality and often impure, as remarked by the French engineer Auguste Fougeroux de Bondaroy who visited the Solfatara in 1768 and observed the way alum was collected there. Insufficiently purified, it still contained traces of ferrous sulphate that could impair the quality of light dyes.[26] The author of the *Memoirs* does not indicate any price for this alum: either he does not use it, or it is included among the alums used in the few mordanting and saddening processes for which he does not specify which sort of alum is to be employed.

The following section in the first *Memoir*, on alum from the Levant, is largely borrowed from the *Dictionnaire universel de Commerce*.[27] That which is described as coming from Smyrna is obviously alum produced at the mines of alunite of Phocaea, some 45 km to the north-west of Smyrna, which had been massively exploited and imported into Europe in the Mediaeval Ages by the Genoese till the capture of the mines by the Turks in 1455.[28] Other production sites, some of which were already mentioned in mediaeval merchants' books, have recently been positively identified in various parts of Asia Minor thanks to geoarchaeological surveys; these probably are the alums mentioned as being exported via Constantinople.[29] In this part of the *Memoirs*, largely borrowed from Savary des Bruslons, it is said that these alums are not much employed in the European textile industries any longer, which is a politically correct statement in the context of the Popes' ban on "buying alum from the Turks".[30] However, the following memoirs contradict this statement, and show that the author actually uses alum from the Levant quite regularly: it is the second most often recommended kind of alum. It may even have been used in greater quantities than Rome alum in the author's manufacture at certain periods, since several recipes of the second memoir indicate that alum either from the Levant or from Spain can be used instead of Rome alum. It certainly appears to be the cheapest (Table 7.2) and this of course would be a very good reason for using it, although under this common name, very different and unequal kinds and qualities of alum must have been sold. In the author's case, however, this problem could be overcome thanks to his close links with the Pinel group of Carcassonne and its network of factors in Marseilles and the Levant, who could select good lots of alum from Phocaea for him, the quality of which would be on a par with that of Rome alum.

The case of "alum from Spain" is different and intriguing because the alum mines of the kingdom of Murcia – Mazarron, exploited since 1462, and Lorca and Cartagena, exploited since 1525 – had definitively closed down in 1592, while they had been producing up to 2600 metric tons of alum per year around 1562.[31] Spain does not appear any longer as alum producer, either in the *Dictionnaire universel de Commerce*, published in 1748, or in the first edition of the *Encyclopaedia* of 1751. This is confirmed by the Irish naturalist William Bowles, noting in 1775 that the Murcia mines are abandoned; the only active site of alum production he mentions in Spain is the vein of native alum close to the town of Alcaniz, in Aragon, 105 km to the west of Tarragona.[32] This certainly is the alum from Spain used in the author's manufacture. Its main quality seems to be that it is cheaper than Rome alum, even when bought legally, via Marseille (Table 7.2). It must therefore be even cheaper when bought via the well-organised smuggling route mentioned by the author.[33] How much cheaper, he cannot, for obvious reasons, take the risk to precise but this is probably the way by which he mainly obtains it. The same applies to the "alum from France", which does not refer to any of the best known contemporary productions of the west of the Lyons region and of the Beaujolais, or of the present department of Aveyron, closer to Bize,[34] but exclusively to a small alum vein then in exploitation in nearby Roussillon; it is described in the *Encyclopaedia*, which locates it more precisely near the town of Prades.[35] This alum is probably used for the mordanting and saddening processes where the author does not recommend any special kind of alum. He does not give any price for it either, only remarking that it "turns to better account for dyers because of the vicinity of the mine". It must have been at least as easy to smuggle it into Languedoc as the alum coming from Aragon by coastal transport, maybe on the same type of "small flat boats".

The two other kinds of alum mentioned in the first *Memoir* offer remarkable examples of the consequences of political circumstances on textile production in Languedoc, and further confirmation of the fact that the *Memoirs* are written at the time of the Seven Years' War. Alum from England cannot but be mentioned by the author, it being one of the "three main sorts of alum" according to Savary des Bruslons' *Dictionnaire universel de Commerce (*Encyclopedic Dictionary of Commerce).[36] It is also described as such in Diderot's and d'Alembert's *Encyclopaediea*, from which most of this section is copied.[37] However, although the author adds that it is commonly used by Languedocian dyers in the same manner as alums from Spain or Sweden, he does not mention any price for it, nor does he recommend it specifically for any of his recipes. He may have used it in the processes for which any available kind of alum will do, but it is more probable that he could not get any at the time of writing, or not at a reasonable price, because of the war between England and France. The renewed importance of the exports of Yorkshire alum to the south of France in the period after the war has recently been highlighted by the English historian David Pybus' research into series of data preserved in archives: from 1785 on, yearly exports to Marseille amount to between 152 and more than 305 metric tons – with a record of 325 tons in 1791, two years after the beginning of the French Revolution. In 1788, 61 tons of alum left the harbour of Littlebeck directly bound for the Languedocian harbour of Sète, from which it was only a short way to the many cloth manufactures of the region: part of it may well have ended in the author's Royal Manufacture of Bize.[38]

Conversely, imports of alum from Sweden, one of the allies of France in the Seven Years' War, must have been encouraged, and they may have contributed to the impressive development of alum production there during the second half of the 18th century. The first alum factory, from pyritous shales, had been created in 1583 at Dylta, in the parish of Axberg (province of Närke). Around 1765, just after the end of the Seven Years' War, several other alum factories have opened, in the same province and in the south of the country; at Andrarum, in Scania, 288 tons of alum are being produced per year.[39] This alum from Sweden had already been known in France for some time: it is mentioned in the *Encyclopaedia.*[40] The author of the *Memoirs* buys and uses some: he has been impressed by the size of some pieces of this crystallized alum, weighing up to 80 kg. He buys it via Marseilles at prices comparable to the official prices for alum from Spain. The nine cloth samples which illustrate recipes of lemon, golden yellow, golden wax (three samples, of different shades), *festiqui*, King's colour, dark green and dark grass green in the fourth Memoir, may be the oldest positively identified evidence of textiles mordanted with alum from Sweden.[41] This alum, mostly produced by advanced techniques, developed by Swedish scientists and engineers, has a double advantage: it is cheaper than Rome alum (Table 7.2) for a comparable quality. This is shown by a comparison between two recipes of *couleur de Roy sans bruniture* (King's colour without saddening) in the *Memoirs*, and the corresponding samples. The first process figures in the first *Memoir*, which gives orthodox recipes in conformity with the regulations, and the second is described in the fourth *Memoir* where the author proposes ways to obtain equally good results at lower cost. The two recipes only differ in the type of alum used for mordanting, and the similarity of the results obtained with equal quantities of alum from Sweden or of alum from Rome, confirmed by the colorimetric data of the two samples, offers a brilliant demonstration of the quality of this alum from Sweden.[42] Finding it better than alum from the Levant, moreover, the author feels he can use lesser quantities of it for mordanting yellows and greens (3 pounds per cloth piece, instead of 4 pounds of alum from the Levant), thus obtaining absolutely identical colorimetric results, as exemplified by two "full grass greens".[43]

The significance of this issue of quality-price ratio concerning the alum supply in the author's dyeworks becomes evident when one calculates that 158 kg of alums of diverse provenances have been employed in the 99 processes in the *Memoirs* where alum is used as mordant or saddening agent. An estimation of the yearly consumption of alum in the Royal Manufacture of Bize gives even more impressive results. Several documents allow to calculate it with a fair degree of probability. A list of ten bundles of Londrins Seconds cloths from Bize, shipped to the Levant on 16 July 1751, indicates the colour assortment in each bundle of ten pieces.[44] Since nearly all the colour names correspond to recipes described in the *Memoirs*, it is possible to know how many of the pieces have been dyed into colours needing a pre-mordanting or saddening process using alum, and to evaluate how much alum was used in each case. The result is that dyeing of the cloths in this consignment has required an estimated total of 96.255 kg of alum. Of course, different colour assortments would imply some variations in the quantities of alum necessary. Fortunately, other similar lists of consignments of cloths for the Levant, mentioning the colour assortments in each bundle, are preserved in the archives of Languedoc. Dated from 1749 to 1775 and concerning different manufactures, these lists allow to calculate that the average consumption of alum per 100 cloth pieces is in the order of 80 kg.[45] Based on this average figure, the quantity of alum used at the Royal Manufacture of Bize in a year of record production such as 1764, when 2510 Londrins Seconds and 240 pieces of superfine Londrins Premiers and Mahoux cloths were dyed, would amount approximately to 2.2 metric tons of alum.[46]

The *Memoirs*, therefore, do demonstrate the strategic importance alum supplies keep having for textile industries in the 18th century: only a reliable supply of tons of alums of reasonably regular qualities can enable dyers to confidently prepare their successions of mordanting or saddening baths for the crimson, purple, amaranth and wine colours, the mauves and lilacs, the yellows and cinnamons, the greens, olives and hazels requested by customers in the Levant. Regularity of supply, both in quantity and quality, justify importing alum from such remote places as the Levant, Italy, England or Sweden. When trade routes are disrupted by political circumstances, tapping nearer resources gets indispensable and may prove economical, but it implies some risks, adjustments of the processes, and new challenges for the master dyers.

Tin liquor

At the time when the *Memoirs* are written, the technique of using a tin salt to mordant cloths still is a comparatively recent innovation, resulting from a serendipitous discovery by a Dutch inventor at the beginning of the 17th century: tin dissolved in *aqua fortis* turns the purplish red of cochineal into a fiery orange red. This colour reaction would never have had such considerable economic and technical consequences had it not concerned the most prestigious of all colours, scarlet. Scarlet, at the time, was the colour most in demand by the elites of all countries in the world, but it could now be obtained from a new, exotic and comparatively cheaper dye source. For millennia, this slightly orangey, very luminous red had only been provided naturally from dyers' kermes, the scale insect that gave its name to the kermes oak, the Mediterranean tree on which it lives as a parasite.[47] Since the discovery of America, and the colonial exploitation by the Spaniards of the colouring resources of the vast Empire they had conquered there, a new source of red, a cochineal raised on some species of cacti by indigenous people, has been imported regularly into Spain and sold to users in other European countries. The new mordant elaborated by the Dutch scientist Cornelis Drebbel (1572–1633) now allows to shift its naturally crimson dye towards more orange shades and thereby to use it as an alternative to kermes for scarlet dyeing.

The process, basically consisting of dissolving granulated tin in nitric acid (*aqua fortis*), was first kept a secret by Drebbel's sons-in-law, the brothers Abraham and Johannes Sibertus Kuffler who emigrated to England and specialised in the new scarlet dye, called *color Kufflerianus*, in their dye works at Bow, to the east of London.[48] Before long, however, various versions of the recipe spread among European scarlet dyers, who characteristically call the new mordant "tin liquor for scarlet", or in French, *composition d'écarlate* (scarlet spirits). It is one of the pillars of 18th century dyeing, particularly in dyeing of cloths for the Levant, where bright colours are relished.

No wonder, then, that this is the first mordanting process described, both in the second and fourth *Memoirs* (Table 4.3). It reveals the author, not only as an experienced dyer, but also as a practical chemist, enjoying giving detailed descriptions of the minutiae of the process, in the first person – "I take… I put in… I let the boiling… I have employed some… I have experienced that..." One can imagine him in a rudimentary laboratory, busy adjusting the proportions of ingredients and watching the progress of the chemical reaction between water, sal ammoniac, nitric acid and tin, according to the seasons, their influence on ambient temperature and quality of water, and to the various concentrations of commercial *aqua fortis* he can find. The only ingredient of regular quality in this reaction is tin, imported from Cornwall which was by far the main producer in Europe at the time. In this the author agrees with Partridge's description on how "to make tin liquor for scarlet": "the tin is always obtained from Cornwall in stamped blocks, and none other can be depended upon as being genuine."[49] It was not a cheap product, hence the importance not to waste it and to reach an optimum dissolution, in spite of the variations in the quality of nitric acid used. The author and later, Partridge, are both in the same situation of having "a lot of work to dispatch", which makes it unpractical for them to prepare their own *aqua fortis,* following Jean Hellot's prescriptions. It is remarkable, in such conditions, that the author's empiric understanding of chemistry helps him consider that sal ammoniac can be substituted for the cheaper common salt, in agreement with Partridge's comment on this point: "in France, they add to the aqua-fortis, previous to adding the tin, some sal-ammoniac in place of the salt, which practice is also followed by some of the English dyers; but if this be added to the least excess, the colours will incline to a pink."[50] More importantly, his repeated experiences lead him to find such sound proportions – given in his fourth Memoir – that they match those still in industrial practice at the end of the 19th century, in the large dye works of Théophile Grison, another genius of practical dyeing chemistry (Table 4.3). In the *Memoirs on Dyeing*, tin liquor is used not only for the whole range of orangey reds to yellows, pinks and pinkish beiges, but also to give additional vividness to purples, purplish reds and wine colours which the water of the Cesse would tend to sadden too much. In the fourth *Memoir*, the author even reports his experiments to add tin liquor to the dye baths for a Saxon green and a celadon green. As a total, for the 43 recipes in which tin liquor figures, he has prepared 259 pounds (107.226 kg) of it, using between 7 and 9 kg of pure tin.

By-products of wine making: tartar, wine lees ash

A large part of the information on tartar and cream of tartar in the first *Memoir* is compiled from the entries on these products in the *Encyclopedic Dictionary of Commerce* by Savary des Bruslons and in the popular *Dictionnaire oeconomique contenant divers moyens d'augmenter son bien et de conserver sa santé* (Economic Dictionary including diverse means of increasing one's property and of preserving one's health) by Noël Chomel; it is from the latter that the author of the *Memoirs* has borrowed the picturesque expression: "Tartar has the sap of the grapes for a father, fermentation for a mother, and the barrel for womb."[51] However, his remark that tartar is collected from winegrowers in the countryside, by the same people who collect linen rags for paper making, obviously reflects local practices and personal experience. He is quite right to state that both white and red tartar "are of great use" in dyeing. In one form or another – crude red or white tartar, cream of tartar – for mordanting or in the dye baths, or at both stages, tartar figures in 129 out of the 206 dyeing processes described in the *Memoirs*. The total quantity used in these processes amounts to 253 pounds, i.e. 104.742 kg of tartar, the red one being used in the highest number of recipes and in biggest quantities (Table 4.4). This is easily explained by the abundant supply and comparative cheapness of this by-product of wine production, an activity so widespread in Languedoc and the neighbouring regions. From a technical point of view, since tartar deposited in vats and barrels consists mostly of potassium bitartrate, but with varying admixtures of calcium tartrate and of different colorants, depending on the varieties of grapes vinified, purified cream of tartar or crude white tartar were preferred for "high priced, delicate and bright colours", as explained by William Partridge, while red tartar or argol – as he calls crude tartar – "is always preferred for dark colours, such as bottle greens, dark browns, blacks, etc."[52]

In the *Memoirs on Dyeing*, tartar – both white and red – is commonly used in mordanting baths – "boilings" – both with alum and with the tin liquor, in proportions of 6.6 to 8 % of the dry weights of the cloth pieces. The liberal use of tartar for dyeing the cloths made in Languedoc for export to the Levant certainly contributed to the beauty of the colours obtained and to the long-lasting success of these cloths among Oriental customers. Conversely, William Partridge attributed "the principal deficiency in the production of furnace colour" in his country of adoption "to the want of employing these salts [cream of tartar and argol] more generally."[53] Although the usefulness of tartar in mordanting baths and in the dye baths for certain colours was commonly acknowledged by dyers, its role was not completely understood from a scientific point of view until Valery Golikov's recent research into cochineal dyeing.

Golikov, a Russian scientist, found that adding cream of tartar to mordanting and dye baths produced several different beneficial effects: it allows to obtain better, more beautiful results with waters not particularly good for dyeing, because it prevents some undesirable insoluble compounds (of calcium, magnesium or iron, according to the water composition) from precipitating onto the textile fibres; it also improves the extraction and dissolution of colorants in the dye bath and their fixation onto the cloth; lastly, it allows a fine-tuning of the shades of cochineal reds and of other dyes sensitive to the pH of the dye baths.[54] The dye baths in which the author of the *Memoirs* uses tartar – of the same or of another sort as that used in the mordant bath – are nearly all cochineal dye baths meant to give colours with a purplish/pink tinge: purples, violets, *soupevins* (wine colours), amaranth, *vessinat* (colour of vetch flowers), *griotte* (morello cherry), cherry, pink, peach flower, lilac, mauve, dove, linen gray; however, he also adds tartar to a dye bath of young fustic to obtain a bright daffodil yellow and to a dye bath with young fustic and a little madder to make a "chocolate with milk" colour.[55]

Wine lees ash (*cendre gravelée*), of complex composition, mainly consists of soluble potassium carbonate, sulphate and chloride. As opposed to tartar and tartaric acid, it is strongly alkaline. The author of the *Memoirs on dyeing* gives the best description of its preparation I have come across so far, and the only one mentioning the offensive smell caused by this production. Mordanting and nuancing woollen as well as silk textiles with wine lees ash was already a popular practice among Italian dyers in the Middle Ages. They called it *allume di feccia* (lees alum) in Toscan, *lume de feza* in Venetian,[56] using it in the same way as the author here, to give a crimson bloom to madder reds and to make the crimsons of cochineals (extracted, in their case, from *Porphyrophora* spp.) more violet.[57] They also used it as an alkaline agent in the woad vats for wool, and in the indigo vats for silk; finally, they brightened their yellows and greens with it, "*che ffa i piu vaghi colori che ssieno*" ("for it gives the most delicate colours that may exist").[58] In the *Memoirs*, however, wine lees ash is only employed in three recipes: two of wine soups and one of amaranth.[59]

Vitriols and verdigris

Green vitriol or copperas, iron(II) sulphate heptahydrate, and blue vitriol or Cyprus vitriol, copper(II) sulphate pentahydrate, are called vitriols from the Latin name for glass, *vitrum*, because their respectively blue-green and light blue crystals look like glass fragments. Verdigris is mostly copper(II) acetate as used by the author, since it is produced in large quantities in Languedoc, particularly in the region of Montpellier, using by-products of wine making – grape pomace, tartar, wine lees – to corrode thin copper plates. These iron or copper salts act as mordants, binding with colorants and fibres, but, as opposed to alum and tin mordants, they also modify the hue of the dye, acting in some measure as mineral colorants.

Green vitriol – or "copperas" – figures among the drugs that were described and whose provenance and price were discussed in the part of the first *Memoir* that is now lost. In the two memoirs giving dye recipes, it is the main ingredient used to obtain darker shades, in ranges of reds (King's colour), purples (plum colours), browns (cinnamons, cloves, hazelnuts, the different shades of tobacco and coffee), brownish greens (olives), all greys – including a "*poudre à canon des Anglais*" (English gunpowder) colour – and all blacks. This process is called *bruniture* in French, saddening in English. It involves heaving the cloth out of the dye bath when it is judged that it has taken enough colour, adding a small quantity of copperas, mixing it very well in a bath that must not be too hot, the door of the furnace being closed, no more fuel added, and the heat kept moderate. The cloth is then plunged back into the dye bath, turning it constantly.[60] If the colour is not saddened enough, the process is repeated after adding another small quantity of copperas. The quantity may therefore vary according to the dyer's taste, like salt or spices in a chef's recipe. Consequently, the author seldom gives more precision than "saddened with a little copperas", "with very little copperas", "with very little copperas and with caution". Exceptions are an ash grey in the fourth *Memoir* for which he recommends using 1 ounce of copperas; a saddened red in the same *Memoir* and the colour "rat on hazelnut ground", for which he gives the quantity of ¼ lb of copperas; mole greys, which, he says, need 1 lb of copperas; and of course, blacks, which require between 1 and 3 ½ pounds of copperas per piece, depending on the particular black process carried out. In grey and black-dyeing processes, iron sulphate does not only act as a postmordant, but actually contributes to forming a black colorant, through a chemical reaction with gall-nuts or other sources of tannins in the dye-bath, which produces a black ink. Other French sources on dyeing, like Jean Hellot and Antoine Janot, similarly do not give definite quantities of copperas to be used.[61] William Partridge does, but only in his recipes for dyeing wool in the fleece, or for dyeing cloth black, like the author of the *Memoirs*. For other colours, he adopts the same method – visual estimation – as the French authors, for instance to dye cloth a fawn colour: "boil the ingredients two hours, and the cloth two, heave out and sadden to pattern with copperas."[62]

As opposed to the section concerning green vitriol or copperas, the sections on blue vitriol and verdigris in the first *Memoir* have been preserved. The first part of the section on blue vitriol is entirely borrowed from the relevant entry in the *Dictionnaire du citoyen, ou abrégé historique, théorique et pratique du commerce* (The Citizen's Dictionary, or historical, theoretical and practical compendium on commerce) by Honoré Lacombe de Prézel, published in 1761, which gives another clue for dating the writing of the *Memoirs* towards the end of the Seven Years' War.[63]

Blue vitriol and verdigris, however, only figure in very few processes, described in the second *Memoir*. They include three recipes of celadon greens: in the first one, copper sulphate employed in proportions of 8 to 10 % of the cloth weight, is the unique source of green colorant; in the second recipe, it is employed in proportions of 10 to 12 %, dissolved in a dye bath of water to which have been added two bucketfuls of a welding bath. In the third, verdigris is used as the only colorant, in proportions of 16 to 20 % of the dry weight of cloth.[64] These recipes of greens entirely based on the colouring power of copper salts are an absolute heresy in a treatise dedicated to "the best mode of dyeing". Being well aware of it, the author has chosen not to waste any of his superfine cloths with such dyes of "no fastness whatsoever", so that there is no sample in the *Memoirs* to illustrate these processes. At the moment when he is writing, in any case, the sweeping infatuation for celadon green receding, "this colour is not in demand any longer".

However, during the previous years, the sudden and irresistible fashion of celadon green which had appeared among the elites of the Ottoman Empire in the years 1745–1750, had created technical challenges for European dyers. Celadon, a very pale, slightly blue green, inspired by the colour of the glaze on some Korean and Chinese ceramics very much admired at the time, happened to be "one of the most delicate colours in dyeing" but also one of the most difficult to obtain in the best mode of dyeing. As Antoine Janot explained in 1746, in a small memoir dedicated to this topic, it was classified as the seventeenth and palest shade of the stepped gradation of woaded greens obtained following the regulations, and its welding, particularly, required great dexterity to avoid giving the cloth "a yellowish tinge that is not at all what is wanted". He is one of the Languedocian dyers who routinely managed to make perfect celadon greens in the orthodox manner, as demonstrated by the beautiful pattern glued in his memoir, and he would relentlessly express his contempt for the cheap substitutes based on blue vitriol or verdigris that some chemists, such as the "Mr Albert" often mentioned in the *Memoirs* and other contemporary documents, try to present as wonderful improvements.[65] From the substantial files dedicated to this "celadon affair" preserved in the archives in Languedoc, it transpires that the English had started flooding the Levantine markets with cloths dyed in celadon green by new processes. Investigations into these and the ingredients employed were immediately commenced in Languedoc.

In August 1749, an inspector on broadcloth production writes: "we must try and imitate the English taste, from the points of view both of the quality and the colours of our broadcloths, in order to reduce their outlet by a greater consumption of our cloths."[66] Moved by a desire to save on production costs and find easier technical solutions, a group of local clothiers and dyers mount a kind of lobby to persuade the Intendant and inspectors of the province that "celadon green is a colour that it is impossible to make with a ground of woad blue and weld or another yellow dye plant" and, they argue, "that can only be done with verdigris without woad. Mr. Albert has developed a process for this dye that makes it very fast, so that its colour loses none of its beauty from staying exposed to the heat of the sun or to outdoor conditions." They produce patterns that "look rather well imitated when compared with the English cloths of that same colour". They further argue that the possibility of making "celadon green with *verdet* or verdigris" had already been considered in a regulation on dyeing, the *Instruction générale pour la teinture des laines de toutes couleurs, & pour la culture des drogues ou ingrediens qu'on y employe* of 18th March 1671.[67]

As a result, the Intendant allows the inspectors to let cloths dyed in celadon with verdigris to be exported to the Levant, under the condition that the clothiers first must ask for a permit, declaring the number of cloths they intend to dye in this manner and providing a sample of the results they obtain. Very soon, though, new processes based on "blue Cyprus vitriol" appear, either combined with a previous soap bath or without it. The group now asks for the "permission to dye some pieces of cloths celadon green with Cyprus vitriol, because celadon is a fashionable colour about which it is of utmost importance for our trade to satisfy the Orientals."[68] Unsurprisingly, one finds the same Mr. Albert busy again advertising for another of his wonderful new processes and performing a full-scale demonstration of it, invited by Jean Marcassus, director of the Royal Manufactures of La Terrasse and Auterive, to the south of Toulouse. It is only in the arguments presenting the advantages of this new recipe of celadon green that the serious defects of the processes using verdigris are now revealed for the first time: "Clothiers and dyers prefer blue vitriol to verdigris 1° because they find the former easier to employ 2° because the resulting colour is brighter and more fast to sunlight and to pressing."

A letter by a Mr. Cazaban, inspector of broadcloth manufactures, written in Saint-Chinian in August 1749, is even more telling: the celadon green obtained from verdigris is being abandoned, not only because it results more expensive than the process using blue vitriol, "but even without taking this into account, the new system that consists in adding two pounds of soap per piece of cloth, assuredly makes it very dangerous to use it 1° because it discredits our cloths in the Levant, since all those dyed in this manner are greasy and stinking and true dust traps; 2° if any water happens to splash onto the clothes, they are stained nearly irreversibly; 3° when one beats these clothes with rods, off comes a white powder as fine as that used for wigs, together with the verdigris."[69]

Considering that the author of the *Memoirs* is as capable as Antoine Janot of making a beautiful celadon green on a woad ground, as illustrated by a nice sample on p. 60 of the manuscript, it is probably only for the sake of completeness that he has chosen to include these recipes of celadon based on copper salts.

Celadon green does not figure once among the colour names found in the many pattern books of Gloucestershire and Wiltshire clothiers and dyers I have been able to consult. William Partridge includes blue vitriol in only one recipe of green on a woaded ground, besides using it and or verdigris in all his recipes for blacks.[70]

Notes

1 Partridge re-ed 1973, p. 224 and ff.
2 Partridge re-ed 1973, p. 87.
3 Hellot 1750, p. 277.
4 Ponting's notes in Partridge re-ed 1973, p. 237.
5 Pages 83 and 96 of the *Memoirs on dyeing*.
6 Hellot 1750, pp. 325–326.
7 P. 80 of the manuscript.
8 Defoe 1962, vol. 1 p. 282.
9 Mann 1971, pp. 10–11.
10 Partridge re-ed 1973, p. 88.
11 *Ibid.*, p. 92.
12 Rudder 1779, p. 711.
13 Partridge re-ed 1973, p. 7.
14 Cardon 2007, pp. 4–6, 20–48.
15 Singer 1948, pp. xvii-xviii.
16 *Règlement pour la teinture des étoffes de laine et des laines servant à leur fabrication* (Regulation on dyeing of woollen cloth and wools for their fabrication) of 15th January 1737, An.1746, p. 236, art. 36.
17 Hellot 1750, pp. 345–346, 348–350; p. 88 of the *Memoirs on dyeing*.
18 P. 82 of the manuscript.
19 Savary des Bruslons 1748, p. 639.
20 Delumeau 1962; Singer 1948, pp. 139–157.
21 Savary des Bruslons 1748, p. 639; Fougeroux de Bondaroy 1769, p. 19.
22 Roux, Aubert de La Chesnaye-Desbois and Goulin 1762–1764, t. 1, pp. 53–54.
23 Delumeau 1962, pp. 172–175, diagram 4 and table 10.
24 It is the Roman steelyard pound of Marseilles, used for commodities weighing more than 20 pounds; 1lb = 0.403 kg, Charbonnier *et al.* 1994, p. 124.
25 Ancien Régime France had a system of custom duties between the different parts of the kingdom. Languedoc was one of the "*provinces reputées étrangères*" (provinces considered as foreign from the taxation point of view) where high duties were fixed both for import into and export to neighbouring provinces, such as Provence. Marseilles was in still another category "*provinces à l'instar de l'étranger effectif*", free of taxes on trade with foreign countries. But high custom duties had to be paid to import products from there into Provence and other regions of France.
26 Fougeroux de Bondaroy 1768, p. 277; Cardon 2007, p. 15, 39.
27 Savary des Bruslons 1748, p. 639.
28 Singer 1948, pp. 89–94.
29 Çolak, Thirion-Merle, Blondé and Picon 2005; Singer 1948, p. 141.
30 *Ibid.*, p. 144.
31 Cordoba de la Llave, Franco Silva and Navarro Espinach 2005, p. 128.
32 Bowles, transl. de Flavigny 1776, pp. 387–390.
33 P. 5 of the manuscript.
34 Picon 2005.
35 Diderot and d'Alembert 1751, vol. 1, p. 309.
36 Savary des Bruslons 1748, p. 639.
37 Diderot and d'Alembert 1751, vol. 1, pp. 308–309.
38 David Pybus, personal communication; see also Pybus 2009.
39 Aki Arponen, "Alum in Sweden", paper presented at the meeting Dyes in History and Archaeology, 20.
40 Diderot and d'Alembert 1751, t. 1, p. 309.
41 Pp. 89–91, 93 of the manuscript.
42 Pp. 52–53, 91, see colorimetric data in appendix 3, samples n° 43 and 151; the difference between the two samples is of 2.8 units (under 2, differences are not perceptible by the human eye).
43 Pp. 55–59, 89, 93; colorimetric data, appendix 3, samples n° 58 and 159, difference = 0.
44 AD11 3J 342. Each bundle contains 10 cloths, each of which corresponds to one half of a piece as woven on the loom (see chapter 1, n. 45). These "half-pieces" are the lengths of cloth considered as pieces or "ends" by dyers; they measure 15 to 17 ells of France (17.82 to 20.19 m) and weigh around 25 pounds (10.35 kg).
45 AD34 C2160; AD 11 3J 342 (12 bundles, Royal Manufacture of Pennautier, 24 July 1749; 2 bundles, same manufacture, 3 November 1767; 6 bundles, same manufacture, 8 October 1775; 6 bundles, Manufacture of Germain and Jacques Pinel in Carcassonne, 22 April 1748; 2 bundles, Royal Manufacture of Saint-Chinian, 30 June 1750).
46 Production statistics for the Manufacture of Bize, AD 11 9 C 31.
47 Cardon 2007, pp. 607–619; Cardon 2014, pp. 593–602.
48 Cardon 2007, pp. 47–48.
49 Partridge, re-ed 1973, p. 101.
50 *Ibid.*, p. 103.
51 Savary des Bruslons 1726, t. 2, p. 1711; Chomel 1732, t. 2, p. 1192.
52 Partridge re-ed 1973, p. 100.
53 *Ibid.*
54 Golikov 2001.
55 Pp. 85 and 87 of manuscript.
56 *Arte della Lana*, dated 1421, Biblioteca Riccardiana, Florence,Codex 2580, folios 151–152 verso: "… *sappi chell' alume della feccia sie feccia di vin bianco arsa che ss' inpasta sopra cierti fuscielli e ardesi ogni cosa, tosi poi detta feccia arsa ed'è l'allume detto*" ("… you must know that lees alum is the calcined lees of white wine that are pasted onto some twigs and the whole is burnt; then take such calcined lees and this is the said alum"); Rebora 1970, p. 66–67, 69–70,102, 105, 112, 115, 119–120, 128, 130.
57 Cardon 2014, pp. 616–633.
58 *Arte della Lana*, Bibl. Riccardiana, Cod. 2580, folio152 verso.
59 Pp. 83 and 93 of the manuscript.
60 The way to proceed is described p. 53 of the manuscript, where it is said that this description, concerning the saddening of madder reds, is valid for all following processes ending in a saddening of the colour.
61 Hellot 1750, pp. 434–435, 439–443.
62 Partridge re-ed 1973, p. 194.
63 Lacombe de Prézel 1761, t. 2, p. 253.
64 Pp. 65–67 of the manuscript.
65 Janot's notes on celadon, AD34 C 2240.
66 AD34 C 2240, other document.
67 An. 1730, p. 434, art. 55.
68 AD34 C 2240.
69 *Ibid.*
70 Partridge 1973, pp. 136–141, 178.

Table 4.1: Importance of alum and tin mordants, tartar and iron sulphate in the 206 dyeing processes described in the Memoirs

	Number of recipes	*Percentage of total number of recipes (206)*
No premordanting or nuancing with alum or tin	64	31
Alum in premordanting or saddening bath	99	48
Tin liquor in premordanting and/or dye bath	43	21
Tartar in premordanting and/or dye bath	129	63
Copperas in saddening bath	54	26

Table 4.2: Different kinds of alum employed: respective importance, price range

Kinds of alum mentioned in recipes	*Number of recipes*	*Percentage of total weight of alum used (158 kg)*	*Range of buying prices in Marseilles (percentage of maximum price mentioned for Rome alum = 40 pounds/quintal)*
Rome alum	33	35	100
Alum from the Levant or Spain	24	24	25–85
Alum from the Levant	19	20	25–67
Alum from Spain	9	10	42.5–85
Alum from Sweden	8	7	42.5–85
Alum without mention of provenance	6	4	Price not specified

*Table 4.3: Tin liquor for scarlet (*composition d'écarlate*) in the 18th and 19th centuries according to French sources and one Anglo-American source*

Ingredients	*2nd* Memoir *p. 42 (quantities for 5 pieces)*	*4th* Memoir *p. 81*	*Janot 1744 f°7v°-8*	*Hellot 1750 p. 281–4*	*Partridge 1823 re-ed 1973 p. 101*	*Grison 1884, v. 2 p. 217*
Water	30 lb = 12.42 kg 100 %	12 lb = 4.968 kg 100 %	4 lb = 1.656 kg 100 %	8 oz = 0.244 kg 100 %	2 pints = 1.134 kg 100 %	20 kg 100 %
Aqua-fortis	17 lb = 7.038 kg 56.6 %	9 lb = 3.726 kg 75 %	2,5 lb = 1.035 kg 62.5 %	8 oz = 0.244 kg 100 %	4 pints = +/−2.268 kg 200 %	15 kg 75 %
Tin	2 lb = 0.828 kg 6.6 %	1 lb = 0.414 kg 8 .3 %	6 oz = 0.155 kg 9.3 %	1 oz = 0.030 kg 12.5 %	8 oz = 0.226 kg 19.85 %	1.750 kg 8.7 %
Sal ammoniac	¼ lb = 0.103 kg 0.8 %	3 oz = 0.077 kg[1] 1.5 %	1 oz = 0.025 kg 1.5 %	½ oz = 0.015 kg 6.25 %		
Common salt	-	3oz = 0.077 kg 1.5 %		-	1 handful = +/− 0.070 kg 6.17 %	0.750 kg 3.7 %
Saltpetre	-	-		2 *gros* = 2/128 lb = 0.007 kg 3.1 %		

1 The author remarks that "it can be replaced by common salt". Quantities for sal ammoniac and common salt in this table are therefore not to be added: either one substance or the other is used.

Table 4.4: Tartar in the 206 dyeing processes described in the Memoirs

Kind of tartar	*Number of recipes*	*Weight (kg)*	*Percentage of total weight for all tartars mentioned (104.742 kg)*
White	39	24.633	23.5
Red	81	63.756	60.9
Cream/crystal of tartar	3	4.14	4
Non-specified	11	12.213	11.6

5

Blues

Blue first

At the time when the *Memoirs on dyeing* were written, professional dyers in Europe had for centuries been using an increasing number of dye sources, while their way of using them had constantly been based on the same basic principle. Their dye baths were composed with the aim of obtaining one of the three primary colours: blue, red and yellow. A fourth group of colorants adds a range of beige and brown colours, the *fauve* (tan) or *couleur de racine* (root colour), to sadden the three first primary colours or to obtain black by various successive combinations of colorants and mordants.

This traditional understanding of the art is clearly presented by Jean Hellot at the beginning of his own *Art de la teinture des laines et des étoffes de laine en grand et petit teint* (Art of dyeing wool and woollen textiles in the best mode and in small dyeing).[1] Colours resulting from a combination of blue and red (all purples and mauves), or blue and yellow (all greens), and all sorts of shades requiring some degree of blue, are obtained by submitting the cloth or wool previously dyed blue to a mordant bath and then to a second dye bath with one or several red, yellow or brown colorant(s), because the chemical structure of the colouring molecules present in most natural dye sources for these ranges of colours did not allow combining them satisfactorily with the most important natural blue dye – indigo – in one dye bath. This is due to the fact that indigo, from whatever plant source, is deposited onto textile fibres by the precipitation of its soluble colourless form into its insoluble blue form under the action of the oxygen in the air. It forms microlayers on the fibres, the accumulation of which by successive dippings into the dye bath gives more and more saturated and dark blues. These progressive degrees of blues served as a kind of scale to dyers, who evaluated them by visual comparison with a standard range of cloth swatches dyed in the successive shades of blues.[2] These degrees of blue, in turn, determined the different shades of purples and greens that can be obtained by dyers. This is exceptionally well demonstrated and illustrated in the present *Memoirs*.

However, these degrees cannot be accurately assessed if the cloth or wool has already received another colour. This is why the author began his first memoir of dye recipes with a presentation of the method of dyeing into fast blues: "Blue being a primary colour, and the ground for the greatest part of fast colours, it is necessary to start with it to establish some order in this memoir". This order is the traditional order based on the very ancient practice of wool dyers. It is not surprising, therefore, to find that all other French authors of treatises on wool and cloth dyeing written in the 17th and 18th centuries begin their descriptions with the blue-dyeing process of the woad-and-indigo vat.[3]

English wool and cloth dyers were following the same order, imposed by the chemical nature of the dyes: in the dyeing book of a firm of Gloucestershire clothiers who did not dye blue themselves but commissioned indigo dyeing to two different specialised vat dyers, the recipes for all "oaded" (woaded) colours always start with the mention that the wool has first been sent to the woad dyer: "Oaded Slate – To Timbrell for 3 d blue, July 13, 1797" – and only then comes the recipe of the following process(es), performed in the dye-house of the firm.[4]

Indigo and the wool vat

The only source of blues really fast to light and washing ancient dyers possessed was the blue pigment known as indigo. For ages, all over the world, and in different forms (compost of leaves, thick slur of "mud indigo", cakes of dried indigo precipitate), indigo has been extracted from many plants belonging to several different botanical families.[5] For a European wool dyer of the 18th century, dyeing a strong and fast blue cannot mean anything else but dyeing in the woad vat. Woad is used in it in the form of couched woad, a mass of round lumps concentrated in indigotin, the blue component of indigo, extracted by a double fermentation from the leaves of the woad herb, *Isatis tinctoria* L. (Brassicaceae), the only indigo-producing plant growing in Europe.[6] To this woad vat, the wool dyers who specialised in blue dyeing – called *guéderons* in French (from the name of the plant, *guède*, derived from the same Germanic root as woad) – had started to add some proportion of indigo pigment extracted from the leaves of bushes of different species of *Indigofera* (Fabaceae). This happened gradually during the 16th and 17th centuries, beginning with the Age of Discovery, when the opening of direct sea routes to the indigo-producing regions of India, the Caribbean and Mesoamerica made exotic concentrated indigo widely available to European dyers, at much lower prices than during the Middle Ages, when it was imported from, or via, the Eastern Mediterranean.

Why, then, was woad still commonly used to dye wool and woollen cloth blue, not only until the 18th century but well into the 19th century?[7] The reason was that exotic indigo, imported as dry cakes of compressed indigo pigment, did not ferment and get into the soluble reduced form in which it could dye wool as easily as did couched woad, which still contained bacteria, enzymes, and sugars favouring the right sort of reducing fermentation in the vat: in the new, hybrid "woad vat" reinforced with indigo, common to all European wool dyers, woad was acting both as colouring and fermenting agent, even though indigo contributed most of the blue colouring power of the vat, being more concentrated in indigotin. On the other hand, the indigo vats without woad that had been developed for cotton and silk in Europe since the Middle Ages were not suitable for wool, because in such vats the solution and reduction of indigo was produced by chemicals in a strongly alkaline medium which would have damaged the wool or cloth.

The precise description of the process of the woad-and-indigo vat given at pages 39–40 of the *Memoirs*, is one of the earliest based on daily practice in a big broadcloth manufacture. Even though the author does not work this woad vat himself and does not pretend to do it – very honestly, he does not once write in the first person in this part – he obviously daily watched the progress of his vats and conferred with the "vatman" he surely employed. In Languedoc, these specialised dyers were among the few workers in broadcloth production that were employed permanently and paid – fairly good salaries – on a monthly basis.[8] By a fortunate circumstance, two other descriptions of this type of vat, by commission dyers of the same region and period, have been preserved. They allow very interesting comparisons, not only between the three of them, but also with the only known equivalent description of this process as practised in the West of England, published by the Anglo-American dyer William Partridge at a later date, but based on earlier experience in his country of origin, and occasionally using woad obtained from there, from his brothers (Table 5.1).[9] Partridge also provides an essential information for such comparison, lacking in the French sources: "the contents of the vessel" which, he rightly points, "should be given with it, when receipts are given for woad dying, otherwise an artist ignorant of the business, might fail for want of being put in possession of this very simple fact, for" – he goes on to explain – "were the same materials to be used [in the French vat and in the English vat that have different contents], one of them would be too strong, when the other would have only its proper quantity."[10]

As a working hypothesis for the comparison proposed in Table 5.1, it is fairly safe to use the data on the contents of the French vat provided by Partridge and by Thomas Cooper, because the woad-and-indigo vat was an industrial process, standardised in each country. The volume they mention for the French vat, moreover, is within the same order of magnitude as the woad-and-indigo vats still in operation around 1870 in Verviers, an important centre of broadcloth manufacture in the east of Belgium.[11] The enormous contents of these vats were devised to ensure an optimal thermal inertia, helping maintain the initially boiling water at the right temperature – around 50°C – for the proliferation of the indigo-reducing bacteria. With the same aim, the wooden vats were partly sunk into the ground and kept tightly covered with plank lids and blankets between dyeing sessions.[12] They were deep, Partridge explains, to ensure that "there be room enough to hold all the sediment collected during one working, so as it shall not interfere with the goods dyed."[13]

In the Royal Manufacture of Bize, three such vats for blue-dyeing were constantly kept in operation, as shown by the itemised inventory established in 1774, following the death of Germain Pinel, owner of the Manufacture. They are described as "made of oak wood, with their lids of fir wood".[14] Having at least two vats working simultaneously was the only way for a big manufacture to always have one "in working state", in Antoine Janot's words.[15] In such vats whose colouring power depends on the reducing power of bacterial fermentation, it is only possible to dye blue when all the indigotin is in solution in the reduced state, which needs a perfect balance of pH (around 9) and temperature (around 50 °C). This balance must be restored after each session of dipping pieces of cloth into the vat, by adding just the right amount of slaked lime, and letting the bacterial fermentation set off again. It can take a few hours before the vat can be used again. Having several vats in operation allows alternate and nearly continuous working sessions to take place. It also allows dyers to constantly be able to dye the whole range of blues, from the darkest, obtained from newly set vats, to the lightest, easier to manage in nearly exhausted ones. Partridge, too, considers that "in dying with woad, there should always be two vats in operation at the same time, one that has been worked for one or two months, and a new vat. The wool to be coloured, should be primed in the new vat, and finished in the older one."[16]

The process he is describing differs from the Languedocian vats in several interesting aspects. The first concerns the proportions of ingredients. Not those for woad, though, which are very similar in all vats, confirming the validity of the content mentioned by Partridge for the French vat. Even Janot's maximum quantity of four bales of woad does not make a big difference. It might be a traditional mnemonic way of knowledge transmission among vat dyers: 4 bales of woad, 4 pounds of indigo, giving 4 pieces of cloth of each of the degrees of the scale of vat blues... The main differences between the French vats and the vat described by Partridge are that he does not add weld to the boiled water he sets the vat with, and adds much less bran and madder than the Languedocian dyers, who use all three ingredients as complements for the cellulose in couched woad, providing carbohydrates and nutrients for the bacterial fermentation and reduction of the indigotin of woad and indigo. On the other hand, he adds much more indigo into the vat – between more than double and nearly double the proportions used by Languedocian dyers – and about ten times more lime. This is all due to the fact that his vat combines two reduction processes: through the bacterial fermentation of the traditional woad vat, and through the chemical reaction between lime and copperas. Characteristically, this last ingredient does not figure at all in the three other recipes.

The reasons for such differences may be that it was easier and cheaper for Partridge in America, and even for English dyers of the 18th century, to get good indigo from South Carolina, than to have a regular supply of good woad, rich in indigotin. Therefore they mainly relied on the addition of indigo for the colouring power of their vat; but to reduce such high proportion of indigo, they combined two reduction processes, both needing lime. In Languedoc, by contrast, dyers were very close to the biggest region of production of best quality woad, which they could buy easily and comparatively cheaply. Moreover, they were obliged by regulations to stick to the traditional process of the woad vat as the only method for reducing the indigo they had eventually been allowed to add into it "as per the dyer's judgment."[17] They, too, had access to very good indigo, massively produced in the French colonies in the Caribbean, first in Guadeloupe since the 17th century and later on in Saint-Domingue.[18] Particularly in the case of the Royal Manufacture of Bize, it could be obtained at the best possible price, since indigo was one of the commodities in which the Pinel group, who owned the manufacture, was trading.[19]

Differences in processes, therefore, depended also on the output that was expected from the vats. In Partridge's example, a vat is worked during six months, the liquor being reheated and replenished with smaller quantities of ingredients twice a week. "There can be obtained from it, during the working down, four hundred pounds of dark blue wool, two hundred of half blue, and two of very light. This is the calculation in all well regulated English dye-houses."[20] By contrast, Languedocian dyers appear more interested in producing light and very light blues, on which we shall see they had built whole ranges of fashionable colours, with their Oriental customers in mind. According to Antoine Janot, each working of a vat, set as summarized in Table 5.1, lasts for one week. It allows dyeing 12 pieces of cloth into dark blues, 12 pieces into medium blues and 16 pieces in light to very light blues.[21] According to Hellot, the darkest blues are obtained first, usually in several dips, and the lightest are done at the end of the week, when the vat is getting less concentrated in indigotin. Antoine Janot mentions the possibility "to reheat the same vat several times, each time adding new woad and indigo that produce the same effect as the first reheating, this is what is called *anter* (to graft) the vat."[22]

However, beautiful, bright light blues are easier to obtain in new vats, Hellot observes.[23] In the *Memoirs*, indigo blues or colours that need a blue ground represent one-third of the processes described. Among them, there are 17 dark blues, 33 medium blues and 20 light blues.

The stepped gradation of indigo blues

Page 41 of the *Memoirs on Dyeing* is a particularly precious and rare document. For the 18th century, I know only one equivalent, giving the names of the entire range of blues that were commonly obtained from a woad-and-indigo vat and illustrating all of them with cloth samples. This is folio 2 of the unpublished *Mémoire* of recipes to produce the colours that were in demand in the Levant, written by Antoine Janot, a master dyer in Saint-Chinian, which he finished and signed on 31 March 1744.[24] Another, later, manuscript from the same region, the *Livre de Teinture à l'usage de Casimir Maistre* (Casimir Maistre's Dye Book) includes 15 cloth samples of vat blues but they mostly represent the darkest shades of the scale, from Persian blue down to azure blue, obtained from two vats worked in the author's manufacture of Villeneuvette, near Clermont-L'Hérault, from 1 June to 3 June 1817.[25] Still later, Théophile Grison's invaluable book, *La Teinture au dix-neuvième siècle en ce qui concerne la laine et les tissus où la laine est prédominante* (Dyeing in the nineteenth century as regards wool and fabrics in which wool predominates), includes a complete stepped gradation of vat blues on broadcloth, classified however in five degrees only, without names, and designated as "n°1" for the darkest to "n°5" for the lightest.[26] There exist several other sources, mediaeval ones inclusively, giving the names of the successive degrees in the scale of blues obtained from the woad vat.[27] Unfortunately, they are not illustrated with samples. Several of these names have been handed down through the centuries. However, some are placed differently on the scale, in different sources (Table 5.2). Persian blues (*persi*) stayed right at the top of the scale, where they already figured in the 15th century in documents from Venice and Florence. Similarly, the *bleu de lait* (milky blue) of the *Memoirs* still figures among the lightest shades, as already did the milky blue (*alatado*) of the mediaeval Venetian scale of blues. By contrast, azure blues (*azuri*) and celestial blues *(celestri*) that used to be classified as darker than Turkish blues (*turchini*) in mediaeval Italy, are lighter than *bleu turquin* (Turkish blue) for 18th century Languedocian dyers.

It cannot be overemphasised that these names are not merely poetical descriptions: dyers and their clients did assess the exact match of a freshly dyed piece of cloth by visual comparison with sets of standards of dyed cloth; moreover, each degree of blue was evaluated in terms of production costs (amounts of woad and indigo consumed, number of dips necessary to achieve the desired intensity of blue), and was priced accordingly. This is why, in some important broadcloth producing regions, degrees of blues in the scale of vat blues were described, not by names, but by monetary values. This is particularly clear from a series of 15th-century Catalan and Balearic regulations on dyeing, where the degrees of blue for any colour needing a woad ground are expressed in *sols* (pennies).[28] In Mallorca, this value was stamped onto a lead seal attached to standard samples of each of the degrees in the stepped gradation of vat blues, a set of which was kept in the clothiers' hall.[29] The same system is used by William Partridge to specify the blue ground needed for each shade of green, "to be first dyed blue in the vat, such as can be done for ten cents per pound [of wool]". The range of other blues he mentions goes from "a full 20 cent blue" down to "a light 7 cent blue", other degrees specified being the 15 cent, 14 cent, 13 cent, 11 cent and 9 cent blues.[30] For lack of corresponding samples, there is no way one could figure out the colorimetric values of these blues.

However, I have recently been able to ascertain that Partridge's system was an adaptation to American currency of the system which the English vat dyers and clothiers of the West of England were using in the 18th century. In dye books and pattern books of different firms, preserved in archives in Gloucester and Chippenham, I found numerous references to "3 d blue" (3 pennies blue) or "3 blow" and other values of blues, found in the same documents, range from a "12 d blue" down to "lighter blues" than the "3 d blue", with mentions of "10 d", "9 d", "4 d" and "3 ½ d" blues.[31] What is even more interesting in these sources is that, just like in the *Memoirs on dyeing* and in Janot's manuscript, samples of these degrees of blue are preserved. It made it possible to relate the French and English scales of vat blues by visual matching. This was done by using a complete set of samples of the shades of blue obtained on woollen cloth from a pure woad vat, reconstituted following a mediaeval recipe described in a Florentine manuscript.[32] I used this set of samples as common standard for visual comparisons with blue samples in both French and English documents.

This has revealed that the "3 d light blue" of English dyers was equivalent to the *bleu d'azur* (azure blue) of French dyers and to Grison's n°4. The French *bleus d'azur* are remarkably consistent between themselves: the sample of azure blue on page 41 of the *Memoirs on Dyeing* is identical to the *bleus d'azur* on pattern sheets of the Royal Manufactures of La Terrasse and Auterive and nearly identical to Janot's *bleu d'azur.*[33] By contrast, English "3 d" blues are slightly greener and greyer, perhaps reflecting differences in the vat process, such as have already been evidenced by the comparisons proposed in Table 6.1. Some of the English "3 d" blues, a little darker than the rest, are closer to the French *bleus d'azur foncés* (dark azure blue) of the Royal Manufactures of La Terrasse and Auterive and of Casimir Maistre's *Dye Book*. The English "3 ½ d" blues are equivalent to Janot's *bleu céleste* (celestial blue) and to Grison's n°3, and the "4 d" blues are equivalent to the *bleus de Roy* in the *Memoirs* and in Janot's manuscript.

In the range of light blues, the "lighter blues" in the pattern book of two clothiers from Trowbridge, Thomas and Josh Clark, are equivalent to Janot's *bleu de ciel* (sky blue), to the blue ground for agate or dove colours in Casimir Maistre's *Dye Book*, and to Grison's n°5.[34] A "sky blue" is recorded in May 1759 in a later pattern book of Thomas Clark but the corresponding sample is missing.[35] In the 18th century French dyers' scales, there are between three (in the *Memoirs on dyeing*) and five (according to Hellot) degrees of lighter blues below sky blue. In pattern books of clothiers from the West of England figure beautiful light to very light blues, unfortunately not qualified, either by name or value.

In the upper degrees of the scale, from the "9 d" blues upward, all English dark blues preserved in the documents from the West of England consulted, are darker and blacker than the darkest French blues, even the "dark Persian blue" in Casimir Maistre's *Dye Book* and the earlier Persian blue on a pattern sheet from the Royal Manufacture of La Terrasse, both obtained purely by dyeing in the woad-and-indigo vat. This observation corresponds rather well with Partridge's statement, mentioned above, that English dyers were mostly interested in producing dark blues. It is also possible that they were using an afterbath of logwood with alum to intensify and sadden their dark blues. This process will be discussed further down, in the section dedicated to this dye wood.

Saxon blue

Saxon blue, now more commonly called indigo carmine, is the dye called *bleu de Prusse* (Prussian blue) by the author of the *Memoirs on dyeing*. It has nothing to do with the blue pigment today called Prussian blue, which is a hydrated iron hexacyanoferrate complex compound discovered around 1706 by Johann Jacob Diesbach in Berlin. Since it is insoluble, it took years of research and experimentations before the intricate process allowing to dye wool blue with it was developed. It involved producing a chemical reaction to encourage the formation of the colorant directly on the fibre in the dye bath. This discovery took place much later than the period when the *Memoirs* were written: under the Napoleonic Empire, supplies of indigo being disrupted by the naval blockade, research into all other possible sources of blue dyes were encouraged by the French government. However, it was not until 1822 that the first process valid at industrial scale for dyeing broadcloth into Prussian blue was developed, and patriotically called *bleu de France*.[36]

It is not surprising that, around the middle of the 18th century, the author should call *bleu de Prusse* a totally different dye, because this was one of the names then given by French dyers to Saxon blue. It is a dye obtained from a solution of indigo in sulphuric acid, a process invented par the German chemist Johann Christian Barth, at Grossenhain, in Saxony, around 1743.[37] In his *Éléments de l'art de la teinture* (Elements on the Art of Dyeing), published in 1804, Claude-Louis Berthollet still remarks that "in dyeworks, this solution is called '*composition*', and Saxon blue is often called Prussian blue".[38] Sulphuric acid, also called oil of vitriol, used to dissolve indigo to prepare Saxon blue, could be sold in various concentrations, hence the variations in the proportions of sulphuric acid and indigo that are found in the numerous recipes of *composition* – or "chemick", as William Partridge calls it – published during the second half of the 18th century, and reviewed by Berthollet. The first proportion proposed by the author of the *Memoirs* (8 parts sulphuric acid for 1 part indigo) is exactly the same as the proportion recommended by the Swedish chemist Torbern Olof Bergman, according to Berthollet.[39] The second preparation he describes, devised to dye five pieces of Londrins Seconds cloth, is a little more concentrated in indigo (6 parts sulphuric acid for 1 part indigo), but less so than the proportion given by William Partridge in his recipe of chemick, which uses only 4 parts of oil of vitriol for one part of indigo because, he argues, "it is worse than useless to employ more than is sufficient, for in all cases the action of the vitriol is injurious to the goods dyed".[40]

The author of the *Memoirs*' remark that "these colours [Saxon blues and greens] are not often requested, particularly nowadays" does not only confirm the date proposed for the period when he was writing but also shows how fast technical inventions were spreading and fashions were changing in the 18th century. As just mentioned, the process and the new possibilities it opened for dyeing had been discovered in Saxony around 1743. After being kept secret for a few years, it was sold to a prospective user during the 1748 Easter Trade Fair in Leipzig.[41] By 1749, in Languedoc, several chemists and dyers are already at work to develop "improved" versions of the recipe, which has spread like wildfire. The same Mr. Albert as was already involved in promoting celadon greens with verdigris (see previous chapter) is very busy with Saxon blues and greens too. Full-scale demonstrations of the allegedly improved processes are organised in different Royal Manufactures. One takes place in October of that year in Bize, whose manager at the time is not yet Paul Gout, but a man called Jean Mailhol. Another demonstration is staged at La Terrasse, to the south of Toulouse, and the inspector of manufactures invited to watch the process and its results soon writes to the Intendant of Languedoc, extolling "the freshness of the colours in the fashion of Saxony" obtained.[42]

It is not surprising that dyers should have been instantly seduced by this new, semi-synthetic dye: being soluble, it can be treated like any other of their mordant dyes, by contrast with the other ways of using indigo, all involving reduction processes. The difference between the two dye recipes for Saxon blue given by the author of the *Memoirs* is in the number of steps in the process: in the former, the cloth is first mordanted in a boiling bath, with proportions of alum and tartar similar to those used for mordanting yellows and greens (13 to 16 % of alum and 6.6 to 8 % of white tartar by weight, based on the dry cloth weight). This is quite close to the proportions still recommended by Théophile Grison, more than a century later, to dye cloth into sky blue with indigo carmine: 13.6 % of alum, 4.5 % of white tartar in crystals. The author of the *Memoirs* is obviously aiming at ensuring the best possible fastness to this "problematic blue dye".[43] After being mordanted for a long time in the boiling bath, the cloths are heaved out of the bath, part of which is taken out to make room for cold water, because the ensuing process requires the bath to be just warm, "*tempéré*"; with indigo carmine, the degree of blue obtained depends entirely on the quantity of dissolved colorant then added to the bath – in this case, the sample joined shows that a light blue was wanted.

For the second process "without boiling" a mordant bath from the former recipe is used. It still contains some alum and tartar, but 5 to 6 % of the very concentrated sort of calcined alum which was called *farine* (flour) by French dyers are added to the bath, together with indigo carmine. This is therefore a one-bath process at comparatively low temperature, in which the cloth must not remain for long, which corresponds rather well to these comments by

Théophile Grison: "to obtain a fresh sky blue, the cloth must not remain in the dye bath for more than twenty-five to thirty minutes, the bath not heated to more than 65 to 70°C. If it boils or if the process is continued for too long, the blue gets green or markedly duller, due to the formation of a sulphuric derivate of the indigo blue which is a green product. This is why the precise moment when the degree of blue is high enough, and the dye uniform, must be watched, and the cloth immediately heaved out of the bath."[44]

Characteristically, no sample is joined to illustrate this process, in the *Memoirs*. The author obviously did not want to waste one of his superfine cloths with a dye whose poor lightfastness and washfastness had soon become apparent. Indigo carmine, on wool or silk, changes from blue to light greenish blue and eventually yellowish, as it fades, and runs in water with a blue colour.[45] The *Memoirs* show that dyers very quickly mastered ways to make the best possible use of Saxon blue, in terms of fastness and beauty of shade, but that in spite of all efforts, it took little more than ten years for Saxon blues and greens to be discredited among the producers of cloth for the Levant, certainly because of the bad comments they must have received from factors in the Eastern Mediterranean.

The same must have happened in England, and for the same reasons. The process for dyeing Saxon blue was bought by the wool dyers of Norwich and patented in England as early as 1748.[46] However, by 1773, it apparently was little used in the West of England for dyeing fine cloths. I could only find four mentions of Saxon blue in the pattern books of Gloucestershire and Wiltshire clothiers I consulted: they figure in a pattern book of Thomas Clark of Trowbridge.[47] Later still, Saxon blue is mentioned only once in 1804 and once in 1805, in a stock book/account book of the Hawker family of Stroud.[48] Partridge gives a very detailed recipe "to mix oil of vitriol and indigo to make chemick", but he does not give any recipe to dye blue with it and only mentions its use in connection with Saxon greens, discussed below.

Logwood: "dyeing blue in the furnace"[49]

Logwood, also called blockwood in the 16th–18th centuries, was called *bois de Campêche* (Campeachy wood) or *bois d'inde* (Indian wood) by French dyers of the same period. It consists of the heartwood of a Mesoamerican tree, *Haematoxylum campechianum* L. (Fabaceae, Caesalpinioideae). Its description, distribution, chemistry and the history of the exploitation of its wood have been very much studied and it is one of the sources of natural dyes that may become of major economic importance again.[50] The author of the *Memoirs* could obtain it quite cheaply because the tree grew abundantly on several Caribbean islands that had become French colonies. Until the middle of the 18th century, exportations of logwood from Martinique are recorded separately for each year: in 1752, 717 *quintaux* or hundreds (= 35 metric tons) of logwood are shipped to Bordeaux, and the price indicated corresponds to the range of prices mentioned by the author of the *Memoirs* – taking into account that the highest prices he mentions must correspond to the period of the Seven Years' War and the disruptions it occasioned in commercial maritime routes.[51] For the second half of the century, it is in the registers of *Entrées des Isles françoises de l'Amérique* (Imports from the French Islands of America) preserved in Bordeaux, that the amount of logwood and of Santa Marta wood (a red dye wood discussed below) are recorded: in 1771, 73,161 pounds of logwood (35.812 metric tons) arrive in Bordeaux, corresponding to the pre-war high levels of imports, at prices which have started decreasing again after the war.[52]

Supplies of logwood and its price, however, are not a major concern for the author of the *Memoirs*, because he is supposed to only make a very limited use of it. Long before the invention of Saxon blue, the dark blue dye obtained by simply boiling rasped logwood in water had, in the 16th century, already given European dyers hopes to be able to "dye blue in the furnace" after a mordant bath, in a similar process as those they used for most other dyes. But they were soon disappointed: the blue and violet dyes from logwood were not lightfast. Consequently, regulations were issued to prohibit them for high quality textiles and particularly for broadcloth, in France as in England. In France, the prohibition was maintained for a long time. It is reiterated in 1737, in the *Règlement pour la teinture des étoffes de laine et des laines servant à leur fabrication* (Regulation on dyeing of wool cloth and of the wools used for their making) of 15 January which prohibits the use of logwood to dye any good quality broadcloth into any shade of blue, grey, violet, purple and amaranth.[53] It is only allowed for dyeing black, under certain conditions that will be discussed further down, in the chapter concerning this colour. The *Memoirs on dyeing* show that in spite of such regulations, Languedocian dyers working in or for broadcloth manufactures, have started experimenting with logwood, mainly with the aim to save on the indigo grounds traditionally considered as necessary for a number of compound shades. The application they are mostly interested in, has been evoked earlier in this chapter. It consists in using logwood to falsify the darkest degrees of the scale of blues obtained from the woad-and-indigo vat. At page 40 of the manuscript, the author of the *Memoirs* lets it transpire that, instead of consuming most of the colouring potential of a vat to produce Persian blues, it is possible, "sometimes", to *aviver* (intensify) and *augmenter* (upgrade) a dark blue of slightly lesser intensity by top-dyeing the woad-dyed cloth in a logwood bath with alum added. This actually became a common practice in industrial dyeing during the 19th century, as explained by Théophile Grison: "the fastest vat blues are dyed purely with indigo; but the price of the colouring matter coming up to between 2 and 4 francs per kilogram of wool, according to the higher or lower intensity of the blue shade... the expense can be lessened by only dyeing a vat blue ground of one quarter, one half or three quarters of the desired degree of blue, and by ending with a *remontage* or *avivage* (upgrading or intensifying bath) with logwood."[54]

In the 18th century, nothing prevented English dyers from doing this, since the prohibitions on the importation and use of logwood had been suppressed in 1662.[55] It has been suggested above that the very dark, almost black blues figuring in the pattern books preserved in archives in the West of England may have been obtained by top-dyeing an already dark blue ground obtained in the woad-and-indigo vat with logwood. This is the case for the colour called "raven" in these documents. The recipe for a "dark raven" in the *Wool Dyeing Book* of Marling and Evans, in Stroud, requires a woad ground of a "3 d blue" which is then top-dyed with logwood and indigo carmine.[56] A final saddening with copperas actually makes it closer to black than to very dark blue, which will be discussed further, in connection with black dyes. For many other very dark blues found in pattern books there is no corresponding colour name or recipe joined. Dye-analyses only might detect the use of logwood, but such research remains to be done.

English dyers were just as aware of the poor lightfastness of this blue dye employed alone as were their French colleagues. William Partridge's assertive judgment on the process of "dying blue in the furnace" is that "this is often done for very common purposes, but never on anything like fine goods."[57] However, their experiments with logwood had allowed them to develop alternative processes where it was used with alum or iron mordants, and occasionally with indigo carmine, to obtain not only blacks but also diverse shades of greys, greens, purples and violets, which will be discussed in the following chapters.[58]

In his last *Memoir*, the author reports on his experiments in using small quantities of logwood – in the order of 7 to 8 % of the cloth weights – for top-dyeing some indigo grounds or even madder dyes to obtain some very dark colours at slightly lower costs. He uses even lesser proportions of it (from 0.8 to 1 %) to give their exact shade to some fashionable colours such as *musc* (musk), *noisette* (hazelnut), *gris de rat* (rat grey), *gorge de pigeon* (dove breast). It is only for "English violets and purples" and some blacks presented as new processes "in Albert's manner", that he gives recipes with high proportions of logwood, as shall be seen. All in all, and even including these last recipes, only 63.5 pounds, that is around 26 kg, of logwood have been used in the course of the implementation of the 206 dye processes described in the *Memoirs*. There could not be a better indicator of the minor importance of logwood in the *teinture de grand teint* as it was understood in France in the 18th century.

Notes

1 Hellot 1750, p. 46.
2 Illustrated in Cardon 2007, p. 361, fig. 36 and *id.* 2014, p. 355 fig. 34.
3 *Livret*, 1667; Janot 1744; Hellot 1750.
4 Gloucestershire Archives, D948/1, Wool Dyeing Book of Marling and Evans of Ebley, Stroud, dated 1794–1804, cloth n° 10905.
5 Cardon 2007, chap. 8, pp. 335–408; lesser known exotic indigo plants described in *id.* 2014, pp. 327–396.
6 Illustrated in Cardon 2007, pp. 342–343, *id.* 2014, p. 335, 337.
7 Cardon 1992a, 1992 b.
8 Marquié 1993, pp. 138–139.
9 Partridge re-ed 1973, pp. 156–157.
10 *Ibid.*, p. 153.
11 Van Laer 1874: two of the vats described contain 6000 L, another one 8000 L.
12 Illustrated in Cardon 2007, p. 372 and *id.* 2014, p. 364.
13 Partridge re-ed 1973, p. 125.
14 AD 11 3E 1246.
15 Janot 1744, folio 1 verso.
16 Partridge re-ed 1973, p. 158.
17 Hellot 1750, p. 100.
18 Yvon 2015.
19 Marquié 1993, pp. 186–187, 193.
20 Partridge re-ed 1973, pp. 158–159.
21 Janot 1744, folio 2.
22 Janot 1744, folio 2 verso.
23 Hellot 1751, pp. 76–79, 102.
24 Janot 1744, *Mémoire – on trouvera les opérations de la teinture du grand et bon teint des couleurs qui se consomment en Levant avec la quantité et qualité des drogues qui les composent*, AD 34, C 5569.
25 AD 34 11 J 39, folio 2 verso–3 verso.
26 Grison 1884, vol. 2, p. 125.
27 Cardon 1992a, p. 28; *id.*1992c, pp. 304–307.
28 Ordinance "For broadcloth" of 21 November 1430, de Capmany y de Montpalau re-ed 1962, p. 472, doc. 322 §27.
29 Ordinance of 23 August 1452, AHCB *Ordinacions originals*, §25.
30 Partridge re-ed 1973, pp. 178–81.
31 Pattern book of Thomas Long of Melksham, dated 1698–1728, Wiltshire and Swindon History Centre, 947/1802, patterns n° 194, 214; Wool Dyeing Book of Marling and Evans of Ebley, dated 1794–1804, Gloucestershire Archives, D948/1.
32 *Arte della Lana*, Biblioteca Riccardiana, Florence, Codex 2580, folio 141verso–147, vat set in February 1418. Text and translation into English and French in Cardon 1995/1998, pp. 50–57.

33 Pattern sheets of the royal Manufactures of La Terrasse and Auterive, AD34 C5550; Janot's *Mémoire*, AD34 C5569.
34 Dated 1743–1744, Wiltshire and Swindon History Centre, 927/6.
35 Wiltshire and Swindon History Centre, 927/8.
36 Delamare 2013, pp. 168–171.
37 De Keijzer *et al.* 2012.
38 Berthollet and Berthollet 1804, vol. 2, p. 98.
39 *Ibid.*, pp. 93–94.
40 Partridge re-ed 1973, p. 106.
41 De Keijzer *et al.* 2012, p. S88, based on Kortum 1749, pp. 5–6.
42 Several files are dedicated to Saxon blues and greens in the *Archives départementales de l'Hérault*: AD34 C 2238, C 2544.
43 This is a quotation from the title of the paper on indigo carmine by de Keijzer *et al.* 2012. The authors did not know how fast Saxon blue was adopted by Languedocian dyers working for the Levant and consequently present a Dutch manuscript of 1775 as the earliest recipe for the preparation of, and dyeing with, the Saxon colours, while it actually is at least ten years later than the *Memoirs*.
44 Grison 1884, vol. 1, pp.134–135.
45 De Keijzer *et al.* 2012.
46 *Ibid.*, p. S88.
47 Wiltshire and Swindon History Centre, 927/7, including dates from 1773 to 1778.
48 Gloucestershire Archives, D1181/3/1.
49 This is Partridge's definition of dyeing with logwood, Partridge re-ed 1973, p. 162.
50 McJunkin 1991; Cardon 2007, pp. 263–274; Cardon 2014, pp. 266–273.
51 ANOM C8 B21.
52 AD33 C4390, doc. 21.
53 An. (1746) *Manufacture de Carcassonne*, art. 37, 44, 57.
54 Grison 1884, vol. 1, p. 126.
55 Charles II, 1662: An Act for preventing Frauds and regulating Abuses in His Majesties Customes, in Statutes of the Realm: Volume 5, 1628–80, ed. John Raithby (s.l, 1819), pp. 393–400 http://www.british-history.ac.uk/statutes-realm/vol5/pp393–400 [accessed 25 July 2015], cap. 26, Recital of use of logwood by dyers.
56 Gloucestershire Archives, D948/1,
57 Partridge re-ed 1973, p. 162.
58 *Ibid.*, pp. 135–216.

Table 5.1: Setting of the woad-and-indigo vat: 18th-century Languedoc; Partridge's Practical Treatise on Dying of Woollen *(reflecting West of England's practices, end of 18th c.)*

Ingredients	Memoirs on dyeing *c. 1763 pp.39–40*	*An. AD34 7J7 s. d. 18th c. folio 9*	*Janot 1744 AD34 C5569 folio 1–2*	*Partridge 1823 re-ed 1973 pp. 153–4*
Water (vat content)	not mentioned in source but French vats according to Cooper + Partridge[1] = 1428 ale gallons[2] = 6,597 L 100%	not mentioned according to Cooper + Partridge = 1428 ale gallons = 6,597 L 100%	not mentioned according to Cooper + Partridge = 1428 ale gallons = 6,597 L 100%	1640 ale gallons = 7,577 L 100%
Weld	8–10 lb = 3.312–4.14 kg 0.05% – 0.06%	-	3 lb[3] = 1.242 kg 0.01%	-
Wheat bran[4]	½ *setier*[5] = 35.5 L = 8.875 kg 0.13%	6 *boisseaux* (bushels) = 13.32 L = 3.33 kg + 2 bushels (3 hours before 1st dipping) = 4.44 L = 1.11 kg Total 4.44 kg 0.06%	2 *pugnères* = ½ *setier*[6] = 35.5 L = 8.875 kg 0.13%	1 peck = 9.09 L = 2.27 kg 0.02%
Madder	8 lb = 3.312 kg 0.05%	6 lb = 2.484 kg + 2 lb (3 hours before 1st dipping) = 0.828 kg Total 3.312 kg 0.05%	3 lb per bale for 4 bales: 12 lb = 4.968 kg 0.07%	5 lb = 2.26 kg 0.02%

(continued)

Ingredients	Memoirs on dyeing *c. 1763* *pp.39–40*	*An. AD34 7J7* *s. d. 18th c.* *folio 9*	*Janot 1744 AD34 C5569* *folio 1–2*	*Partridge 1823 re-ed 1973* *pp. 153–4*
Couched woad	2 to 3 bales 3 bales = 540 lb = 223.56 kg 3.39%	3 bales = 540 lb = 223.56 kg 3.39%	up to 4 bales[7] = 720 lb = 298 kg 4.5%	560 lb = 253 kg 3.34%
Indigo	as per the dyer's judgment[8]	12 lb = 4.968 kg 0.07%	4 lb per bale[9] for 4 bales: 16 lb = 6.624 kg 0.10%	15 lb = 6.8 kg + 15 lb (filling up) = 6.8 kg Total 13.6 kg 0.18%
Lime	2 *trenchoirs*[10] per bale for 3 bales: 6 *trenchoirs* = +/- 600 g 0.01%	9 *trenchoirs* = +/-900 g 0.01%	as per the dyer's judgment	¼ peck= 2.27 L = 1.13 kg + 7–8 quarters of peck (filling up) = 15.9- 18.17 L = 7.95–9.08 kg 0.11 – 0.13%
Copperas	-	-	-	4 lb = 1.81 kg 0.02%

1 Cooper 1815, p. 36 and Partridge re-ed 1973, p. 153.

2 1 ale gallon = 4.62 litres, Doursther 1965, p. 155. 1 L of water = 1 kg, which allows to conveniently propose percentages for all ingredients, on the basis of the water content of the vats, and in the very probable hypothesis that all French vats have the same standard content.

3 The pound used in Saint-Chinian is the same as that used by the author of the *Memoirs*, that is the *livre grosse* of Montpellier of 0.414 kg. Should the weight of weld mentioned be understood as the quantity to be used per woad bale – as it is the case for madder – it should be multiplied by four in this recipe.

4 1 litre of bran, as weighed by myself = 250 g.

5 This is the *setier* of Narbonne of 71 litres, see appendix 1.

6 Again the *setier* of Narbonne.

7 Antoine Janot states that a bale of woad weighs about 180 pounds and that "up to 4 woad bales are used when setting a vat" (Janot 1744, f° 1).

8 Allowance dating from the *Règlement pour la teinture des Étoffes de Laine et des Laines servant à leur fabrication* of 15th January 1737 (An. (1746) *Manufacture de Carcassonne*, p. 238, art. 44)

9 It is the same proportion as in the recipe of AD 34 7J7 indicating 12 lb of indigo for 3 woad bales.

10 "This a kind of wooden palette which serves to roughly measure the quantity of lime that is put into the vat. It is 5 *pouces* (inch) wide and 3 ½ *pouces* long: it can contain a good fistful of lime, more or less" (Hellot 1750, p. 61). A good fistful of lime = +/- 100 to 120 g (personal experiment).

Table 5.2: Stepped gradation of vat blues in 18th century French treatises on dyeing, from darkest to lightest

Memoirs *p. 41*	*Janot 1744* *f°2r°*	*Hellot 1750* *p. 168*
		bleu d'enfer infernal blue
Bleu pers en faux False Persian blue obtained with a logwood saddening		*bleu aldego* aldego blue
Bleu pers augmenté à la cochenille Persian blue obtained by reinforcing a lighter blue with cochineal	*aile de courbeau* raven's wing	*bleu pers* Persian blue
		Fleur de guesde Garter blue (*lit.* = woad flurry)
Bleu turquin Turkish blue	*Bleu turquin* Turkish blue	*Bleu de roy* King's blue
Bleu de roy King's blue	*Bleu de roy* King's blue	*Bleu turquin* Turkish blue
Bleu tané Tanned blue	*Bleu céleste* Celestial blue	*Bleu de reine* Queen's blue
Bleu d'azur Azure blue	*Bleu d'azur* Azure blue	
Bleu céleste Celestial blue	*Bleu de ciel* Sky blue	*Bleu céleste* Celestial blue
Bleu mignon Dainty blue	*Bleu mignon* Dainty blue	*Bleu mignon* Dainty blue
Bleu de lait Milky blue		*Bleu mourant* Dying blue
	Bleu déblanchi Off-white blue	*Bleu pâle* Pale blue
Bleu de blanchy Off-white blue	*Bleu déblanchi plus clair* Lighter off-white blue	*Bleu naissant* Nascent blue
	Bleu déblanchi encore plus clair still lighter off-white blue	*Bleu blanc* Whitish blue

6

Reds

Dyeing in the furnace

Unlike indigo blues, prepared in huge wooden vats were dye baths are maintained warm enough for several days without heating, all reds are obtained from mordanting and dyeing baths heated to boiling or simmering point in very large vessels made of tin, copper or brass. This is why this branch of dyeing is defined as "*teinture de bouillon*" (dyeing at boiling temperature) in French, "*bouillon*" also designating more specifically a mordant bath preceding a dye bath, particularly in the *Memoirs*. In English, this mode of dyeing is called by William Partridge "furnace dying" and premordant baths are also called "boiling" in his *Treatise.*[1] The furnaces consist of large metal vessels resting on brick work including a fire-place and a chimney. The reason for heating the baths to high temperatures is to allow the dilatation of fibres, and the even diffusion of the metal ions of the mordants and of the colouring molecules of the dyes into the textile mass.[2] Piece-dyeing of cloths of such large-dimensions as have been described in the first part of this book obviously required vessels that were masterpieces of metal working. At the Royal Manufacture of Bize, the itemised inventory established in 1774, following the death of Germain Pinel, owner of the Manufacture, describes three furnaces in the dye-house. Two of the vessels are made of copper: the first one weighs 6 *quintaux* (hundredweights) and 50 pounds (amounting to a total weight of about 266 kg) and its value is estimated at 780 *livres tournois*; the second one, specified as "serving for welding", weighs 8 *quintaux* (+/- 328 kg) and is estimated at 960 *livres tournois*. The third furnace "used for scarlets", is made of brass, weighs 5 *quintaux* (+/- 205 kg) and is valued at 500 *livres tournois.*[3] In Antoine Janot's dye-house in Saint-Chinian, by contrast, a tin vessel is used for scarlets.[4]

In his *Art de la teinture des laines*, Jean Hellot discusses the issue "of the material in which the caldron used for scarlets should be made"; he remarks that "in Languedoc they use vessels made of pure tin", while in Paris, either block-tin or brass vessels are used by equally renowned dyers. His own experiments using vessels made of these two metals lead him to observe that the length of cloth "dyed in the tin vessel was of a slightly more fiery red" than the other which "was a little more pink". But he argues that, by using a little more tin mordant when dyeing scarlets in a brass vessel, one could obtain the same colour as by dyeing in a tin vessel. He highlights the double disadvantage of using tin vessels: "those of big enough size, cost between three and four thousand pounds, which is a lot, while they can get melted the very first time they are used through the workers' carelessness."[5] This could and did happen, because tin is one of the metals with lowest melting point (232 °C). Emptying and cleaning of the vessels when dye baths had been exhausted required precautions, well described by William Partridge: "When a furnace is made of block-tin, it will have to be very stout, particularly at the bottom, and, when the fire is drawn, after a day's cooling, the liquor in the furnace will have to be cooled down, and the fire drawn some time before it is emptied, otherwise the bottom of the furnace will be liable to fall down and ruin it."[6] From the estimate of prices of the furnaces in the Manufacture of Bize, it appears that the brass vessel used there for scarlets was much cheaper than the prices mentioned by Hellot for tin vessels. However, brass vessels – and copper ones too – needed to be very carefully cleaned, "perfectly scrubbed", as highlighted by the author of the *Memoirs*, "and the liquor must never be permitted to remain in the furnace after a day's colouring is finished," adds Partridge, because otherwise the vessels might get corroded and oxidized, and some copper salts might get dissolved in the bath and sadden the colour.[7]

From scarlet to pale flesh – the exemplary recycling of cochineal

By the 18th century, American cochineal had nearly totally replaced kermes as the main source of scarlet, the most prestigious colour among the hundreds of shades that routinely emerged from the huge furnaces of the large dye-works and broadcloth manufactures flourishing in various regions of Europe. The best sort, the "fine or mesteck cochineal", *Dactylopius coccus* (Costa, 1829), was the only Coccid insect species domesticated by the indigenous people of the pre-Columbian civilisations of Mesoamerica and the Andes. They bred it mostly on the Barbary fig or prickly pear, *Opuntia ficus-indica* (L.) Miller.[8] During the whole period of the Spanish Empire in the New World, this fine cochineal could only be imported legally into Europe through Cadiz or Seville.[9]

The *Memoirs on Dyeing* offer unexpected keys to understanding the extraordinary economic and ecological use of this exotic dyestuff at a Royal Manufacture in 18th-century France. In the international context of intense competition between Dutch, English, and French cloth producers for the markets of the Levant that has been presented in the first part of this book, the beauty and quality of the dyes are an essential asset: "It is the vividness of the colours of the cloth that causes them to be bought."[10] At the top of the list of the "high" or "strong" colors that can win a market are the three that can be obtained from cochineal: *écarlate* (scarlet), *cramoisi* (crimson), and *soupe au vin* (wine soup). The difference between the former and the two others, the author of the *Memoirs* explains, depends on the mordanting process: "The colour that is given naturally by the fine cochineal employed without acid ranges from lilac grey to crimson. There is an infinity of shades between these two [...]. By using acids one makes them vivid, bright and orangey."[11] While crimsons and *soupes au vin* are still being mordanted in most dye-works with a variety of native or manufactured alums, which hardly modifies the colour of the dye, it has by then become possible to create new ranges of more or less orangey scarlets from cochineal, thanks to the invention of the new mordant composed of tin dissolved in nitric acid, described in a previous chapter.

As already mentioned in the first part of this book, the second half of the century brings a situation of "latent permanent crisis" in the trade.[12] After winning the favour of the Eastern Mediterranean elites by including five pieces of "high colours" per bale of cloth, French manufacturers gradually have to reduce their offer to four, three, or only two pieces of scarlet, crimson, or wine soup per bale. Cochineal is the most expensive of the dyestuffs they use, the only one whose price is indicated by the pound and not by *quintal*. The manufacturers are particularly sensitive to the price fluctuations of cochineal, which are largely due to the recurring wars between France and England and their respective allies. They are also strictly bound by the French system of royal regulations. Since April 18, 1713, a decree of the King of France's Council had forbidden manufacturers to dye any piece of good quality cloth into scarlet, crimson, or wine soup with less than 1¾ pound of cochineal. This amounts to 7 to 7.5 % of the weight of dry cloth, as opposed to the 5.3 % described by William Partridge as the usual proportion used around 1793 by his father, broadcloth producer and dyer in the Stroudwater area, one of the regions of England most famous for its scarlets (Table 6.1).[13]

It is very difficult for French manufacturers and dyers to cheat on the quantity of cochineal used, because of the series of quality controls to which cloth pieces are subjected before being allowed to be exported via Marseilles. Standards (*matrices*) of the best qualities of broadcloth dyed with the regular amount of fine cochineal are deposited in all the control centres, allowing the conformity of the colour of each piece of cloth to be visually assessed by comparison with the standard (Fig. 6.1). Numerous sources prove that controllers were quite good at spotting pieces lacking the right intensity of colour, which they describe as "*affamées*" (famished). Frauds are severely punished: not only are the faulty pieces re-dyed into less prestigious colours at the clothier's expense, he has to pay a heavy fine.[14] Some clothiers' accounts preserved in the *Archives départementales de l'Hérault* allow evaluation of the proportional cost of cochineal in the production of a bale of cloth, composed of 20 "half pieces", including five dyed in "high colours." While the cost of dyeing into all the other non-cochineal colours amounts to 5 to 8 % of all production costs – dyestuffs and labour included – the cost of cochineal alone equals 5 % of the total production costs. By comparison, the wages of weavers and their bobbin-winders only account for 7.3 to 8 % of the production cost of a bale.[15]

To ease somewhat the burden of such considerable investments, the government waives part of the import duties normally imposed on cochineal in favour of the cloth producers of Languedoc exporting to the Levant. "Passports of cochineal" are issued to the producers each year, allotting each a portion of the 210 *quintaux* markweight (10,279 kg) of duty-free cochineal, proportional to the volume of their production (Fig. 6.2).[16] But even this demonstration of the government's good will does not prove helpful enough. The manufacturers have to devise other ways to recoup their investment in cochineal, using the precious dyestuff down to the last colorant molecule. How systematically and cleverly they managed to do this is only now fully understood, thanks to the *Memoirs on Dyeing*.

When using fine or mesteck cochineal for scarlets, crimson or *soupevin* (wine soup), a small part of the compulsory amount of 1¾ pound of cochineal is already added to the tin-mordant bath, the *bouillon*, at about 0.3 % in proportion to the weight of the dry cloth piece. The aim is to already give the cloth a slightly coloured ground which will help obtain intense reds from the ensuing dye-bath. The mordant bath, having been used once, is

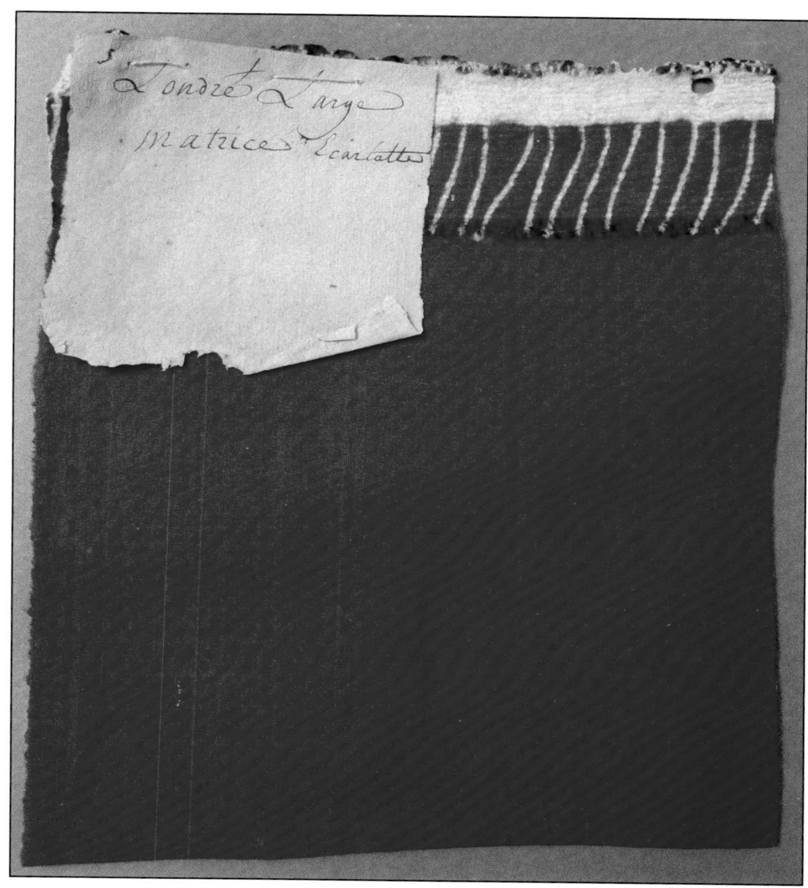

Fig. 6.1. Matrice *(standard) of scarlet for the quality of cloth named "Londre Large" (Broad London). The selvedge has been covered with cotton cloth for economy reasons, well explained by William Partridge on page 123 of his dyeing treatise: "Cloth, intended for scarlet, or any other cochineal colour, is always girt-webbed, to prevent the lists from taking the dye, as it would, being heavy and coarse, absorb much of the cochineal." Archives départementales de l'Hérault, C 5550. Photo: D. Cardon.*

not thrown away. Without any addition at all, the pale buff shade obtained from this mordant bath recycled as dye bath is called *biche* (doe). It can also be reused to compose dye baths with young fustic, *Cotinus coggygria* Scop. (Anacardiaceae), the source of a bright orangey yellow dye. The plant, and the range of colours obtained from it, will be discussed in the next chapter on yellows.

The dye bath for scarlet, crimson or wine colour that follows the mordant bath is called *rougie* (reddening bath) by French dyers. It is prepared with the rest of the 1¾ pound of cochineal and more tin mordant (13 to 16 %). After being used for the first time to dye into regulation-quality scarlet, crimson or *soupevin*, these dye baths are also reused in an extraordinary succession of processes (Table 6.2). First, they can serve as the basis of a mordant bath for another scarlet, in which case no cochineal needs to be added at the mordanting stage, only the tin mordant. They can also be used as a mordant bath for a series of paler red to purplish-pink shades, such as *vessinat* (colour of vetch flowers), *griotte* (morello cherry), and *cerise* (cherry). Lastly, a scarlet dye bath can be re-used to compose a special mordant bath, with varying quantities of tin mordant and young fustic, for the whole range of "flame colours" (*couleurs de feu*). Using young fustic in the mordant-bath gives the cloth an orangey ground. Although fustic was normally classified as *petit teint* (not fast enough for good quality textiles), an exception to the regulations that forbid its use was made for the broadcloth producers of Languedoc, because the brilliant flame colours it gives when associated with cochineal were extremely popular among the Levant elite. Consequently, a mordant bath for a regular scarlet can be recycled as a mordant bath for "flame scarlet," also dyed with 1¾ pound of cochineal, as well as for the range of colours called *fleur de grenade* (pomegranate flower), jujube, *langouste* (spiny lobster), orange, and *abricot* (apricot). These are produced with high proportions of fustic (up to 24 %) in the mordant bath and by decreasing quantities of cochineal (from 5 to 0.25 %) in the dye bath (Table 6.3).

But that is not all: instead of being recycled into mordant baths, a scarlet or crimson dye bath can also be used for dyeing again, once or even twice (Tables 6.2 and 6.4). Recycled for the first time with decreasing amounts of cochineal added, it forms the dye baths of the purples, violets, *vessinat*, *griotte*, and cherry pinks; with no more cochineal but a little gallnut, it produces a *gris-de-prince* (prince's grey); without anything, a shade called *chair* (flesh). Used a second time with minor additions of tin

JEAN-EMMANUEL DE GUIGNARD, *Chevalier, Vicomte de Saint Prieſt, Conſeiller du Roi en ſes Conſeils, Maître des Requêtes ordinaire de ſon Hôtel, Intendant de Juſtice, Police & Finances en la Province de Languedoc.*

LE ROI ayant, par Arrêt du Conſeil du 3 Décembre 1712, accordé aux Marchands de Draps de la Province de Languedoc, qui travaillent pour le Levant, l'exemption de tous Droits d'Entrée, qui ſe levent ſur la Cochenille, juſqu'à concurrence de deux cens dix quintaux poids de Marc, pour laquelle quantité il leur ſera par Nous expédié chaque année des Paſſeports, chacun à proportion du nombre des Balles qu'il a accoûtumé de faire; NOUS avons accordé au Sr. Marcassus de la terrasse

le préſent Paſſeport, pour la quantité de Deux cent soixante huit livres poids de marc de Cochenille pour l'année 1758 au moyen duquel il pourra faire entrer dans le Royaume ladite quantité de Cochenille, par tels Bureaux des Fermes qu'il trouvera à propos, ſans payer aucuns Droits d'Entrée, qui ſe levent au profit de Sa Majeſté, en conformité dudit Arrêt du Conſeil: Ayant ledit Sr. Marcassus fait ſa Soumiſſion en notre Greffe de rapporter

Fig. 6.2. Passport of cochineal granted by the Intendant of Languedoc to Mr. Jean Marcassus of the Royal Manufacture of La Terrasse, for 268 pounds of cochineal duty free, for the year 1758. Archives départementales de l'Hérault, C 2230. Photo: D. Cardon.

mordant, cochineal, and tartar, the dye bath for cherry gives a *rose* (pink), while that for *vessinat* gives a *fleur de pêcher* (peach-flower) colour. And without any addition, the dye bath that has already given a flesh colour produces an even paler shade, *chair pâle* (pale flesh). The liquor of the dye bath for purples and violets can also be reused as basis for the mordant-bath for madder reds. The utmost degree of economy is reached with a range of shades obtained from dye baths that consist of a mordant bath made with an already used dye bath: if this is a flame scarlet with young fustic, additions of small amounts of young fustic, madder, or tin liquor give pale pinkish-beige colours called *café au lait* (coffee with milk), *chocolat au lait* (chocolate with milk), and chamois; if the original dye-bath was for scarlet, crimson or winesoup without fustic, adding various small quantities of gallnuts and copperas produces a range of pale greys – beaver grey, pearl and silver grey.

The purpose of this complex succession of recycling processes (called *suites*) was evidently to exhaust the mordant baths and dye baths, which were initially rich in the costly tin mordant and even more expensive cochineal. Because all the processes must follow each other in quick sequence, this demanded an amazingly clever management of the workshop equipment and space. Jean Hellot explains why: "if the dyers cannot find time to use these baths for the second and third time within the next twenty-four hours, the colour of the bath gets spoilt, and the liquor, which was of the colour of a rose, becomes turbid and entirely loses this colour." On the other hand,

Hellot continues, when one manages a smooth succession of all the possible recycling options, "one saves a lot of fuel and time."[17]

The result of such adroit management of resources is reflected in a list of the colours of 100 half-pieces of Londrins Seconds cloths exported to the Levant in July 1751 from Bize, the very manufacture where the author of the *Memoirs* was to work a few years later. Each of the ten bales sent together includes a half-piece of scarlet. The ten dye baths in which they were produced have obviously been superbly recycled: half of them have been re-used to obtain colours needing the addition of young fustic (five pieces of flame scarlet and five pieces of jujube are distributed among the half bales); the other half were used to obtain a range of colours that must not have any young fustic, including five purples, top-dyed over a woad-and-indigo blue. A further recycling stage may have produced the five pieces of pink and five pieces of linen grey also found in the list, while recycling of the mordant bath with fustic would have given five more pieces, dyed into *cassie* yellow.[18] Altogether, the colour of 45 out of the 100 half-pieces in the lot are derived from a mordant bath or a dye bath for scarlet and the successive recycling of the precious tin and cochineal they contained. In the *Memoirs*, written about ten years later, 37 recipes include some cochineal in varying proportions and their implementation has needed a total amount of 40 ¼ pounds (16.663 kg) of fine cochineal. This is without taking into account the processes in the fourth *Memoir* in which the author reports on his experiments using cheaper sorts of cochineal: in ten processes he has used 11.5 pounds (4.786 kg) of *cochenille avariée*, that has been wetted during the transport by ship, and in five more, he has tried some *garbeau* or *granille*, by-products of the garbling of cochineal, to an amount of 28 pounds (11.592 kg). As a result of the long recycling chain described above, every bit of the colouring power of fine cochineal was fully exploited. However, the commercial success of the fine cloths dyed with the range of colours resulting from this succession of ever more diluted dye baths very much depended on cultural, psychological, and political factors. Had the French Court and the French *élites* not been considered at the time as fashion leaders in many parts of the world, it might have been difficult for the clothiers' factors in the ports of the Levant to convince their customers to adopt these new shades, so different from the intense reds commonly associated with cochineal.

Faced with the same problem of the high cost of the ingredients, English scarlet dyers appear to have adopted a more straightforward policy of daily recycling mordant baths for scarlets – in which turmeric instead of young fustic is used with tin liquor, cochineal and tartar – into dye baths for scarlets, recycled in turn into mordant baths, as described by William Partridge: "it is usual in all establishments where scarlet dyeing is carried on upon a large scale, to colour twenty or thirty pieces in one day, and by this means, much expense is saved. If a furnace is brought on early in the morning, some pieces are boiled in the first operation, this prepares the liquor for finishing; three or four lots that were boiled a day or two before, are then finished in the same liquor, and afterwards three or four lots of white pieces are boiled. The first lot boiled after finishing, needs no cochineal, the others follow it until the whole of this valuable drug is taken out of the finishing liquor – tin liquor, cream of tartar, and turmeric, are added in the quantity prescribed; but no cochineal, excepting for the second boiling[19] – the third and fourth lots are called runs, and are boiled a second time. Sometimes a whole day is employed in boiling."[20]

Red equals madder

In the era of pre-synthetic dyes, for French and English dyers alike, the colour names *rouge* or "red" implicitly imply a madder dye.[21]

This is the very fast blood-red dye obtained, not from an exotic insect like cochineal, but from a Mediterranean plant, *Rubia tinctorum* L. (Rubiaceae), familiar to Languedocian dyers. However, big scale cultivation of madder having much declined in France since the 16th century, by the middle of the 18th century they are mostly importing this dye from the Netherlands. It is a pity that the part of the first *Memoir* concerning madder has not been preserved, for it may already have reflected the revival of cultivation that did happen during the second part of the century. It starts in 1747, with a general enquiry ordered by the Intendant of Languedoc with the aim of identifying all remaining stations of subspontaneous growing of the plant, to evaluate the local resource in relation with the needs of the clothing industry. Reports from all over the province confirm that madder still grows in many parts of Languedoc, particularly in the region of Narbonne and Béziers – close to Bize – and around Montpellier, but raise the point that it is entirely collected for its medicinal applications and sold to apothecaries. For the large quantities they need, clothiers still have to import it from the Netherlands, via Bordeaux. The inspector of manufactures in charge of the diocese of Narbonne reports the annual madder consumption at the Royal Manufacture of Bize as fluctuating between 6 and 7 *quintaux* (248 to 290 kg), consisting mostly of unstripped madder, bought in Bordeaux at prices varying between 60 and 75 pounds per *quintal*.[22]

Although extremely fast to light and washing, madder red was generally considered as "never as beautiful as kermes or cochineal reds".[23] This is why the author of the *Memoirs* uses every opportunity to recycle his cochineal dye baths, particularly for purples and violets, as liquor for "boiling" madder reds. These mordant baths are not prepared with the tin mordant, but with alum and tartar, employed in the same proportions as for purples, violets, and all other colours where the naturally purplish tones of cochineal are wanted. The amount of madder in the ensuing dye-bath is generous, corresponding to 27–32 % of the cloth weight; the author further intensifies the resulting red by adding 3.3 to 4 % ground gallnut which acts as complementary mordant.

Other ways he mentions – but does not advocate – to get more vivid reds are, first, to add 3.3 to 4 % tin liquor to the dye bath. William Partridge is against using tin liquor "in the same liquor with the madder" but he adds 4 to 6 % of it, together with alum and tartar, in the mordant bath and he asserts that "no madder red is ever dyed in England without the cloth being prepared with more or less of the solution of tin."[24] This does not seem to have been as systematic as he claims, at least in the earlier part of the 18th century, judging from the evidence of the Crutchley Archive, preserved in the Southwark Local History Library. This is a collection of dyeing and accounts books dating from the 1720s to 1740s, concerning the Crutchley family dyeing business that existed from 1700 to 1799 in Southwark, Surrey.[25]

An initial assessment by Dr. Anita Quye, lecturer at the Centre for Textile Conservation and Technical Art History, University of Glasgow, with Dr. Patricia Dark, archive manager at the SLHL, has shown that in the period represented by this archive, the Crutchleys specialised in producing red dyed fabrics using madder and cochineal, often together in the same dyeing process, and exclusively with an alum mordant.[26]

Another trick to obtain a brighter red from madder is described by the author of the *Memoirs* as still another of "Mr. Albert's" discoveries. This funny character, who describes himself as a "Doctor in Medicine, from the Academies of Sciences of Montpellier and Toulouse", has already been mentioned in previous chapters, in connection with celadon greens with verdigris and Cyprus vitriol, and with Saxon blues and greens. Always busy perfecting or inventing dyeing processes and attempting to promote them and to obtain money from selling them or for publishing them, he is moderately appreciated by professional dyers.[27] The author of the *Memoirs* knows him and follows his researches with half interest, half amusement. Concerning madder dyeing, he is not at all convinced by Albert's process, involving playing with the pH of the dye bath by adding 13 to 16 % vinegar. When Albert had publicly demonstrated this process in Saint-Chinian, on 30 July 1749, it had already attracted sarcastic comments from Antoine Janot, the most respected master dyer of the place.[28]

In dyeing with madder, the beauty of the red obtained also depends on the temperature of the bath. The author's recommendations not to let the madder dye bath get too hot too fast, and not to let the cloth boil in it, are echoed by William Partridge, who provides precious information on the dyers' way of judging temperatures in this part of his treatise: "the cloth should be had in at a blood heat, and well reeled for six hours; by this time the liquor should just break out to a spring heat, the fire to be then drawn." Spring heat, he indicates, corresponds to about 206°F, that is 96°C or just below boiling point. And he explains why: "those who attempt to dye red on woollen must take care not to let the madder liquor boil, as the yellow of the madder will become fixed on the goods, and spoil the colour."[29]

Bright madder reds, *garance vive*, and *rouges brunis*, madder reds saddened by adding some copperas that reacts with the gallnut in the bath, are in regular demand in the Levant. They figure in assortment lists and pattern sheets of cloths sent not only from Bize, but other prestigious manufactures, such as that of the Pinel brothers in Carcassone, or Pennautier, close to that city.[30]

Although less precious than cochineal dye-baths, madder dye-baths also are recycled in the author's dye works. A first reuse, on a ground of yellow, gives a range of cinnamon colours. The reuse of these gives a wax colour and a whole range of olives on green grounds; they will be discussed in a following section. On white cloth and with additions of gallnut, brazilwood and a little cochineal, madder dye-baths are reused to obtain *prunes* (plum colours); with brazilwood and soot added, they give certain shades of *noisette* (hazelnut).[31]

Red dyewoods were sometimes used to give madder reds a more crimson shade, as in the recipe of *garance cramoisillée*. They are the third source of reds used in the *Memoirs*.

Reds from dye-woods

Since the Middle Ages, European merchants' and dyers' books often mentioned "brazil", a dye-wood so called because of the colour of red embers or glowing coals (*rouge de braise*, *brasa*) it gave to textiles. At the time, it could only refer to sappan wood from the small tree *Caesalpinia sappan* L. (Fabaceae), imported from south-east Asia. The discovery of America soon led to another important discovery, namely that other trees, different species of *Caesalpinia* and a species of the related genus *Haematoxylum*, also sources of red dye woods, were growing in different parts of the continent and on some islands. These trees, the heartwood of which was found to give reds similar to those obtained from sappanwood, were soon generally called "brazilwood".[32] The provenances and descriptions of the different brazilwoods given by the author in his first *Memoir* are directly borrowed from the *Dictionnaire universel de Commerce* by Savary des Bruslons, also copied in Diderot's *Encyclopédie.*[33] However, he adds some useful personal information on the species that were really available to dyers in his days: though he mentions a "brazil from Japan", the *Memoirs* show that sappanwood was not used any longer in wool dyeing. He only actually uses American brazilwoods: his "brazil from Lamon", mentioned by Savary des Bruslons as coming from All Saints' Bay, must be identical with Fernambuco/Pernambuco wood, since only one species of tree giving a red dye wood grows in the Atlantic forest of north-east Brazil.[34] This is the *pau-brasil* tree, *Caesalpinia echinata* Lamarck, whose high content in brazilin, the main source of colorant common to all these dyewoods, has been confirmed by a recent chemical survey.[35] The brazilwood from Santa Marta was extracted from another species rich in brazilin: *Haematoxylum brasiletto* Karsten, only growing in two parts of America: a big region including Santa Marta, on the north coasts of Colombia and Venezuela, and the arid islands of Aruba, Curaçao and Bonaire; and a wide band along the Pacific coast of Mexico and Guatemala.[36] Positively identifying the last sort of dye-wood, called *bresillet* or *bois de la Jamaïque* (Jamaica wood) by the author of the *Memoirs*, remains a conundrum, since it cannot be *H. brasiletto* which does not grow in Jamaica. It could be *Caesalpinia violacea* (Mill.) Standl., the present valid name of *C. brasiliensis* Sw., an indigenous species of Cuba and Jamaica, in the heartwood of which a derivate of brazilin has been evidenced by Witold Nowik.[37]

The range of landed costs of Santa Marta wood in Bordeaux mentioned by the author corresponds to the figures in an official *liste des Entrées des Isles françoises de l'Amérique* (list of entrances from the French Islands in America) for 1771, in which a consignment of 4529 pounds (2.215 metric tons) of Santa Marta wood is valued at 26 pounds per hundredweight, which falls just between the lowest and highest prices quoted in the first *Memoir.*[38]

The problem with these red dyewoods was the same as for logwood: they are not fast to light and consequently, their use was strictly prohibited to dyers in the best mode of dyeing. The *Règlement pour la teinture des étoffes de laine et des laines servant à leur fabrication* (Regulation on dyeing of woollen cloths and of the wools used to make them) of 15 January 1737 forbid them "to have... Brazil wood, Santa Marta wood, brazilwood from Japan and Fernambuco wood within their premises".[39] The author of the *Memoirs* nearly faithfully complies with these prohibitions: it is true that, at page 41, he gives a recipe of false Persian blue in which 2 pounds of logwood and 2 pounds of non-specified brazilwood (that is 6 to 8 % of the cloth weight for each dyewood) are used to top-dye a piece of cloth having received a ground of King's blue in the woad-and-indigo vat, thereby raising it to the intensity and darkness of a Persian blue. But he describes this as an unsatisfactory way to upgrade a vat blue and characteristically, there is no sample corresponding to this recipe in the manuscript. In the only other recipe including some non-specified brazilwood – a recipe of *prune à la cochenille* (damson in grain) – small proportions of brazilwood and cochineal (8 to 10 % of each), with some gallnut added, serve to give a purplish shade to a reused madder dye bath to which a small quantity of madder is further added.[40] Fernambuco wood – the best sort of redwood – is employed in still smaller proportions (1 to 5 % of the cloths weights) to give more crimson shades to a madder dye, an amaranth tinge to cochineal dyes, and reddish shades to some *marrons* (chestnut browns) and *noisettes* (hazelnut).[41] Interestingly, in the cloth sample corresponding to the recipe of *garance cramoisillée* (crimsoned madder), where only 3 to 4 % of Fernambuco wood are reckoned to have been employed, marker compounds for soluble redwoods could still be clearly detected by dye analysis, together with the colorants of madder.[42] All in all, only 8.25 pounds (3.415 kg) of all sorts of brazilwood including Fernambuco, have been used in the course of the implementation of the 206 dye processes described in the *Memoirs*.

Reds on blues versus blues on reds

To the complex succession of recycling processes devised to make the best possible use of cochineal, belongs the beautiful range of purple to pale mauve shades resulting from top-dyeing cloth having already received decreasing degrees of blue grounds in the woad-and-indigo vat, with ever less concentrated dye-baths of fine – or not so fine – cochineal (Tables 6.2, 6.5–6). Their common technical characteristic is that alum and tartar are used for the mordant bath preceding the cochineal dye-bath, by contrast with the tin mordant used by the author for scarlets and even for crimsons and wine soups.[43]

Beside this range of "orthodox" purple to mauve shades produced in accordance with regulations, the author is aware that there exist other processes to produce a similar gamut more cheaply. In his last *Memoir* reporting on some of his experiments, he gives several recipes of "English" purples and violets. These are not dyed on cloths that have received a blue ground in the woad-and-indigo vat but are entirely "dyed in the furnace", on undyed cloths that are first mordanted with various proportions of tin and/or alum with tartar, then dyed with cochineal and then dyed a second time – he says "*brunis*" (saddened) – with logwood. The quantity of logwood is only specified in an intermediary recipe in which the cloth has received a light ground of sky blue. In that case, the proportion is of 20 to 24 % of the cloth weight, so presumably, it is higher when there is no woad ground and the purple colouring from logwood is left to the dyer's appreciation and skill.

The recipe for "a rich purple" given by William Partridge employs very little cochineal (1.7 % of the cloth weight) which is added to the mordant bath combining alum, cream of tartar and tin liquor. In the ensuing dye-bath, more red is provided by 6.6 % brazilwood and the blue component of the colour comes from a combination of logwood (20 %) and indigo carmine (4 %).[44]

Although the "English purples" described by the author of the *Memoirs* did not conform to the regulations, they must have been occasionally tolerated by the inspectors and obviously found customers in the Levant: a piece of *pourpre anglais* figures in an assortment of Mahoux Seconds sent from the Royal Manufacture of Pennautier on 3 November 1767.[45] Much later, a sample of *pourpre anglais* is included on a pattern sheet of Londrins Seconds sent to the Levant by Bernard Darle, a clothier in Carcassonne, in August 1816.[46]

Not only cochineal, but madder, too, can be used to top-dye cloths that have received a blue ground of woad and indigo. Ranging from azure to dainty blue, they are then mordanted with alum and tartar and dyed with 20 to 24 % madder. The resulting shades of maroon called *couleur de Roy* (King's colour) only differ by the addition of various small quantities of gallnut and/or of Fernambuco wood that can be further saddened with a little copperas.[47] They may not have been much demanded, however: there is not one piece of *couleur de Roy* in the assortments and pattern sheets I could consult in archives in Languedoc.

Red on yellow

Within the cycles of reuses of cochineal dye and mordant baths described earlier, a range of reddish orange to orange colours are produced by cochineal dye-baths of decreasing concentrations, on cloths that have already received golden yellow colours in the tin mordant baths, because these are prepared with high proportions of fustic (up to 24 %). This is how the colours called *fleur de grenade* (pomegranate flower), jujube, *langouste* (spiny lobster), orange, and *abricot* (apricot) are obtained.

Madder also is used, in proportions of 20–24 %, to top-dye cloths that have been dyed yellow. In this case, the cloths are first mordanted with Rome alum and red tartar and then dyed with weld. The resulting colour, *maure doré* (golden Moor) is already mentioned among the shades of wools used in tapestry weaving in the *Statuts et Reglemens pour les Teinturiers en Soye, Laine et Fil* (Statutes and Regulations for Dyers of Silk, Wool and Linen) of August 1667.[48] It can be slightly saddened with copperas.

It has been mentioned that madder, although less costly a dye than cochineal, was also recycled. The whole range of *canelles* (cinnamon colours) in the *Memoirs* is obtained by reusing the madder dye-baths for *maure doré*. The mordant bath for these shades, however, is less concentrated in alum. The mordanted cloths are then dyed with weld, and top-dyed with madder added to the reused dye-bath of *maure doré*. The proportion of madder can thus be reduced to only 16 % of the cloth weight. Gallnut is added to all cinnamon shades and can be made to react with some copperas for *canelle brûlé* (burnt cinnamon). A second recycling of those dye-baths that have not been spoilt by additions of copperas can be done, with even less madder added (6.6 to 8%). It gives the colour *cire doré* (golden wax). A piece of cloth dyed *canelle* is included in half the bundles sent to the Levant in April 1748 from the manufacture of the Pinel brothers in Carcassonne.

The two recipes given by Partridge for cinnamon have nothing in common with the recipes for *canelles* in the *Memoirs*, apart from the alum and tartar mordant. One "for a bright cinnamon" only employs redwood, alum and tartar. The other, "for a dark cinnamon", uses alum, tartar, old fustic, barwood and redwood. He does mention another way of proceeding, using madder, but in the reverse order to that described in the *Memoirs*: "cloth for cinnamon is oftentimes prepared with ombre madder, alum, and argol, and then finished with fustic, redwood and barwood in another liquor".[49] The colours cinnamon, cinnamon brown and bright cinnamon are mentioned in two pattern books of the Clark brothers from Trowbridge corresponding to the years 1753–1778, but the corresponding samples are missing and the processes by which they have been obtained are not described, since pattern books do not include recipes.[50]

The *Memoirs* describe other uses of madder on undyed cloth mordanted and dyed yellow with weld. It is the main ingredient of dye-baths that give the different shades of *tabac* (tobacco) (Table 6.8). They only differ from the dye-baths for cinnamons by having slightly lesser proportions of madder, by the addition of 6.6 to 8 % of old fustic and by being saddened with copperas or an uncommon ingredient: soot. The greenish shade of tobacco is done on cloth that has received a woad ground of dainty blue, the same degree of blue as required for light green and two shades of *olive*.

Red on blue and yellow

In the *Memoirs*, all olive shades are obtained by reusing cinnamon dye-baths (themselves recycling a madder dye-bath) to top-dye cloths that have already been dyed into greens on woad grounds of various degrees of blue, mordanted with alum and tartar and dyed yellow with weld. To the recycled madder-bath, only 3.3% to 4% ordinary madder needs be added. The characteristic features of olive dye-baths are the addition of 7 to 12 % soot and their prolonged boiling. This allows the extraction of the brown colorants in madder roots and enhances the brown colour given by soot. Differences in proportions of copperas in the saddening process and the length of time the cloth is exposed to it contribute to the different nuances of olives, but are of less importance than the degrees of the woad grounds, that range from milk blue, needed for both *olive sèche* (dry olive) and light olive, to sky blue, required to obtain parrot greens for the darkest shades of olive (Table 6.7).

In spite of its unappetising name, *olive pourrie* (rotten olive) was a colour appreciated in the Levant: a piece of this colour figures in half the bales of Londrins Seconds sent from the Royal Manufacture of Bize on the 16 of July 1751.[51] A piece dyed in *olive sèche* is included in one in two of the 20 bundles of Londrins Seconds sent from that of François Pinel in Saint-Chinian in June 1750 and 20 more bundles "with the same assortment" are promised for September of the same year.[52]

Olive colours were at least as popular with English dyers and their customers. In the years 1752 to 1755, the Clark brothers of Trowbridge dye various shades of olive, bright olive and green olive, and also "oaded olives" – which implies that the rest have not received a woad ground.[53] Partridge's recipe "for a fine olive green" does not include any madder: it is a true woaded green on a ground of "a nine cent blue", but it is only top-dyed with the yellows from old fustic and weld. Alum and copperas are then added to the dye-bath.[54] By contrast, none of the olive shades produced in Gloucestershire by Marling and Evans of Ebley, in the years 1794 to 1804, are woaded but they do result from a combination of yellow, blue and red colorants. Their olive, dark olive, brown olive, light brown olive and dark green olive are dyed with "chip fustick", logwood, barwood, copperas and alum; in their green olive, barwood is replaced by madder.[55]

Notes

1 Partridge re-ed 1973, p. 133.
2 Explanation of the scientific rationale of mordant dyes in Cardon 2014, pp. 14–15.
3 AD 11 3E 1246.
4 Janot 1744, folio 8.
5 Hellot 1750, pp. 295–303.
6 Partridge re-ed 1973, p. 124.
7 *Ibid.* and page 45 of the manuscript of the *Memoirs*.
8 The species of *Dactylopius* and their plant hosts and distributions have been much revised recently; synthesis on these updates in Cardon 2014, pp. 602–616.
9 Donkin 1977, p. 37.
10 Report of a French merchant in the Levant, quoted by Jacques Savary in his *Le Parfait Négociant*, Savary, re-ed. Richard 2011, p. 1074, 1120.
11 Page 11 of the *Memoirs*; the "acid" mentioned is the dissolution of tin in *aqua-fortis*.
12 Minard 2003, pp. 69–89.
13 Partridge re-ed 1973, p. 184.
14 ADH C 5569, document dated 1734.
15 ADH C 5552, document dated 1751–1753.
16 ADH C 2230, dated 1758.
17 Hellot 1750, p. 320.
18 AD11, doc. 3J 342, dated 1751.
19 This is in agreement with the practice described in the *Memoirs*, as shown in table 4.1.
20 Partridge re-ed 1973, pp. 187–188.
21 Page 52 of the manuscript of the *Memoirs*; Partridge re-ed 1973, p. 165.
22 AD34 C 2229.
23 Hellot 1750, p. 379.
24 Partridge re-ed 1973, p. 165.
25 The Crutchley Archive has been donated in 2011 by the family to the Southwark Local History Library (SLHL), London.
26 Anita Quye, 20 June 2014, "The Crutchley Archive: initial assessment report – Southwark Local History Library, London", personal communication.
27 AD34 D152, document dated 1758.
28 AD34 C2487.
29 Partridge re-ed 1973, pp. 165–167.
30 AD11 3J 342, documents respectively dated 16 July, 1751, 22 April 1748 and October 1775.
31 Pages 69–70 of the *Memoirs*.
32 Cardon 2007, pp. 274–289.
33 Savary des Bruslons 1723, p. 478; Diderot and d'Alembert 1751, t. 2, p. 308.
34 Savary des Bruslons, 1741, p. 34.
35 Nowik 2001.
36 Cardon 2007, p. 264; Cardon 2014, p. 267.
37 Nowik 2001.
38 AD 33 C4390, doc. 21.
39 An. 1746, *Manufacture de Carcassonne*, p. 232, art. 20.
40 Page 69 of the *Memoirs*.
41 Pp. 53, 68, 70, 93 of the *Memoirs*.
42 Dye analyses performed by Witold Nowik; all results are grouped in appendix 2.
43 This, he explains, is a way to compensate for the hardness and alkalinity of his water resource (see chapter above).
44 Partridge re-ed 1973, p. 202.
45 AD11 3 J 342.
46 AD11 3 J 961, folio 27.
47 Pages 53 and 91 of the *Memoirs*.
48 An. 1730, *Recueil des Reglemens*, p. 380, art. 45.
49 Partridge re-ed 1973, p. 193.
50 Wiltshire and Swindon History Centre, 927/7 and 927/8.
51 AD11 3 J 342.
52 AD34 C 2160.
53 Wiltshire and Swindon History Centre, 927/8.
54 Partridge re-ed 1973, p. 181.
55 Gloucester Archives, D948/1.

Table 6.1: Proportions of cochineal in scarlets, in the 18th and 19th centuries. 4 French and 1 Anglo-American sources

	2nd Memoir *c. 1762–1763 p. 42–43*	*Janot 1744 f°8*	*Hellot 1750 p. 290*	*Partridge, recipe from c. 1793 re-ed 1973 pp. 184–185*	*Grison 1884 vol. 1 pp. 190–191*
Cloth weight	25 – 30 lb = 10.35 kg to 12.42 kg 100 %	25 lb = 10.35 kg 100 %	1 lb = 0.489 kg 100 %	117 lb = 53 kg 100 %	100 kg 100 %
Cochineal added to mordant bath	1½ ounce = 0.038 kg 0.3–0.37 %	2 ounces = 0.051 kg 0.5 %	1½ *gros* = 0.005 kg 1.17 %	1 ½ lb = 0,679 kg 1.28 %	0.5 kg 0.5 %
Cochineal added to dye bath	1lb10½ ounces = 0.686 kg 5.5–6.6 %	1 lb ¾ = 0.724 kg 7 %	1 ounce = 0.030 kg 6.25 %	4 ¾ lb = 2.151 kg 4 %	10 kg 10 %
Total weights/ proportions of cochineal	1 lb ¾ = 0.724 kg 5.8–7 %	30 ounces = 0.776 kg 7.5 %	9½ *gros* = 0.036 kg 7.42 %	6 ¼ lb = 2.831 kg 5.3 %	10.5 kg 10.5 %

Table 6.2: Multiple recycling of the cochineal dye baths for scarlet, crimson and winesoup in the Memoirs

1st recycling *of a dye-bath already used for* écarlate, cramoisi *or* soupevin			
*as mordant-bath (*bouillon*)*		*as dye-bath (*rougie*)*	
adding young fustic	without young fustic without a woad ground	on a woad ground	without a woad ground
couleur de feu (flame colour) *fleur de grenade* (pomegranate flower) jujube *langouste* (spiny lobster) orange *abricot* (apricot)	*vessina* (vetch flower) *griotte* (morello cherry) *cerise* (cherry)		*vessina* *griotte* *cerise*
	on a woad ground	on a woad ground	
	pourpre (purple) *violet*	*pourpre* (purple) *violet*	
2nd recycling as mordant-cum-dyebath	**2nd recycling** as mordant-bath	**2nd recycling** as dye-bath	**2nd recycling** as mordant-cum-dye bath
chamois *café au lait* (coffee with milk) *chocolat au lait* (chocolate with milk)	*lilas* (lilac) *gorge de pigeon* (dove) *mauve* *gris de lin* (linen grey)	*lilas* *gorge de pigeon* *mauve* *gris de lin* agate	*roses* (pinks) *fleur de pêcher* (peach flower) *couleurs de chair* (flesh colours) *gris de prince* (prince grey) *gris de rat rougeâtre* (reddish rat grey) *gris de castor* (beaver grey) *gris de perle* (pearl grey) *gris d'argent*(silver grey)

Table 6.3: Recycling of a scarlet dye bath for "flame colours" with young fustic. Proportions of added ingredients. 1st re-use as liquor for mordant-baths

Colour	*tin mordant*	*cochineal*	*young fustic*	*tartar*	*madder*
Ecarlate couleur de feu (flame scarlet)					
Mordant bath	16.7–20 %	0	13.3–16 %	6.6–8 %	0
Dye-bath	13.3–16 %	5–7 %	"a little"	0	0
Fleur de grenade (pomegranate flower)					
Mordant bath	16.7–20 %	0	20–24 %	6.6–8 %	0
Dye-bath	13.3–16 %	4.2–5 %	"a little"	0	0
Jujube					
Mordant bath	10–12 %	0	20–24 %	6.6–8 %	0
Dye-bath	6.6–8 %	3.3–4 %	"a little"	0	0
Langouste					
(spiny lobster)					0
Mordant bath					
Dye-bath	10–12 %	0	20–24 %	6.6–8 %	
	6.6–8 %	1.6–2 %	"a little"	0	0
Orange					
Mordant bath	10–12 %	0	20–24 %	3.3–4 %	0
Dye-bath	6.6–8 %	0.4–0.5 %	"a little"	0	0
Apricot					
Mordant bath	10–12 %	0	16.6–20 %	3.3–4 %	0
Dye-bath	6.6–8 %	0.2–0.25 %	0	0	0
2nd recycling, as dye-bath using the liquor of the mordant-baths above described					
Chamois mordant-cum-dye bath	5–6 %	0	0	0	0
Café au lait (coffee with milk) mordant-cum-dye bath	3.3–4 %	0	3.3–4 %	0	0.8–1 %
Chocolat au lait (chocolate with milk) mordant-cum-dye bath	0	0.4–0.5 %	3.3–4 %	1.6–2 %	0.8–1 %

Table 6.4: Recycling of a dye bath for scarlet, crimson or winesoup. Proportions of added ingredients

Colour	*tin mordant*	*cochineal*	*alum*	*tartar*	*other*
Vessinat (vetch flower)					
Boiling	0	0	16.7–20 %	6.7–8 %	–
(reused dye-bath) Finishing	0	3.3–4 %	0	3.3–4 %	
Griotte (amarelle cherry)					
Boiling	0	0	16.7–20 %	6.7–8 %	–
(reused dye-bath) Finishing	0	2.5–3 %	0	3.3–4 %	
Cherry					
Boiling	10–12 %	0	0	3.3–4 %	
(reused dye-bath) Finishing (re-used dye-bath)	6.7–8 %	1.6–2 %	0	0	–
Flesh (reused dye-bath)	0	0	0	0	-
Prince's grey (reused dye-bath)	0	0	0	0	0.2–0.25 % Nut gall + copperas
		2nd recycling of finishing bath			
Rose (reuse of cherry bath)	3.3–4 %	0.8–1 %	0	3.3–4 %	-
Peach flower (reuse of vetch flower bath)	3.3–4 %	0.4–0.5%	0	3.3–4 %	-
Pale flesh (reuse of flesh)	6.7–8 %	0	0	0	-

Table 6.5: Range of colours derived from a reused cochineal dye-bath on a woad ground. Proportions of added ingredients

Colour	*Woad ground*	*alum*	*tartar*	*cochineal*
Pourpre (purple)	*Bleu d'azur clair* (light azure blue)			
Mordant bath (reused dye-bath)		16.7–20 %	6.7–8 %	0
Dye-bath (reused dye-bath)		0	3.3–4 %	4.16–5 %
Violet	*Bleu d'azur* (azure blue)			
Mordant bath (reused dye-bath)		16.7–20 %	6.7–8 %	0
Dye-bath (reused dye-bath)		0	3.3–4 %	3.3–4 %
2nd recycling: mordant and dye baths reusing a dye-bath for purple or violet				
Lilas (Lilac)	Sky blue			
Mordant bath (reused dye-bath)		11.6–14 %	3.3–4%	0
Dye-bath (reused dye-bath)		0	3.3–4 %	1.7–2 %
Gorge de pigeon (dove breast)	Sky blue			
Mordant bath (reused dye-bath)		11.6–14 %	3.3–4%	0
Dye-bath (reused dye-bath)		0	3.3–4%	0.8–1%
Mauve	Dainty blue			
Mordant bath (reused dye-bath)		11.6–14 %	3.3–4%	0
Dye-bath (reused dye-bath)		0	3.3–4 %	1.7–2 %
Gris de lin (linen grey)	Milk blue			
Mordant bath (reused dye-bath)		11.6–14 %	3.3–4%	0
Dye-bath (reused dye-bath)		0	3.3–4 %	0.4–0.5 %

Table 6.6: Purples and violets in the Memoirs

	pourpre *purple* *p. 51*	pourpre foncé *dark purple* *p. 94*	pourpre violet *violet purple* *p. 94*	*violet* *p. 51*	violet d'évêque *bishop's violet* *p.93*	pourpre anglais *English purple* *p. 88*	pourpre anglais foncé beau *beautiful dark English purple* *p. 96*	violet anglais *English violet* *p. 89*
Woad ground	Light azure	Dark azure	Dark azure	Azure	King's blue	-	Sky blue	-
Mordant bath								
alum	Rome 16.6–20%	Spain 13.3–16%	Spain 13.3–16%	Rome 16.6–20%	-	Rome 3.3–4%	-	Spain 16.6–20%
tin mordant	-	-	-	-	18.3–22%	16.6–20%	13.3–16%	-
tartar	red 6.6–8%	red 6.6–8%	red 6.6–8%	red 6.6–8%	white 6.6–8%	white 6.6–8%	unspecified 6.6–8%	red 6.6–8%
Red dye bath								
cochineal	fine 4–5%	reused bath + *granille* 4–4.7%	reused bath + *granille* 3.3–4%	unspecified 3.3–4%	fine 5–6%	spoilt 4–5%	3.3–4%	spoilt 4–5%
tartar	red 3.3–4%	unspecified 3.3–4%	unspecified 3.3–4%	unspecified 3.3–4%	-	-	-	-
tin mordant	-	-	-	-	-	8.3–10%	10–12%	8.3–10%
larch agaric	-	0.8–1 %	0.8–1 %	-	-	-	-	-
Saddening								
logwood	-	-	-	-	-	+ quantity not specified	20–24%	+ quantity not specified
alum							6.6–8%	
tartar							3.3–4%	

Table 6.7: Correlations between the stepped gradations of blues, purples, greens, blacks and greys, olive and tobacco shades

Degree of woad and indigo ground	*Gradation of purples*	*Gradation of greens*	*Shades of olive and tobacco*	*Blacks and greys*
Bleu turquin Turkish blue		*Vert brun*[1] Dark green *Vert obscur*[2] Obscure green		
Bleu de Roy foncé Dark King's blue		*Vert de canard*[3] Duck green *Herbe très foncé*[4] Very dark grass green		Noir de Sedan Sedan black[5]
Bleu de Roy King's blue	*Violet d'évêque*[6] Bishop's violet	*Vert noir*[7] Black green *Vert d'herbe*[8] Grass green *Vert d'herbe rempli*[9] Full-bodied grass green *Vert bronzé*[10] Bronze green		
Bleu de Roy clair Light King's blue		*Vert de mer foncé*[11] Dark sea green		
Bleu d'azur foncé Dark azure blue	*Pourpre violet*[12] Violet purple *Pourpre foncé*[13] Dark purple	*Vert d'herbe*[14] Grass green *Vert de chou*[15] Cabbage green *Vert de câpre*[16] Caper green *Vert brun*[17] Dark green		
Bleu d'azur Azure blue	Violet[18] *Couleur de roy*[19] King's colour	*Emeraude*[20] Emerald *Vert de mer*[21] Sea green		*Noir au pastel ordinaire*[22] Woaded black oridnary *Taupe*[23] Mole *Poudre à canon des Anglais*[24] English gunpowder
Bleu d'azur clair Light azure blue	*Pourpre*[25] Purple			
Bleu de ciel ou céleste foncé Dark celestial or sky blue				*Taupe*[26] Mole
Bleu de ciel ou céleste Celestial or sky blue	*Lilas*[27] Lilac *Gorge de pigeon*[28] Dove *Couleur de roy*[29] King's colour *Agathe*[30] Agath *Pourpre anglais foncé beau*[31] Beautiful dark English purple	*Vert perroquet*[32] Parrot green *Vert pistache*[33] Pistachio green	*Olive brun* Dark olive Olive[34]	*Ardoise*[35] Slate colour *Gris de plomb*[36] Lead grey

(continued)

Degree of woad and indigo ground	*Gradation of purples*	*Gradation of greens*	*Shades of olive and tobacco*	*Blacks and greys*
Bleu de ciel ou céleste clair Light celestial or sky blue	*Lilas*[37] Lilac			
Bleu mignon Dainty blue	Mauve[38] *Couleur de roy*[39] King's blue	*Vert clair*[40] Light green	*Olive pourrie*[41] Rotten olive *Olive verte*[42] Green olive *Tabac verdâtre*[43] Greenish tobacco	*Gris de rat*[44] Rat grey
Bleu de lait Milk blue	*Gris de lin*[45] Linnen grey *Agathe clair*[46] Light agate	*Vert gay*[47] Gay green *Vert céladon*[48] Celadon green	*Olive claire*[49] Light olive *Olive sèche*[50] Dry olive	
Bleu déblanchi Off-white blue		*Vert naissant*[51] Nascent green *Vert pomme*[52] Apple green		*Gris de rat foncé*[53] Dark rat grey

1 Recipe p. 58
2 Recipe p. 58
3 Recipe p. 58
4 Recipe p. 95
5 Recipe p. 73
6 Recipe p. 93
7 Recipe p. 58
8 Recipe p. 82
9 Recipe p. 58 and 93
10 Recipe p. 60 and 84
11 Recipe p. 95
12 Recipe p. 94
13 Recipe p. 94
14 Recipe p. 58
15 Recipe p. 60
16 Recipe p. 88
17 Recipe p. 93
18 Recipe p. 51
19 Recipe p. 53 and 91
20 Recipe p. 59
21 Recipe p. 60
22 Recipe p. 74
23 Recipe p. 75
24 Recipe p. 85
25 Recipe p. 51
26 Recipe p. 90
27 Recipe p. 51 and 94
28 Recipe p. 51 and 94
29 Recipe p. 53
30 Recipe p. 76 and 82
31 Recipe p. 96
32 Recipe p. 59
33 Recipe p. 59
34 Recipe p. 61
35 Recipe p. 75 and 90
36 Recipe p. 75
37 Recipe p. 85
38 Recipe p. 51
39 Recipe p. 91
40 Recipe p. 59
41 Recipe p. 61
42 Recipe p. 61
43 Recipe p. 72
44 Recipe p. 75
45 Recipe p. 52
46 Recipe p. 76
47 Recipe p. 59
48 Recipe p. 60
49 Recipe p. 62
50 Recipe p. 62
51 Recipe p. 59
52 Recipe p. 60
53 Recipe p. 75

7

Yellows

Yellows of weld and *trentanel*

Like the reds of cochineal and madder, fast yellows are "dyed in the furnace", the cloths having first been mordanted. The mordant bath is prepared with the same ingredients as for the purplish shades of cochineal and for madder reds – alum and tartar – only with lesser proportions of alum (13.3 to 16 %) which, in this case, can be of a cheaper sort than top quality Rome alum. The dye bath for yellows is called *gaudage* (welding) by French dyers because the most commonly used yellow dye plant – not only in France but in all Europe and since prehistoric times – is the herb called *gaude* in French, i.e. weld, *Reseda luteola* L. (Resedaceae).[1] While a wealth of other plants, in all natural environments, will give yellow dyes, weld was preferred for a combination of reasons. One was its high yield in colorants: it is a tall plant, easily sewn and grown, and the yellow colorants (flavonoids) are present in all parts of the plant. Therefore, a field of weld provides a fair quantity of dye. Another important asset of weld is that its main colorant, luteolin, belongs to the chemical group of flavones that gives the most lightfast yellow dyes. Ancient dyers, with their remarkable empirical understanding of the chemical phenomena they could observe, had obviously noticed this.

The description of the welding process given by the author at page 55 of the *Memoirs* is very detailed and useful. As the bundles of weld, pressed down, remain in the furnace all the time the dye-bath is boiling, colorants go on being extracted until the shades obtained get too weak. The usefulness of adding alkaline substances such as lime and ash to weld is confirmed by William Partridge, who uses pearl-ash.[2] The weight of dry weld mentioned in the second *Memoir* can only be an approximation, because in reality series of cloth pieces are passed in the dye bath, one after the other. The technical term employed by the author "*donner 8 à 10 bouts*" to each piece means that the cloth piece is wound through the dye-bath from one end to the other by a large winch or reel placed over the furnace, and then backwards, which allows it to be evenly dyed.[3] The number of times it is passed in the bath in this way – in this case, approximately 8 to 10 times – depends on the dyer's visual assessment of the intensity of colour achieved. The shades described as *jaune*, yellow, need proportions of weld ranging from 53 % to 68 % of the cloth weight, while as little as 40 % can suffice to obtain the less saturated and slightly greener *limon* (lemon), the more brownish *feuille morte* (dead leaf) and the pistachio colour *festiqui*. Two ingredients typical of the practice of Languedocian dyers are used to obtain these last shades: *trentanel* and soot.

Trentanel is the name in the Occitan language of flax-leaved daphne, *Daphne gnidium* L. (Thymelaeaceae), a very common bush of the mediterranean *garrigues*, much used by Spanish, Moroccan, Algerian, Greek and Sardinian dyers for their yellow and green dyes.[4] Their reasons for selecting this plant are partly the same as in the case of weld: the yellow colorants, present in the whole plant, are very fast flavones and flavone C-glycosides.[5] Since the plant grows as dense, fairly high bushes, does not need to be cultivated, and grows again abundantly after the branches have been cut down and collected for dyeing, it is a very good resource. In Languedoc, it made the use of weld seem superfluous to dyers and its cultivation pointless to most farmers, as pointed in 1731 by Inspector of manufactures Batizat in a *Mémoire pour servir au rétablissement de la culture de plusieurs ingrédiens propres à la teinture des Draps pour les Echelles du Levant et pour Celle des Laines pour Ceux de Mélange, dans la Province de Languedoc* (Memoir aiming at reestablishing the cultivation of several ingredients useful for dyeing Cloths for the Levant and Wools for Medley cloths in the Province of Languedoc).[6] Indeed, a shortage of weld supply is occasionally mentioned by Languedocian clothiers, for instance in 1749, when the high prices of wheat in the preceding years has induced farmers to prefer cultivating it instead of weld in the usual regions of production of the dye, around Bouzigues, near Montpellier and Sète.[7] *Trentanel* is a cheap and easy yellow dye source for all dye-houses situated in villages or towns surrounded by large areas of scrublands. This is the case in Bize, where *trentanel*, still growing today within a quarter of an hour's walk from the ancient Manufacture, could daily be brought fresh, full of sap, as the author highlights it must be used at page 36 of his first *Memoir*. He does not conceal that its use had been prohibited since the *Instruction générale pour la Teinture des laines de toutes couleurs, & pour la culture des drogues ou ingrediens qu'on y employe* (General Instruction on Dyeing wools in all colours and for the cultivation of the drugs and ingredients employed in it) of 18th March 1671 because the plant also contains a toxic compound, which does not get fixed on the cloth but which, the regulation denounces, "smells foul and damages the eyesight of those who use it".[8]

However, no prohibition of this dye – the earliest being regulation 163 of Ibn Abdun's Treatise of *Hisba*, end of 11th–beginning of 12th centuries – ever managed to discourage Mediterranean dyers from using it.[9] At the beginning of the 19th century, in another cloth manufacture not so far from Bize, Villeneuvette, the owner Casimir Maistre indifferently uses the words *gaudage* (welding) or *trintanellage* ("trentanelling") for yellow dye-baths into which the same quantities of either fresh *trentanel* or dry weld are put – which testifies to the excellent colouring power of flax-leaved daphne.[10] In the *Ancien Régime* days of the author of the *Memoirs*, regulations on dyeing being still enforced, he has to be very cautious to let it understand that it is a good idea to use *trentanel* for yellows with a greenish tinge, without positively admitting that this is what he does.

He seems not to have been too systematic about it, though: while in both the second and fourth *Memoirs* he recommends using a combination of weld and *trentanel* in the dye-baths for lemon yellow, dye-analysis of the sample of *limon* in the fourth *Memoir* has shown that only weld and no flax-leaved daphne has been used.[11] On the other hand, analyses have revealed that the two samples of *festiqui*, on pages 56 and 90, were dyed with flax-leaved daphne as the only yellow dye, without any weld. This colour name, derived from the name of the pistachio fruit in Greek (*fistiki*) is not mentioned in French regulations on dyeing, but it has been transmitted among Mediterranean dyers since the Middle Ages. In a Venetian dyer's manual of the 15th century, "*festechin* green is the lightest and it is done on a ground of milk blue" (*si è nome festechin che è el piui chiaro et questo se fa de guado alatado*).[12] Later on, in Languedoc, Antoine Janot also mentions *festequet* as a shade of very light green requiring a ground of dainty blue and he illustrates it in his dye book.[13] Dye-analysis of the sample of *festiqui* on page 56 of the *Memoirs* has detected indigo, together with flax-leaved daphne, although no woad ground is mentioned in the corresponding recipe: maybe this is just an oversight by the author or his copyist.

Another unusual ingredient figures in the recipes for slightly brownish yellows, that is, soot. "Soot from chimney" was actually one of the sources of *fauve* (light brown, or literally wild animals' hair) colours permitted by the *Instruction générale pour la teinture des laines* of 18th March 1671, particularly for "all the nuances of *feuilles-mortes* [dead-leaves] and *couleurs de poil de boeuf* [cattle's hair colours]".[14] However, it was later classified in another regulation of 1737 as an ingredient that only dyers in common colours could have in their workshop.[15] Jean Hellot, in his *Art de la Teinture des Laines*, nevertheless finds it a good ingredient, cheap, efficient to give brown colours or darken other shades, and of nearly equal fastness to the other sources of *fauves*. His only reproach is that "it makes the wool harsh, and it gives an unpleasant smell to the cloths".[16] To my knowledge, the *Memoirs* are the technical text on dyeing that mentions soot in connection with the greatest variety of recipes.

Among the shades of yellows that could be obtained from weld and/or flax-leaved daphne, light acid yellows such as *limon*/lemon appear to be the most popular in the Levant around the middle of the 18th century. They regularly figure in assortments sheets with samples of the colours required, and in invoices sent by different manufactures.[17]

"Lemon" was one of the shades of yellows also proposed by English dyers, as shown by a sample in the Wool Dyeing Book of Marling and Evans, of Ebley, in Stroud, Gloucestershire, dated March 1798; however, it was obtained differently, purely with old fustic and alum.[18]

Yellows of young fustic

The heartwood of young fustic or smoke tree, *Cotinus coggygria* Scop. (Anacardiaceae), is rich in yellow and orange colorants, as well as in tanins.[19] The precise and accurate description of the plant and of its uses in the first Memoir shows that the author is familiar with this Mediterranean small tree. Not very common in Languedoc, it extends on whole hillsides and low mountain slopes in the north of Provence, on the eastern side of the Rhône valley, where it was exploited as a source of both dye and tanin from the Middle Ages until the end of the 19th century. It is from Provence and – a little surprisingly – not directly but through his business contacts in Marseille, that the author of the *Memoirs* buys the not so big quantities of young fustic he needs. A total of 175 pounds (= 72.45 kg) of young fustic are used in different recipes in the second and fourth Memoirs.

Young fustic may be employed in the author's manufacture because, as mentioned in the preceding chapter, an exception to the regulations that forbid the use of this colorant, normally classified as *petit teint* (not fast enough for good quality textiles), had been made for the broadcloth producers of Languedoc.[20] The author justifies this tolerance by the vividness of the colours obtained, as does Jean Hellot, who explains that "if young fustic is tolerated for dyeing cloths in Languedoc, to make the *langouste* (spiny lobster) colours that are so popular in the Levant, it is because old fustic never makes this colour as beautiful as does young fustic".[21]

It has been seen that special mordant baths with varying quantities of tin mordant and young fustic are prepared for the whole range of "flame colours" (*couleurs de feu*) dyed with cochineal as the main ingredient. These mordant baths can be recycled by adding more young fustic, in proportions ranging from 28 to 20 % of the dry weight of cloth, and after a prolonged boiling, and further additions of tin mordant (from 24 to 13%), the resulting dye baths produce a range of warmer, more golden yellows than weld or flax-leaved daphne. Adding "a dash" of madder gives a gold colour (*couleur d'or*). Without madder, but adding a tiny amount of cochineal (0.5 %) a *cassie* yellow is obtained, so named after the colour of the mimosa flower of *Acacia farnesiana* (L.) Willd.), which had been introduced to Europe as an ornamental bush by Odoardo Farnese in the 16th century. With only young fustic and tartar (around 4%) added, the result was a daffodil yellow (*jonquille*). In the last *Memoir*, the author gives a recipe of *jonquille vif* (bright daffodil) in which young fustic is used in proportions of up to 40 % of the cloth weight.

These yellows went to the Levant as complements to the range of flame colours. In the ten bundles containing each ten Londrins Seconds sent from the Royal Manufacture of Bize on 16th July 1751, three of these shades are present: five pieces of flame colour, five pieces of jujube and five pieces of *cassie* yellow. From the corresponding recipes in the *Memoirs*, it can be calculated that for these 15 pieces of cloth, 40 pounds (16.560 kg) of young fustic had to be used.[22]

Yellows on blues: fast greens on a woad ground

For 18th century Languedocian dyers, the obvious way to obtain fast greens is to top-dye cloth, already woaded to various degrees of woad-and-indigo blue, in a yellow dye bath.

This welding is prepared by the same mordant bath as for yellow shades, except that in these cases, cheaper alum smuggled from Spain can be used. For greens, the dye-bath is more concentrated in yellow-giving plant, up to an equal weight of plant and cloth being used. Weld is still the preferred source of fast yellow for greens, according to regulations.

However, the author of the *Memoirs* now does not hesitate to recommend the use of *trentanel*: together with weld and in equal amount in the second Memoir; in higher proportions, up to five times more flax-leaved daphne than weld, or even without weld at all, in the fourth Memoir reflecting the daily practice in his dye-works. Dye-analyses of some of the green samples in the manuscript have shown that indeed, he systematically uses it for his greens, particularly the dark ones. In the *vert d'herbe* (grass green) of page 82, in the fourth Memoir, indigo, weld and flax-leaved daphne have been identified, which perfectly matches the corresponding recipe. The *vert brun* (dark green) sample on page 58, which corresponds to a recipe involving a saddening process with logwood, has actually not needed it; its very dark, nearly black green has been achieved with indigo and flax-leaved daphne only, no other colorant and particularly, no trace of logwood, being detected.[23] This is due to a group of dark colorants present in flax-leaved daphne, that get progressively extracted in the boiling bath of *trentanellage.* This characteristic was precious for dyers who could thus obtain dark greens much more cheaply with this local weed than by using imported logwood. As this dye, moreover, has a strong covering and leveling effect, it was also very useful for hiding any transverse stripes produced by an uneven beating of the weft in some pieces of cloth, as well as the accidental spots on cloths that had previously been dyed lighter greens. Accordingly, another very dark *vert d'herbe*, p. 87 in the fourth Memoir – not analysed – is an example of "failed pistachio green" that could be recuperated in this manner.

Flax-leaved daphne therefore stands out as one of the key assets of Languedocian dyers. Indeed, in the context in which the *Memoirs* are written – fierce international competition for the markets in the Levant, Seven Years' War – *trentanel* could be viewed as one of the "secret weapons" that ensured the success of "French colours" and of Languedocian cloths in the Levant where green, as will be discussed, was so important a colour.

In green dyeing, however, other factors were equally crucial: the degree of blue of the woad and indigo ground given to a particular cloth; and the dyers' skill in managing the optimal order, number and length of time for passing the mordanted, half-mordanted or unmordanted pieces through the yellow dye-baths (Table 7.1). At page 57 of the second *Memoir*, the author gives by far the most detailed description of this art of dyeing green I know of. It shows that the large quantities of yellow-dye plant(s) that are put into the baths are planned so that the dye, very concentrated at first, gets gradually exhausted, the order in which the cloth pieces are passed in it determining the subtle balance between blue and yellow which distinguishes each shade of green.

In his dye works, the author adopts a personal order, different from that followed both by Antoine Janot at about the same period, and by Casimir Maistre in Villeneuvette, half a century later.[24] This is because his chosen policy is to save on the woad ground of the darkest shades of green which, according to the regulations, should be at least a

Turkish blue. He dyes his *vert noir* (black green) and his *vert brun* (dark green) of page 58, on a ground of the next lower degree of blue, King's blue, in a special dye-bath that he mentions as known also by other Languedocian dyers, under the name of *musique*: it includes copperas or logwood and of course, *trentanel* as source of yellow dye; and this can give such a dark colour that logwood, meant to compensate for the deficiency of the blue ground, only needs to be used in small proportion (6.6 to 8 % of cloth weight) or is not needed at all.

Not including these very dark greens in the sequence of regular welding baths, allows to first pass the cloths that must be dyed middle greens in the fresh yellow bath. These include *vert canard* (duck green), *vert d'herbe* (grass green), *émeraude* (emerald green) and *perroquet* (parrot green). By then, the dye bath is getting less concentrated and close to the concentrations mentioned earlier for yellows. Therefore, it can now be used for undyed cloths that must be dyed yellow, or for cloths that have received a ground of dainty blue and are to be dyed a light green. At the end of this first sequence, two categories of cloths are passed through the bath: cloths that have received a woad ground but have only been mordanted with half the usual proportions of ingredients, that is, with only 6.6 to 8 % alum and 3.3 to 4 % tartar; and cloths that are to be dyed different shades of olive. After that, all pieces to be dyed and remain green – olives therefore excluded – are passed for a second time in the same bath. Lastly come some shades of "blue greens, in which the blue has the ascendency", as William Partridge would describe them.[25] To that effect, the woaded cloths are not mordanted at all, which must have made these colour become less and less green from being exposed to the sun and air, and more and more blue, since their woad ground was quite fast. The samples of "cabbage green" and "sea green" on page 60 now certainly look rather blue, with only a faint greenish tinge.

The poetry of the names defining the different shades of green is in harmony with the beauty of the corresponding colours. The author of the *Memoirs* definitely stands out as a "master of greens", both by the number and range of shades of this colour he proposes, and by the exactness with which he is able to reproduce a certain shade, as illustrated by the identical colorimetric data of the samples of two different pieces of *vert d'herbe* (grass green), one described as *rempli* (full-bodied) in the first Memoir page 58 and the other described as *foncé* (dark) in the fourth Memoir, page 93 (see also Table 7.2).[26]

The second sample is presented as having been mordanted with alum from Sweden and dyed with five parts flax-leaved daphne, one part weld.

Other Languedocian dyers also, clearly put all their know-how and creativeness to develop ranges of beautiful greens and answer the expectations of their Muslim clients in the Eastern Mediterranean. Green, allegedly the favourite colour of the prophet Muhammad, had a strong symbolic power in all parts of the Levant and this must be the main explanation for the unfailing demand of green cloths evidenced in documents. Whatever their date and destination, all the assortments of cloths exported from Languedoc to the Levant in the 18th century include between 20 and 40 % of pieces dyed various shades of green.[27] In Smyrna, between 1759 and 1761, the most popular shades are *vert brun* (dark green), *émeraude foncé* (dark emerald) and *vert naissant* (nascent green).[28]

This was in no way a new trend. Current international collaborative research into the archives of the Salviati, a family of Florentine clothiers and businessmen, has recently shown that in the end of the 15th century, after the fall of Constantinople, and the triumph of the Muslim Ottoman Empire in the Eastern Mediterranean, the colour assortments in the exports of cloths from Italy to the Levant change considerably. Among the 261 cloth pieces exported by the Salviati to Constantinople from 1491 to 1493 for which a colour is mentioned, 35 % are dyed various shades of green, and only 8.9 % are colours "dyed in grain", while a few years before the fall of Constantinople, in 1436 and 1437, the Venetian merchant Giacomo Badoer was only selling 13 % green cloths there, against 51 % of scarlets, crimsons and other "grained" purplish reds. According to the Salviati registers, the most popular greens in Constantinople in the 1490s are pistachio greens (*festichini*) and the very dark green *verdebruni*.[29]

Dark greens on a woad ground are also produced by English dyers. At the end of the 18th and beginning of the 19th centuries, for their dark greens, Marling and Evans of Stroud, in Gloucestershire, first send their wool to be woaded to "a 3 ½ d blue", then prepare a dye-bath concentrated in chipped "fustick" (65 % of wool weight) boiled for two hours, add 15 % logwood, boil again, add 9 % "allom", heave in the wool, boil it for 2 hours and let it "lye in".[30] It has been noted in a preceding chapter on blues that the "3½ d blue" is slightly lighter than the King's blue of the *Memoirs* and is equivalent to Janot's *bleu céleste* (celestial blue). Hence the use of logwood, needed to compensate for a woad ground that would not otherwise allow to obtain the nearly black greens desired. They also dye wool "bottelle", a very dark bluer green, on a ground of "5 d blue", with less old fustic (53 %), the same proportion of logwood, alum (7 %) and tartar (2 %), and they further add some indigo dissolved in vitriol, which makes this green dye a hybrid between a woaded green and the Saxon greens discussed in the next section.[31]

Besides plain "green", other shades of green mentioned in pattern-books of clothiers from the West of England include pea green, "green dyll", "parsley", all mentioned in the years 1753–1754 in a pattern book of the Clark family of clothiers of Trowbridge, unfortunately without accompanying recipes or samples that would allow to characterize these shades.[32] Few names correspond to those employed in French sources. Exceptions are a mention of "festeguy" – probably corresponding to the *festichi* of contemporaneous Languedocian dyers – in the same book, and "grassy", "a colour which is much approved of at London" in January 1732, one learns from a letter sent from Trowbridge by John Jeffries.[33] It may have looked like any of the nuances of *vert d'herbe* illustrated in the Memoirs.

Blue on yellow: Saxon greens

The use of indigo dissolved in sulphuric acid – chemick, as William Partridge calls it – is characteristic of what the author of the *Memoirs* and other Languedocian dyers call Saxon greens, *verts de Saxe*. The difference with Saxon blues is that they involve a yellow dye. In this case, it is neither weld nor *trentanel*, nor even the more golden young fustic, but old fustic, the *bois jaune* (yellow wood) of French dyers. This is the heartwood of Dyers' Mulberry, *Maclura tinctoria* (L.) D. Don ex Steudel (Moraceae).[34] The name *fustok* found in the first Memoir is a distorted form of "fustick", the name of the dye in English contemporaneous dye books. It occurs in a section copied word for word by the author from the *Dictionnaire de la langue françoise, ancienne et moderne* (Dictionary of the French language, ancient and modern), by Pierre Richelet, published in 1732.[35] English dyers, who since the Middle Ages had used the wood of young fustic, which they imported from the Mediterranean regions of Europe, called the new yellow dye wood imported from Central America and the Caribbean "old fustic", because it came as trunks or big logs, which made them think they came from a tree similar to young fustic, only cut down older.

In contrast with logwood, old fustic is one of the colorants permitted by French regulations to dyers in fast dyes.[36] Like logwood, it was imported to Bordeaux from the French colonies in the Antilles, and Tobago may be specially mentioned in the first Memoir as one of the main provenances of *bois jaune* because the island had been ceded to France by the Treaty of Utrecht in 1713 and its resources were being intensively exploited. In spite of being cheaper than logwood, old fustic is not much employed in the recipes described in the *Memoirs*. Apart from its use in combination with indigo carmine to obtain Saxon greens, it figures in the recipe presented as the process for Sedan black, where it is used in proportion of 10 to 12 % of the cloth weight; and in very small quantities – corresponding to a proportion of about 1 % – in nuances of some fancy colours such as *noisette* (hazelnut), *tabac* (tobacco) and *rat*, discussed in the next chapter. A total of only 67.75 pounds (28 kg) of old fustic are used in different recipes in the second and fourth Memoirs.

It is interesting to compare the basic recipe for Saxon green described at page 64 of the *Memoirs* with the equally detailed one given by William Partridge sixty years later for dyeing cloth "a full bodied true green" with chemick.[37] On one hand, they are very similar in the proportions of ingredients used and on the other hand, they differ in a very important aspect, i.e. in the sequence of operations. Basically, Partridge dyes the cloth Saxon blue in a mordant-cum-dye bath, with 17 % alum and 2 % chemick. Then, after heaving the cloth out of it, into the same bath he adds fustic (65 %) and alum (11 %) and extracts the colorants of fustic by prolonged boiling, before adding more chemick (6.5 %), then adding cold water and entering the cloth into it again to dye it green. The author of the *Memoirs* proceeds differently, in that he starts as if he was going to dye a fast yellow with old fustic by preparing a mordant bath with Rome alum, used in proportion of 13 to 16 % and white tartar, at a ratio of 6.6 to 8 % of the cloth weight. The cloth is "boiled" and mordanted in it for two hours before being washed in the fulling mill. Old fustic, used in proportions ranging from 56.6 to 72 % is added to the mordant bath. Its colorants are extracted by a more prolonged boiling than in Partridge's recipe. He, too, adds cold water to cool down the bath to the point at which "one can put one's hand in it". Then only, is the *composition* of indigo carmine added to the bath, and the cloth passed through it turning the reel "briskly". The author of the *Memoirs* adds less indigo carmine than Partridge – only 4.4 to 5 % in total – and does it in two operations, heaving out the cloth, adding more chemick, passing the cloth through the bath again and heaving it out just when the bath starts boiling. This he presents as a personal solution he has adopted, "because this colour is difficult to get even" and he mentions that a similar proceeding is also recommended by "Mr. Albert" who had from the start presented himself as the local expert in Saxon blues and greens. Partridge's advice is to "enter the cloth and rattle over the reel as fast as two men can open it, for unless this be rapidly done, the colour will be uneven, as the blue strikes instantly". He well defines the technical problem dyers are facing with this dye: "in colouring of green, when chemick is used, it is essential to know that the goods take the blue first and then the yellow, and that the longer they are boiled the yellower the colour will be".[38] Two experts' ways of addressing the same difficulty…

To help dyers overcome it, in the first period of enthusiasm at the discovery of Saxon blues and greens, a pamphlet entitled *Manière de teindre un drap blanc en verd, nommé Verd de Saxe* (Manner of dyeing white cloth into a green, called Saxon Green) had been printed in 1750 at government's expense in the *Imprimerie royale* in Paris and it was distributed to the managers and master dyers of the manufactures producing cloth for the Levant.[39] Cloths dyed Saxon greens were included in bales sent to the Levant in order to test the reactions of customers there to these new colours: in July 1751, half the bundles of Londrins Seconds sent from the manufacture of Bize contain one piece of Saxon green, beside pieces dyed in the fast woaded *vert brun* (dark green), *vert d'herbe* (grass green), *vert clair* (light green) and *olive pourrie* (rotten olive).[40]

Since this is the only assortment sheet I found in which Saxon greens are mentioned, it is difficult to know whether, and how long, the fashion lasted.

Notes

1 Cardon 2007, pp. 169–177.
2 Partridge re-ed 1973, p. 169.
3 Good descriptions of this apparatus and the way it is used to evenly dye cloth pieces are given, in French by Jean Hellot (Hellot 1750, pp. 18–20) and in English by William Partridge (Partridge re-ed 1973, pp. 131–133). It is illustrated in Cardon 2007, p. xxii and in the plates of the *Encyclopédie* that illustrate the dye works of the Gobelins Manufacture in Paris, http://www.planches.eu/planche.php?nom=TEINTURE&nr=3 and http://www.planches.eu/planche.php?nom=TEINTURE&nr=8.
4 Cardon 2007, pp. 181–186.
5 Cardon 2014, pp. 195, 659–660.
6 AD34 C 5569, folio 2.
7 AD34 C 2229.
8 Anon. 1730, *Recueil des Reglemens,* p. 427, art. 28.
9 Cardon and Andary 2001.
10 AD34, 11J39, folio 5 recto.
11 All the results of the dye-analyses performed by Witold Nowik, then in charge of the Analytical Department of the Laboratoire de Recherche des Monuments Historiques, Champs-sur-Marne, are grouped in annex 2. Flax-leaved daphne can be clearly distinguished from weld by HPLC-PDA analyses, as evidenced in Nowik 2005.
12 Rebora 1970, p. 97, cap. 68.
13 Janot 1744 folio 3.
14 Anon. 1730, *Recueil des Reglemens,* p. 427, art. 26.
15 An. 1746, *Manufacture de Carcassonne*, p. 232, art. 20.
16 Hellot 1750, p. 419.
17 Several occurrences in AD34 C 2514 and C 2076.
18 Gloucester Archives, D948/1, ninth double page.
19 Cardon 2007, pp.191–195; Cardon 2014, pp. 664–666.
20 The prohibition was maintained on paper, in the *Règlement pour la teinture des Étoffes de Laine et des Laines servant à leur fabrication* of 15 January 1737, An. 1746, *Manufacture de Carcassonne*, p. 232, art. 20.
21 Hellot 1750, p. 472.
22 AD11 3 J 342.
23 Dye-analyses by Witold Nowik, all results grouped in appendix 2 p. 149.
24 Janot 1744 folio 2 verso; Maistre 1817, folio 5 recto.
25 Partridge re-ed 1973, p. 174.
26 Colorimetric analyses of all samples in the *Memoirs* have been performed by Iris Brémaud, Research Scientist, CNRS, Wood Science Team, Laboratoire de Mécanique et Génie Civil, CNRS/University of Montpellier 2; results grouped in appendix 3.
27 A wealth of examples can be found in AD34 C 2076, C 2160, AD11 3 J 342.
28 AD34 C 2537.
29 Ingrid Houssaye-Michienzi and Dominique Cardon "Colour fashions in Constantinople in the light of some unpublished archives of a Florentine Company (end of XVth century)", paper presented during the 33rd Annual Meeting Dyes in History and Archaeology, at the University of Glasgow, UK, October 2014, based on Registers 397–399, Salviati Archives, Scuola Normale Superiore, Pisa, Italy. Research programme ENPRESA, ANR (National Research Agency, France) http://salviati.hypotheses.org.
30 Gloucester Archives, D948/1, 44th double page, samples on the left page, descriptions of processes on the right.
31 *Ibid.*, last double page.
32 Wiltshire and Swindon History Centre, 927/8.
33 Mann 1964, p. 17, letter n° 95.
34 History of uses and chemistry in Cardon 2007, pp. 196–199.
35 Richelet 1732, t. 1, p. 740.
36 *Règlement pour la teinture des étoffes de laine et des laines servant à leur fabrication* of 15 January 1737, An. 1746, *Manufacture de Carcassonne*, p. 231, art. 19.
37 Partridge re-ed 1973, pp. 174–175.
38 *Ibid.*
39 AD34 C 2160.
40 AD11 3 J 342.

Table 7.1: Key factors in green dyeing: woad ground, mordanting process, sequence of weldings

Degree of woad blue	*Degree of green*	*Mordant**	*Running-order in welding bath, saddening or no saddening*
Turkish blue *Bleu turquin*	Obscure green *Vert obscur*[1]	regular	1st in welding
Dark King's blue *Bleu de Roy foncé*	Duck green *Vert de canard*[2]	regular	1st in welding
	Very dark grass green *Herbe très foncé*[3]	regular	2nd in welding
King's blue *Bleu de Roy*	Black green *Vert noir*[4]	regular	*musique*
	Dark green *Vert brun*[5]	regular	*musique*
	Full-bodied grass green *Vert d'herbe rempli*[6]	regular	3rd in welding
	Grass green *Vert d'herbe*[7]	regular	2nd in welding
	Bronze green *Vert bronzé*[8]	0	end of welding
Light King's blue *Bleu de Roy clair*	Dark sea green *Vert de mer foncé*[9]	0	end of welding
Dark azure blue *Bleu d'azur foncé*	Grass green *Vert d'herbe*[10]	regular	3rd in welding
	Cabbage green *Vert de chou*[11]	0	end of welding
	Capre green *Vert de câpre*[12]	0	6–7 turns in welding
	Dark green *Vert brun*[13]	regular	*musique*
Azure blue *Bleu d'azur*	Emerald *Emeraude*[14]	regular	4th in welding
	Sea green *Vert de mer*[15]	regular	end of welding
Celestial blue or sky blue *Bleu de ciel ou céleste*	Parrot green *Vert perroquet*[16]	regular	5th in welding
	Pistachio green *Vert pistache*[17]	half	7th in welding
Dainty blue *Bleu mignon*	Light green *Vert clair*[18]	regular	6th in welding
Milky blue *Bleu de lait*	Gay green *Vert gay*[19]	regular	6th in welding
	Celadon green *Vert céladon*[20]	0	end of welding
Off-white blue *Bleu déblanchi*	Nascent green *Vert naissant*[21]	regular	6th in welding 7th in welding
	Apple green *Vert pomme*[22]	half	

*regular mordanting process: alum 13.3–16 %, tartar 6.6–8 %; half: alum 6.6–8 %, tartar 3.3–4%; no mordant: 0. *Musique*: final saddening bath with 6.6–8 % logwood per end, and a bundle of flax-leaved daphne together in the bath.

1 Recipe p. 58
2 Recipe p. 58
3 Recipe p. 95
4 Recipe p. 58
5 Recipe p. 58
6 Recipe p. 58 et 93
7 Recipe p. 82
8 Recipe p. 60 et 84
9 Recipe p. 95
10 Recipe p. 58
11 Recipe p. 60
12 Recipe p. 88
13 Recipe p. 93
14 Recipe p. 59
15 Recipe p. 60
16 Recipe p. 59
17 Recipe p. 59
18 Recipe p. 59
19 Recipe p. 59
20 Recipe p. 60
21 Recipe p. 59
22 Recipe p. 60

Table 7.2: Colorimetric data of two samples of grass green, in the 2nd Memoir, p. 58, and in the 4th Memoir, p. 93

	*L**	*a**	*b**	*C**	*h(°)*
Full-bodied grass green	26	10	11	15	134
Dark grass green	26	10	11	14	133

8

Browns, Blacks, Greys

Browns

In the classification recognised by dyers in 18th-century France, the fourth group among the *couleurs primitives* (primary colours) encompasses a wide range of browns under the common name of *fauve* (the yellowish browns of wild animals' hair) or *couleur de racine* (root colour), or *couleur de noisette* (colour of hazelnut).[1] Technically, their production differs from that of the colours discussed in the previous chapters because the main sources of these dyes are plants rich in tannins that act as mordants as well as colorants. Most nuances of browns are therefore obtained in one dye bath only. It may be followed by a separate saddening bath, most often consisting of copperas – iron (II) sulphate – but more usually, the cloth is taken out of the dye bath, the saddening and/or mordanting ingredients are added to the same bath and the cloth is plunged back into it for a final nuancing. The brown colorants used by Languedocian dyers at the time of the *Memoirs on dyeing* are different from those used by dyers in northern France. For dyeing broadcloth in Paris and in the northern provinces, walnut husks and walnut tree roots are the most common ingredients for all kinds of russet to brown shades, while for the equivalent colours, the fine cloths for the Levant are dyed with a combination of two main dyes, one exotic and the other typically local: sandalwood and *redoul*, respectively. Sandalwood – also called red sanders or simply sanders by 18th-century English dyers – is the heartwood of the high tree red sanders, *Pterocarpus santalinus* L. (Fabaceae), currently listed as an endangered species by the International Union for the Conservation of Nature.[2] In the 18th century, its red dye wood is easily available to Languedocian dyers, since the tree is endemic to the southern Eastern Ghats, within the zone of French influence in south-east India. Sandalwood therefore is regularly imported to Marseille from the Coromandel Coast by the *Compagnie française des Indes orientales*, the French East India Company. The dye is classified as *faux teint*, prohibited for fine cloths, in the *Règlement pour la teinture des étoffes de laine* (Regulation on dyeing of wool cloth) of 1737 but this is probably only because it tends to roughen the wool when too much of it is used in the dye-baths, Jean Hellot warns.[3] He considers nevertheless that it should be "tolerated in the best mode of dyeing because of the fastness of its colour, which naturally is of a dark yellowish red".

The Languedocian dyers' other favourite ingredient for browns, is *redoul*, *Coriaria myrtifolia* L. (Coriariaceae), a very common bush to the present day, thriving in ditches, on damp roadsides and in humid underbrush in many parts of Languedoc and Catalonia.[4] In Bize, it grows right on the bank of the river Cesse, directly opposite to the ancient Manufacture. As all parts of the plant are very rich in tanins, it is much used locally, not only for dyeing but also to tan hides. It further contains yellow colorants. Easily obtained and cheap, *redoul*, like *trentanel* discussed earlier, is a great asset for clothiers and dyers of the province.

A third ingredient is mentioned by the author of the *Memoirs* in his recipes for browns. This is gallnut, a concentrated source of colourless tannins, very much used by all European dyers, either of wool, silk, linen or cotton. Gallnuts are imported in big quantities from the Levant, where they are formed on the young branches of the Aleppo oak, *Quercus infectoria* G. Olivier (Fagaceae).[5] The two sources of tanins in these recipes, *redoul* and gallnut, react with iron sulphate to form a black colorant that allows to obtain the darker shades of browns. This is a process permitted by the *Instruction générale pour la teinture des laines de toutes couleurs, & pour la culture des drogues ou ingrediens qu'on y employe* of 18th March 1671, in which the use of *redoul* is allowed, not only for dyeing black but all other "colours done with gallnut and copperas, according to the shade that will be desired, and to the Dyers' skill and convenience".[6] The five recipes in the section on *marrons* (chestnut browns) in the second Memoir show that the different shades obtained mostly depend on the varying quantities and proportions of sandalwood and *redoul*, since the proportion of nutgall remains the same – between 1.6 and 2 % of the cloth weight. A predominance of sandalwood will give reddish brown shades while a higher proportion of *redoul* will react with copperas and give darker grey to black colours. Indeed, the samples of *marron* (chestnut brown) and *gerofle remply* (full-bodied cloves), the latter using 13.3–16 % sandalwood and 20–24 % *redoul*, show that large amounts of *redoul* react with copperas to give colours very close to black. Beside sandalwood, other dye woods, used in small amounts, help obtain either redder shades of brown (3.3 to 4 % brazilwood give the *marron rougeâtre*), or bluer and greyer shades, such as *musc* (musk) obtained by adding 6.6 to 8 % logwood to the dye-bath. Unfortunately, no sample illustrates this colour in the manuscript. A sample of musk figures in Antoine Janot's *Mémoire* but the corresponding recipe mentions different ingredients: it is mordanted in the same way as yellows and greens, dyed yellow with weld, then passed in a madder bath with a little gallnut and finally saddened with copperas.[7]

The different shades of *prune* (plum colour) are obtained using the same ingredients as for browns but in lower concentrations, particularly of *redoul*. What distinguishes them from *marrons*, browns, is that, whenever available, mordant baths for scarlets are reused as liquor for the dye-baths. As they still contain traces of tin liquor and cochineal, they contribute to give slightly brighter and redder shades to these plum colours than the preceding browns. This is even more evident for *prunes à la cochenille*, plum colours in grain, which are mordanted with alum and tartar and dyed in a reused dye bath of madder with three red colorants added in very small proportions: 1 % cochineal, 1 % brazilwood, 2 % more madder, with 2 % gallnut to react with copperas and give the required blueish grey tinge. A recipe for the colour amusingly described as *prune sur truffe* (plum on truffle) in the fourth Memoir is actually identical to the recipe of *prune cramoisillée* (crimsoned plum) in the second, where sandalwood and madder are combined as sources of red. This name is a creation of the author of the *Memoirs*, maybe in response to a fashion for new nuances of plum colours that seems to have been spreading in the Levant around the time when he was writing. Samples of *prunes* figure in a pattern sheet sent with a consignment of 12 bundles of Londrins Seconds from the Royal Manufactures of La Terrasse and of Auterive, dated 10 April 1759.[8] The assortments, different for the two manufactures, include pieces of *prune modeste* (modest plum) in one bundle out of two.

Colours described as damson, dark damson, light damson and damson mixtures, and also a colour called "prune", begin to appear in pattern books of clothiers from the West of England in the years 1751–1755, unfortunately without corresponding samples and without recipes.[9] Recipes for damsons in the later dye book of Marling and Evans, of Stroud, are quite different from the recipes of plum colours in the *Memoirs*, only involving barwood and logwood, with a little brazil wood for some nuances, and a final addition of copperas.[10]

Like browns and plum colours, the different shades of *noisette* (hazelnut colour) in the *Memoirs* are obtained in one bath to which all the ingredients are added. While for some, fresh water is used, others reuse a madder dye bath or a dye bath for hazelnut already prepared with a reused madder bath. For one shade only, *noisette biche* – the doe shade of hazelnut – a recycled mordant bath for scarlet is used. Colorants are used sparingly for all shades. They consist of madder and old fustic, a little logwood and, in one case, fernambuco wood, and a source of tannin that may be either *redoul* or gallnut. All the shades of hazelnut are saddened, either with soot or copperas, or both. The author's comment on the great variety of nuances of hazelnut and of processes developed by dyers to obtain them, is still valid more than a century later: in Théophile Grison's book, *La Teinture industrielle au dix-neuvième siècle*, (Industrial Dyeing in the 19th century), figure 9 samples of broadcloth dyed different shades of *noisette* using various proportions of the same ingredients as in *the Memoirs on dyeing* – *redoul* and soot excepted.[11] Curiously, however, hazelnut does not figure once in any of the numerous pattern sheets and lists of cloth consignments sent to the Levant that are preserved in archives in Languedoc. Nor is a colour with that name to be found in any of the dye books and pattern books of English clothiers I have been able to consult, and it is not mentioned either in William Partridge's *Treatise*.

Blacks

"*Noir* (black) is the fifth primary colour of Dyers": Jean Hellot's discussion on this colour is enlightening in that it shows that for 18th century dyers, the notion of black is more technical and consequently, encompasses a much wider colour range, than is the case today. Black, he argues, "encompasses a prodigious quantity of nuances, starting with *gris-blanc* [greyish white] or *gris de perles* [pearl grey], up to *gris de more* [Moor's grey], and ending in black. It is because of these nuances that it is counted among *couleurs primitives* [primary colours]; for most browns, whatever their shade, are finished with the same dye which, on white wool, would produce a grey of lighter or darker intensity. This process is called *Bruniture* [saddening]."[12]

William Partridge's vision of black, albeit somewhat simpler, is technical and quite subtle, too: "There are four different and distinct colours in black, the blue, the red, the yellow and the jet black [...] The terms I have used to distinguish the different colours require no explanation, the blue, red, and yellow being the dyer's primary colours; everyone must know, that when either of these predominate, the colour assumes that name. A jet black is that happy mixture of the three, in which neither of them is in excess so as to be visible."[13]

In order to give a deep black colour to an undyed woollen cloth, ancient dyers would first dye it into as deep a blue as possible, one of these nearly black blues that are obtained from repeated dippings in indigo vats and on which is built the extraordinary prestige of indigo in all civilisations.[14] Mediaeval European dyers of fine broadcloth would then mordant it with alum and tartar and top-dye it with madder, thus producing the very prestigious black called *brunette*, *brunet*, *bruneta* in their respective languages.[15]

However, in the context of ruthless international economic competition in which the author of the *Memoirs* and his Languedocian colleagues are working, this manner of grounding a black dye, in accordance with the French regulations on dyeing into fast dyes, has started to be problematic from an economical point of view because it is getting too costly in woad and indigo, in relation to the selling prices of Londrins Seconds in the Levant. Clothiers therefore try to find solutions to maintain the quality of their dyes and to essentially keep respecting regulations, while introducing some small changes of proportions or even of ingredients that may lower the production cost of their blacks. The three recipes for blacks proposed in the *Memoirs* (Table 8.1) perfectly illustrate the conjuncture: the first one allegedly presents the most prestigious, top quality black dye of the time; the second reflects a phase of experiments to improve logwood blacks and make them acceptable from the point of view of fastness; and the third process is the result of the most cost-effective approach, ensuring best value for money, that Languedocian dyers have managed to find, building on ancient local traditions.

They encounter difficulties, however, in getting their views accepted by a royal administration which, albeit favourably disposed, sticks to the conviction that maintaining high quality dyeing standards is an essential condition for ensuring customer loyalty in the Levant and even conquer new markets for the broadcloths produced in the province. The archives of the Intendance of Languedoc in Montpellier have preserved voluminous files dealing with the "case of black dyes", beginning in 1745, when an increasing infatuation for black cloths in the countries of the Levant starts being reported by factors there, making the cost of their dyeing a matter of major concern. The degree of blue of the ground is the first issue that comes up. According to the *Règlement pour la teinture des étoffes de laine* (Regulation on dyeing of woollen) of 1737, Persian blue is still the required ground for dyeing superfine cloths black; Turkish blue is only allowed for middling fine qualities and celestial blue for common cloths.[16] However, controls in Montpellier soon reveal that 20 pieces of superfine black cloths, out of 200 examined, have received a blue ground inferior to Turkish blue, some pieces not having been woaded at all. The Intendant of the province then decides to make an example of government firmness and distributes a deluge of sanctions, ranging from one of the guilty master dyers being prohibited from practising his trade, to the seizure of faulty pieces, all this accompanied by heavy fines. Two Royal Manufactures are involved – not that of the author, however.[17] The wealth of pleas and explications sent by the offenders and their advocates – among whom Jean Hellot, consulted as expert – reveals that it had become a "common practice among the majority of dyers to give a ground of celestial blue only to black cloths for the Levant and that no complaints had even been sent in return." After more than one year of negotiations and haggling, only the penalties for the most serious offences – the cases of the cloths not woaded at all – are maintained. Moreover, a consensus is eventually reached that "a ground of King's blue is sufficient, considering that these cloths do not sell at high prices and besides, that they are thin and light". It is suggested that, should the Intendant accept such arguments, it would be enough to "inform the controllers, both of dyers and of manufacturers, without the need of any decree or ordinance" that henceforth, they may turn a blind eye on cases of cloths having received woad grounds of one or two degrees of blue immediately below the Turkish blue stipulated by regulations. Several documents in the file show what the defective cloths look like, before and after the *débouilli* – that is the special testing procedure devised to strip the other ingredients of the black dye off from the cloth. From then on, moderate saving on the blue grounds becomes less risky, and it makes it easier for clothiers to respond to the continuing upward trend of the fashion for black cloth. In July 1755, the manager of one of the Royal Manufactures writes that black cloths are "very much in demand in Marseilles, not only for Constantinople, but also for India, since among one hundred pieces which the East India Company has ordered from me, there must be forty that are dyed black."[18] It is in this context that the author's mention of a woad ground of "full-bodied King's blue" in his recipe for a top quality black dye takes its full significance: it is a comparatively recently secured tolerance, obtained with difficulty.

This is not the only way in which this recipe of black dye "in the manner of Sedan" – the only one illustrated by a sample – differs from the original process of Sedan black. This is described in details in a *Mémoire sur la teinture du Bleu et du Noir de Sedan par le Sieur Delo, ancien Elève des Manufactures* (Memoir on dyeing in Sedan Blue and Black by Mr. Delo, former Student of the Manufactures), written in Sedan in 1761, more or less at the same time as the *Memoirs on Dyeing*.[19] According to Delo, Sedan black is obtained on a woad ground of Persian blue or at least of King's blue, by a simple chemical reaction between the gallotannins of Sicilian sumac, *Rhus coriaria* L. (Anacardiaceae) and blue to purple colorants of logwood, and high doses of copperas (19 to 22.5 % of the cloth weight). The author of the *Memoirs*' version, more complex, is also less corrosive to wool. It combines greater proportions of colorants, obtained from two distinct dye woods, and two good sources of tannins – one local and very cheap, *redoul* – with lower proportions of iron(II) sulphate. These differences illustrate the author's skill at using local resources and at adapting processes to the nature of his cloths, much thinner and lighter than fine strong Sedan broadcloth. Otherwise, both Delo's and the author's recipes are remarkably excessive in more than one aspect (Table 8.1). Even taking into account the international renown of black-dyed Sedan cloth, one cannot help wondering at the quantities of ingredients, number of operations, length of time needed and amount of timber consumed, only to intensify and darken the colour of these already dark blue cloths to a true jet black. According to the author of the *Memoirs*, when dyeing Sedan black the cloth piece should be boiled three times in the dye bath containing two dye woods and two sources of tannins. These ingredients must have been boiled together previously during six hours and the cloth is boiled in the resulting liquor for six more hours, but every two hours it must be heaved out and aired and heaved in again. William Partridge has left the best description of the hard work involved in this "black dying", the "broads-men", as he calls the workers, "keeping the cloth open all the time, moving the reel with considerable rapidity", "one person being employed to push the cloth under the liquor on one side, and another on the other side to keep it open; the reel being kept turning during the whole operation"[20] – and everybody having to take care and avoid being scalded while manipulating the hugely heavy hot cloth. In Sedan, the process is called *engallage* (because gallnut was originally employed instead of sumac as source of tannin): according to Delo's description, sumac and logwood are boiled together even longer, for eight hours. However, the cloth is dyed in the resulting bath only once, but it stays boiling in it for four hours, during which workers do not stop moving the piece with the reel, arranging it in folds in the vessel, keeping it open, etc. After the dye-bath, the cloth still needs to be saddened by adding copperas to the bath, toward the end of the process. The black colour is produced by the reaction of gallic acid impregnated into the cloth and ferrous sulphate. The author of the *Memoirs* and Delo both indicate that this is done in two operations, the cloth being heaved out of the bath and aired between the two additions of copperas. In Languedoc, the saddening process is followed by a dye-bath with weld. This final welding is common to the three recipes of black in the *Memoirs*. It was equally popular among black dyers of various regions of Europe. It has a double function: "it makes black faster" and "softens" the cloth, Hellot explains, which is confirmed by Partridge, who considers that "it imparts a softness to all woollens coloured in it, which no other colouring matter does in the same degree."[21] The final repeated streamings and washings at the fulling mill described by the author of the *Memoirs* – which also contribute to the high cost of the dye – are indispensable to prevent the cloth, saturated with colorants and metallic salt, from staining the white linen of the future clients' shirts once it has been tailored into a coat. In total, after the repeated dippings in the woad and indigo vat, each piece of cloth will have received three *feux* (passages in the boiling dye bath) in Languedoc, or a very prolonged *engallage* in Sedan, at least two dippings in a saddening bath; and a final welding, in Languedoc. All these operations will have needed a minimum of fifteen hours of strong fire in the furnace and even more hours of work force, before, during and after the dyeing and saddening processes. It is difficult to imagine how a dye needing such input in raw materials and energy could possibly be cost effective for Londrins Seconds cloths, the selling price of which is valued at 9 pounds 10 *sols* per ell in a document of 1769, that is half the price of the best Sedan cloths.[22] It must have been reserved to dyeing the Mahoux and Londrins Premiers, meant for elite customers such as the Ottoman Sultan and his court.

The second recipe for black in the *Memoirs* is a *noir au tartre* (black with tartar), erroneously attributed by the author to the Mr Albert already mentioned in earlier chapters, in connection with celadon greens with verdigris and Saxon greens. It is one among many such processes for dyeing black without a woad ground and purely with logwood saddened with iron and/or copper salts, in total disregard for the regulations on fast dyes, that were being experimented in Languedoc in the years 1750–1760 to imitate the logwood blacks of English cloths exported to the Levant. Descriptions of some of these processes were printed at the government's expense and distributed to dyers and clothiers, which may seem surprising, in the light of the controls and fight against infringements of regulations on dyeing that have been mentioned in connection with "the case of black dyes", a few years earlier. The author of the *Memoirs* got confused because two of these leaflets were published nearly simultaneously. The first one is anonymous and undated, but documents preserved in the same file in the *Archives départementales de l'Hérault* show that the process had been developed by a master dyer in Montpellier, Fernand Paquier, who actually did not want it to be divulged in that way.[23] This is a similar recipe to the one described in the *Memoirs*, for a black dye with logwood on cloth previously mordanted with tartar and copperas and finally saddened with verdigris. The second leaflet proposes a different process, described in the title as *Procédé de*

M. Albert, Docteur en Médecine, des Académies des Sciences de Montpellier et de Toulouse, pour teindre en noir, sans aucun pied de bleu ni de racinage, une pièce de drap, ou telle autre étoffe de laine du poids de vingt-cinq livres (Process of Mr Albert, Doctor in Medicine, of the Academies of Sciences of Montpellier and Toulouse, to dye black without any blue or tanin ground, a piece of cloth or whatever other woollen of a weight of twenty-five pounds). Printed in Paris at the *Imprimerie royale* in 1758, it describes a black dye using logwood and blue vitriol (copper sulphate) in the same bath, in which the cloth is boiled and passed three times for one hour and a half, being heaved out and aired between each dipping to allow an allegedly better fixation of the colorant and mordant.[24] Documents in the same file in the Archives show that Albert's process was diffused very shortly after the first leaflet. Black dyes based on logwood and tartar actually went on being produced until the end of the 19th century but their lack of fastness was denounced by the expert chemist and dyer Théophile Grison: "blacks dyed after being mordanted with iron and copper sulphates and tartar, get paler from rubbing. This comes from the fact that the coloured lacs produced are not intimately fixed onto the wool, but simply juxtaposed on it [...] Tartar is not a mordant, but simply an intermediary agent which prevents the iron sulphate from getting transformed into ferric sulphate and iron oxide, by keeping it in solution in the bath, which allows it to imbibe the wool."[25]

Of the three processes for black described in the *Memoirs*, it is the last one, presented as "the ordinary manner", that offers the best compromise to obtain a fast fine black, not too corrosive to the wool, at a reasonable cost. It is based on a woad ground of azure blue, two degrees below King's blue, which would not have been tolerated a few years before for dyeing Londrins Seconds. From a technical point of view, however, the deficiency of the woad ground is compensated by the generous use of two distinct sources of gallic tannins and of two sources of colorants, the dark blue colorants of logwood, applied first, with the tannins and copperas, and the red colorants of madder, added to the saddening bath. The final welding bath must have fixed still more colorants, increasing the richness of the black, while giving the cloth a soft hand. The trick consisting in saving on the woad ground, top-dyeing the cloth dyed a medium blue with tannins, and saddening with an iron salt, had already been developed in Languedoc and Catalonia in the Middle Ages and it had finally been accepted in regulations for all but the top qualities of broadcloth: the black so obtained was described as *negre*, as opposed to the prestigious *bruneta* black.[26] In 18th century Languedoc, the author's recipe of "woaded black in the ordinary manner" therefore includes elements of these two different traditional ways of dyeing black, with the important innovation introduced by the addition of logwood, which contributes its dark purplish-blue colour.

Although William Partridge's seven recipes for dyeing cloth blue, yellow, red and jet blacks "are the best that England afford" – he claims – they are puzzling because none of them involves a woad ground, while he does mention the need of a woad ground in his recipes "to dye black wool".[27] For piece-dyed blacks, all his recipes are based on a combination of logwood and sumac, saddened with one, two, and up to three additions of copperas and blue vitriol to the same bath. Among his recipes for dyeing wool in the fleece, figures a "receipt for colouring a raven black for a mixture or for a wool colour" on a woad ground "of a light blue", top-dyed with logwood and saddened with copperas. The resulting colour is "a rich blue black, or what is called a raven, being the hue of the wing of that bird."[28] In the dye book of Marling and Evans of Stroud, who dyed wool in the fleece for medley cloths, there is no recipe nor any sample of wool described as black, but a "dark raven" and a "raven grey" that both look blue-black to a modern eye and have both been obtained on a woad ground of "3 d blue", mordanted with alum, and then dyed with a combination of "spirits" – that is indigo carmine – and logwood, and finally saddened with copperas, in different proportions for the two nuances[29]. These "raven" colours, also mentioned in pattern books of clothiers from Trowbridge[30], are therefore at the border between true blacks and false Persian blues, as their Languedocian colleagues would consider them. One mention of a "star black" however figures in one of these books, on a page dated May 1755, but in that book, unfortunately, mentions of colours are only found when the corresponding pattern is missing, to replace it, and pattern books do not include dye recipes.[31]

Greys

The three last pages of the second Memoir offer a series of recipes for a refined range of greys. They are all based on subtle balances between three colours: blue, black, red. For the colours *taupe* (mole), *ardoise* (slate), *gris de plomb* (lead grey) and *gris de rat* (rat grey), it is the degree of the woad ground that mostly determines what shade can be obtained (Table 6.7). In the case of four of the shades of the rat colour, however, a second blue colorant, logwood, is added to the gallnut bath. The degree of woad ground also distinguishes between nuances of agate, a colour that differs from all other greys in requiring no final saddening bath with copperas. The intensity of the black component contributed by the reaction of copperas with the tannins of the galling baths can be controlled for each particular shade by the choice of the source of tannin used (gallnut or *redoul*) and the proportions in which it is employed. Red colorants, used in small quantities, are either added to the same galling bath as the tanins (this is the case for sandalwood and madder in the recipes of mole, slate, lead and rat greys). Or they are present already in reused dye baths, recycled into some greys: cochineal dye-baths or their saddening baths with alum, for the shades of agate and Prince's grey; mordant baths of scarlets, still containing a little cochineal and tin mordant and giving very soft nuances of greys by small additions of gallnut and copperas, such as the beaver, pearl, and silver greys of the second Memoir and the *gris cendré* (ash grey) of the fourth Memoir. These "*petites nuances de gris*" are doubly advantageous because they cost very little to produce, while being "*couleurs de gout*", elegant and fashionable.

The woad grounds of other greys – mole, slate, lead and agate – makes them quite fast, but more costly. This is also true of one of the author's creations, the colour *poudre à canon des Anglais* (English gunpowder), described and illustrated by a sample in the fourth Memoir. Technically, the recipe is quite similar to a mole grey, with the same ground of azure blue, however it includes no red component. The colour name is unique, although pieces of Londrins Seconds dyed *poudre à canon* figure in different consignments of cloths sent to the Levant. It may only be coincidental that they all correspond to periods of war. The *Memoirs*, it has been explained, are written at the end of the Seven Years' War. Cloths dyed a gunpowder grey are sent from the Royal Manufacture of Pennautier, to the north-west of Carcassonne, in November 1775, the year of the outbreak of the American War of Independence.[32] More pieces of the same colour, made in the manufacture of the firm Besaucèle-Darle in Carcassonne, leave Marseille on different boats, bound for Smyrna and Egypt, between 1807 and 1816, that is, during the period of the Napoleonic Wars.[33] Pattern sheets for the latter consignments show the gunpowder colour as a very dark, nearly black grey, both darker and less blueish than the "English gunpowder" in the *Memoirs*.

The description of the processes for mole grey and slate grey in the fourth Memoir shows how a clever management of the sequence of operations in the dye works of the Manufacture could lessen the costs of woaded greys. Just before these colours are done, several pieces of cloth dyed various nuances of sky blue have come out from the vat. The darkest is dyed into a mole grey by the same process as described in the second Memoir, although its woad ground is one degree lower. By immediately dyeing six more pieces of sky blue cloth in the same dye bath, the author manages to completely exhaust it and obtain good slate colours, only adding a small amount of gallnut (0.3 % of the cloths weight).

In spite of their being more costly, woaded greys are more in demand in the Levant than the other greys: mole greys figure in bundles sent from the Royal Manufactures of Pennautier in July 1749, of Saint-Chinian in July 1750, of La Terrasse in 1759.[34] *Ardoises* (slate greys) are sent from those of Bize in July 1751 and Saint-Chinian in 1754; *gris de plomb* (lead greys) are sent from Saint-Chinian in 1758.[35] That same year, Londrins Premiers and Londrins Seconds dyed *agathe* and Prince's grey are sent from the Royal Manufacture of Montoulieu.[36] Londrins Seconds dyed *castor* (beaver) are sent from Saint-Chinian in 1754.[37] Five bundles with Londrins Seconds *gris de cendre* (ash greys) leave the Royal Manufacture of Auterive in April 1759.[38]

While similar names for greys – slate, lead, silver, ash, beaver, pearl, rat – also appear in English pattern books and dye books, mentions of "oaded slate" or "oaded ash" are very rare; the latter is dated 1751.[39] All other processes for these shades only include various combinations and proportions of dye woods and copperas. A "French grey" and a "rich French grey" are described and illustrated in the Wool Dyeing Book of Marling and Evans of Ebley, in Stroud. They are produced in a completely different way, since this firm dyes wool in the fleece for medleys. The technique, then, consists of dyeing lots of wool into different colours and mixing them in different proportions to obtain either homogeneous or marbled colours at will. French greys are obtained by mixing a portion of wool woaded at another dyer's, a portion of wool dyed in grain, i.e. with cochineal, and some undyed wool.[40]

As far as can be judged from the documents examined so far, it may be that the technique of piece-dyeing blacks and greys was more common among Languedocian dyers and that dyers in the West of England preferred to dye wool in the fleece for this class of colours. Another explanation may be that cloths meant to be piece-dyed into blacks and greys were sent to London factors and entrusted to specialized black dyers in London.

Notes

1 Hellot 1750, p. 407.
2 Cardon 2007, pp. 289–300; Cardon 2014, pp. 264–291.
3 An. 1746, *Manufacture de Carcassonne*, p. 232, art. 20; Hellot 1750, pp. 409–410.
4 Cardon and Pinto 2007.
5 Cardon 2007, pp. 414–418.
6 Anon. 1730, *Recueil des Reglemens*, p. 431, art. 44.
7 Janot 1744, sample on folio 4, recipe on folio 5.
8 AD34 C 5550.
9 Wiltshire and Swindon History Centre, pattern book of the Clark family of Trowbridge, 927/8.
10 Wool Dyeing Book of Marling and Evans of Ebley, dated 1794–1804, Gloucestershire Archives, D948/1.
11 Grison 1884, vol. 2, pp. 87–92.
12 Hellot 1750, pp. 423–424.
13 Partridge re-ed 1973, pp. 135–136.
14 Balfour-Paul 2011; Cardon 2014, pp. 327–394.
15 Cardon 1991.

16 An. 1746, *Manufacture de Carcassonne*, 1746, p. 240, art. 56.
17 AD 34 C 2241.
18 AD34 C 2238 , n° 56.
19 Delo 1761, AN F/12/737.
20 Partridge re-ed 1973, pp. 136–137.
21 Hellot 1744, p. 428; Partridge re-ed 1973, p. 112.
22 Marquié 1993, p. 105.
23 AD34 C 2246.
24 AD34 D 152.
25 Grison 1884, vol. 2, p. 111.
26 Cardon 1991.
27 Partridge re-ed. 1973, p. 142, 145.
28 *Ibid*., p. 147.
29 Wool Dyeing Book of Marling and Evans of Ebley, dated 1794–1804, Gloucestershire Archives, D948/1.
30 Wiltshire and Swindon History Centre, pattern book of the Clark family of Trowbridge, 927/7.
31 Wiltshire and Swindon History Centre, pattern book of the Clark family of Trowbridge, 927/8.
32 AD11 3 J 342.
33 AD11 3 J 961, folio 14, 20, 29.
34 AD11 3 J 342, AD34 C 2160, C 5550, doc. 44.
35 AD11 3 J 342, AD34 C 2076, AD34 C 5550, doc.42.
36 *Ibid*., doc. 84.
37 AD34 C 2076.
38 AD34 C 5550 n° 44.
39 Wool Dyeing Book of Marling and Evans of Ebley, dated 1794–1804, Gloucestershire Archives, D948/1; Wiltshire and Swindon History Centre, pattern book of the Clark family of Trowbridge, 927/8.
40 Gloucestershire Archives, D948/1, eighth double page.

Table 8.1: Blacks: true Sedan, Languedocian woaded blacks, a logwood black

Ingredients[1]	*Sedan black,*[2] *4/3 type, 2nd grade, and 5/4 type, 1st grade*	*Sedan black,*[3] *5/4 type, 2nd grade*	*Black in the manner of Sedan*[4]	*Woaded black in the ordinary manner*	*Black with tartar*
Woad ground	Persian blue	King's blue	Dark King's blue	Azure blue	-
Boiling bath	-	-	-	-	tartar 3.3–4 % copperas 3.3–4 %
Logwood	galling: 9.9% 1st saddening: 12.4%	galling: 6.3% 1st saddening: 10%	23.3–28 %	11.6–14 %	50–60 %
Sumac	galling: 61% 1st saddening: 13.6%	galling: 50% 1st saddening: 12.5%	-	-	-
Gallnut	-	-	3.3–4 %	6.6–8 %	-
Old fustic	-	-	10–12 %	-	-
Redoul	-	-	content of 2 sieves	5–6 %	-
Copperas	1st saddening: 15 % 2nd saddening 7.4%	1st saddening: 12.5 % 2nd saddening 6.3 %	1st saddening: 3.3–4 % 2nd saddening 3.3–4 %	1st saddening: 10–12 % 2nd saddening 1.6–2 %	-
Verdigris	-	-	-		0.8–1 %
Madder	-	-	-	10–12 %	-
Weld	-	-	10–12 %	-	-

1 Proportions of ingredients calculated in relation with dry cloth weight.
2 Delo 1761, f° 15; cloths weights from data in Roland de la Platière 1785, pp. 325–326.
3 Delo 1761, f° 15; cloths weights from data in Roland de la Platière 1785, p. 327.
4 Recipe p. 73 of manuscript.

Epilogue

The last Memoir in the manuscript, *Annotations on the colours made for the Levant*, records a succession of dyeings performed in the author's manufacture over what was probably a brief period of time. In this, it is comparable, but not quite similar, to the contemporaneous dye books and pattern books of firms of clothiers and dyers from the West of England including patterns of cloth or samples of dyed wool. While the latter most often offer detailed mentions of the clients for whom the colours were made, dates when dyeings were done, dye books further mentioning the ingredients used and in what quantities, as each colour has just been made, the fourth of the *Memoirs on Dyeing* is only meant to show how colours already described in the second Memoir can be achieved in an alternative, more economical way, examples being presented in the order in which the daily routine of the dye-works provides them. Economy is the author's key word in that part of the manuscript.

However, it is possible, from the implicit content of the text, to propose another reading of this Memoir, particularly of the last pages, focusing on the author's way of managing his dye-works, and highlighting an effect of his cost-saving policy that is of even greater relevance today than in his time.

The four last pages of the manuscript actually describe the processes implemented during two days of work in the author's dye-house. During that period of time, two parallel dyeing cycles are conducted: the first concerns the range of purples, violets and wine colours; the other, the gradation of greens. They are all meant to make the best possible use of the weekly output of one of the woad-and-indigo vats of the Manufacture. It was planned to produce a sequence of medium blues, of lower degrees than Persian and Turkish blues: as a result, four pieces are dyed a King's blue – one slightly darker than the standard, two exactly matching the standard, and one slightly lighter, described as *petit bleu de Roy* – three more pieces are of dark azure blues, and the last two have come out as sky blues. These different degrees of blues are going to serve as grounds for the corresponding shades of purples and violets, on one side, and of greens, on the other side. Other colours, such as wine colours, that do not need a woad ground but require the same mordanting and dyeing processes as these purples and greens, are integrated to the sequence of operations. In a first step, all cloths receive appropriate mordant baths, respectively to fix the colorants of cochineal and enhance their purplish colour; or to fix and develop the yellow colorants of weld and *trentanel*. Since each of these types of mordant baths can work efficiently for several different colours, the corresponding cloth pieces are mordanted together in the same bath, which saves on drugs and on time and energy. The next step is that of dye-baths: for the range of purples, the recycling system discussed in the section on cochineal dyes is still improved here, in terms of cost-effectiveness, by the use of lower-grade forms of cochineals. For the greens, different "welding" baths cover the whole range between a bath purely with weld to obtain an exquisite nascent green and a bath purely with *trentanel*, perfect for several shades of dark greens. Lastly come the saddening baths with the various ingredients required to reach the exact nuance of either purple or green wanted. These baths too can be used or reused for several cloths.

To achieve the smooth succession of so many different operations implies such an intricate and precise organisation, that it is best understood with the help of diagrams (diagram 1 for purples and wine colours, diagram 2 for greens).

As a result of the multiple reuses described, every particle of mordant and colouring substance was fully used. It also meant that the effluents from the manufacture were fairly innocuous, especially compared with the effluents from dye plants in many countries today.

The last part of the *Memoirs on dyeing* thus offers an impressive manifestation of the author's extraordinary talents for the organisation of tasks and for the management of space, equipment, fuel, water and raw materials – and not least, of his labour force. Here is another illustration

Purples, violets and wine colours of pages 93 and 94 of the *Memoirs on Dyeing* : sequence of operations

BOILING (= MORDANT BATH)
FINISHING (= DYE BATH)
SADDENING
King's blue
alum 16.6-20%
+ tartar 6.6-8%
BOILING N°1
dark azure
dark azure
alum 13.3-16%
+ tartar 6.6-8%
BOILING N°2
1st reuse
2nd reuse
sky blue
no addition
+ alum 10-12%
+ tartar 6.6-8%
+ salt 10-12%
+ fine cochineal 5-6%
1st reuse
+ cochineal 1-1.2%
+ tartar 6.6-8%
2nd reuse
+ cochineal 4.8-5.8%
3rd reuse
+ granille 3.3-4%
+ tartar 3.3-4%
+ larch agaric 0.8-1 %
4th reuse
+ granille 4-4.7%
+ tartar 3.3-4%
+ larch agaric 0.8-1 %
5th reuse
no addition
6th reuse
+ garbeau 4.8-5.8%
+ tartar 6.6-8%
+ larch agaric 3.3-4%
+ brazilwood 6.6-8%
+ lees ash 3.3-4%
+ lees ash 3.3-4%
+ brazilwood 3.3-4%
bishop's violet
winesoup
amaranth
violet purple
dark purple
lilac or dark dove breast
royal wine soup

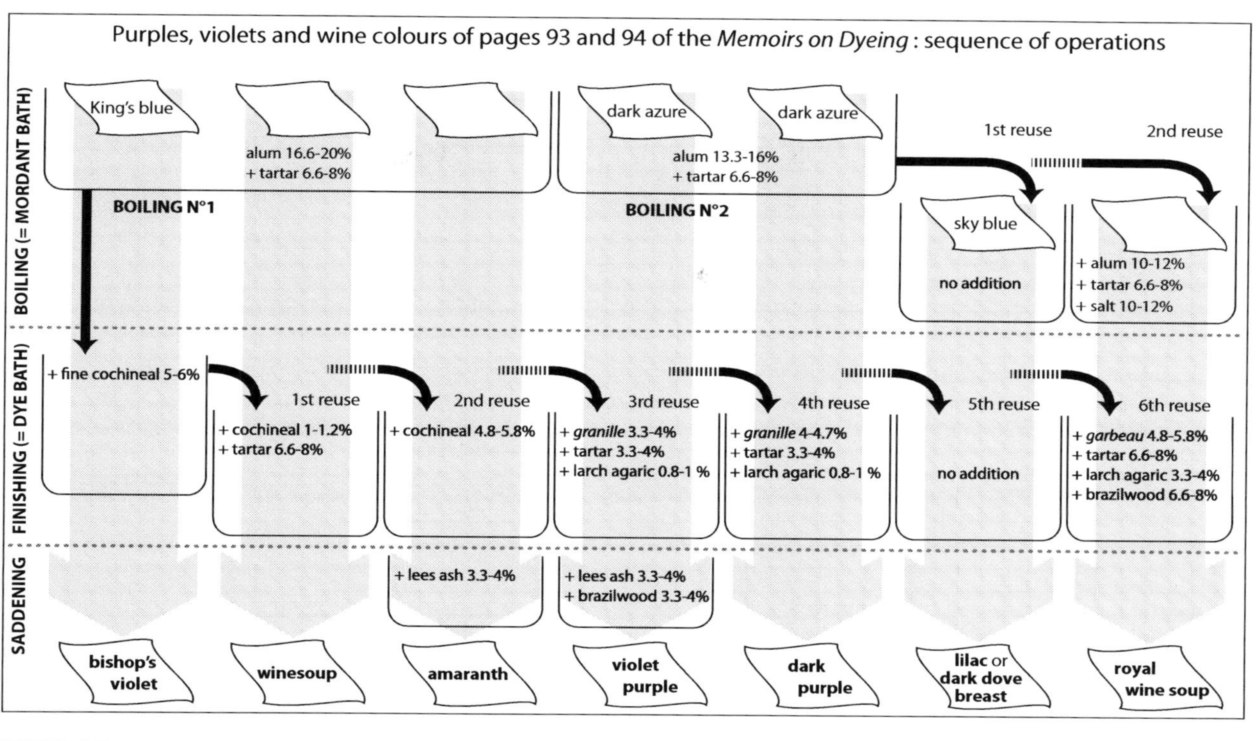

Greens in pages 93 and 95 of the *Memoirs* : sequence of operations

BOILING (= MORDANT BATH)
FINISHING (= DYE BATH)
SADDENING
WASHING
King's blue
dark azure
alum 10-12 %
+ tartar 6.6-8 %
BOILING N°1
dark King's blue
no woading
alum 13.3-16%
+ tartar 6.6-8%
BOILING N°2
weld
8.8 - 10.6%
+ flax-leaved daphne
44.4 - 53.3 %
WELDING N°1
flax-leaved daphne only
53.3 - 64 %
WELDING N°2
light King's blue
reused welding n° 3
+ flax-leaved daphne
53.3 - 64 %
WELDING N°4
chemick
(= indigo carmine)
1.25-1.5 %
+ weld
53.3 - 64 %
WELDING N°3
with copperas
with logwood
3.3 - 4 %
reused saddening
dark grass green
dark grass
very dark grass green
sea green
nascent green

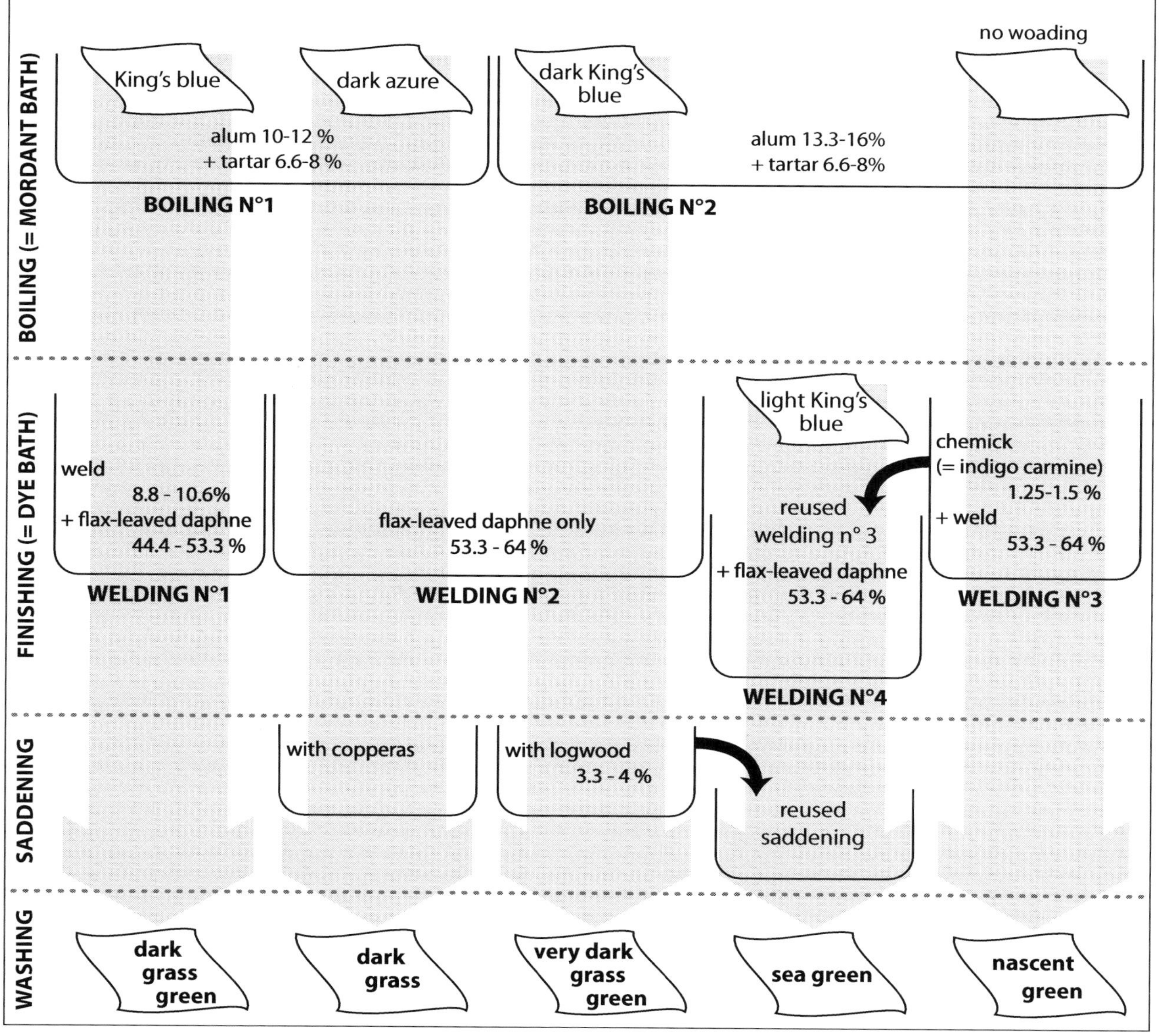

of this "entrepreneurship of exceptional quality", which so much impressed James Thomson when he was appraising the performances of clothiers in the cloth-making region of Clermont-de-Lodève, not far from Bize.[1]

The fascinating tour of his world of colours through which the author has guided his future readers ends abruptly, like an interrupted conversation.

He gives a last recipe, ultimate lesson of economy, for a delicate "*couleur de goût*", a light buff with the name of an elegant animal, the chamois. It costs next to nothing to make, using a small amount of a wild indigenous dye plant. Then, he ends his *Memoirs on Dyeing*, at the bottom of the last page of the booklet he had prepared, without even a final dot.

The task he had set for himself, to write the *Memoirs* in order to share his knowledge and experience and be "useful" to other dyers and clothiers, is accomplished.

Business goes on as usual.

Note

1 Thomson 1982, pp. 8–9.

Appendix 1

Metric equivalents to measurement units mentioned in the *Memoirs on Dyeing*

Length units

The *palme* is called *empan* or *pan* in Languedoc. Originally based on the length of the hand's palm, it is a subdivision of the *canne*. In Languedoc 1 palm = 1/8 of the *canne*. In Bize, the *canne* used is the *canne* of Narbonne, measuring 1.967 m. Hence 1 palm = 24.6 cm.[1]

Weight units

The pound used in Marseille in wholesale trade for commodities weighing more than 20 pounds is the steelyard pound.[2]
1 steelyard pound of Marseille = 0.403 kg

The *quintal* (hundreweight) mark weight = 48.95 kg[3]
It contains 100 pounds mark weight
1 pound mark weight = 0.489 kg[4]
For 1 *quintal* mark weight to be equivalent to 118 pounds, as mentioned at page 6 of the manuscript, 1 pound must equal 0.414 kg. This equivalence proves that it is the pound table weight of Montpellier which is mentioned and used by the author, and not the pound table weight of the department of Aude the mass of which was recorded at the time of the adoption of the metric system, around 1800.
1 pound table weight of Montpellier = 0.414 kg[5]

1 pound table weight of the Aude = 0.407 kg[6]

1 pound is subdivided into 16 ounces, hence 1 ounce = 25.92 g
1 *gros* or dram = 1/8 of ounce, hence 1 *gros* = 3.24 g[7]

The *arroba*, *arrove* or *arrobe* = 11.512 kg
Used in all parts of the kingdom of Castilla, it corresponds to ¼ of *quintal* or hundredweight.[8]

Volume units (solids)

1 *setier* of Narbonne = 71 litres[9]
It is used both in Bize (Aude) and in Saint-Chinian (Hérault).

Currency

At the time when the Memoirs are written the *livre tournois* can be simply mentioned as *livre* since the *livre parisis* had been abolished in 1667.
1 *livre tournois* = 20 *sols tournois* = 240 *deniers tournois*
During the second half of the 18th c. its value is generally estimated as equivalent to about 0.312 g of gold.[10]

Notes

1 Abbe in Charbonnier, Abbe, Brunel *et al.*1994, p. 90-91.
2 Brunel in Charbonnier, Abbe, Brunel *et al.*1994, p. 124.
3 Doursther 1965, p. 459.
4 Machabey 1962, p. 359.
5 Machabey 1962, p. 333, 390; Hélas in Charbonnier, Abbe, Brunel *et al.*1994, p. 191.
6 Abbe in Charbonnier, Abbe, Brunel *et al.* 1994, p. 92.
7 Hélas in Charbonnier, Abbe, Brunel *et al.* 1994, p. 191.
8 Doursther 1965, p. 28.
9 Abbe in Charbonnier, Abbe, Brunel *et al.*1994, p.86; Hélas in Charbonnier, Abbe, Brunel *et al.*1994, p. 205.
10 Marquié 1993, p. 13.

Appendix 2

Dye-analyses of some of the samples of dyed broadcloth in the second and fourth of the *Memoirs on Dyeing*

Dr Witold Nowik
Centre de Recherche et de Restauration des Musées de France (C2RMF), Paris

---

Introduction

Dye analyses of 12 of the samples of Londrin Second cloth glued on the pages of the manuscript of the *Memoirs on Dyeing* could be performed. The samples were selected for various reasons, the main aim being to check how closely the results of the dye analyses of the samples corresponded to the recipes. Other issues concerned the possibility of identifying minor components of complex mixtures and the detection of uncommon and/or more or less forbidden colorants, including some of allegedly poor fastness.

Some potentially interesting samples were excluded because of their small size, our main concern being not to visibly alter the appearance of the document. Only a few fibres or tiny ends of threads were taken from the selected samples.

Experimental

Extraction of dyed textiles

Colorants were extracted from dyed wool fibres with 20μL of a solution of MeOH/H_2O/Oxalic acid 0.2 M (8-:2-:1 v/v/v) for each sample. The solutions were sonicated during 15 min at 35°C, then evaporated to dryness in a vacuum dessiccator overnight. The extracts were dissolved in 20μL of DMSO. The obtained solutions were filtered through a Millex®PTFE microfilters with 0.2μm pores.

High-performance liquid chromatography

Separation of the different dyestuffs was performed on an Agilent HP 1100 system made by (Walbronn, Germany) composed of an in-line degasser (G1322A), a quaternary pump (G1311A), an autosampler (G1329A/G1330A), a column oven (G1316A) and a diode detector (G1315A). Software used was ChemStation version B.03.01. Chromatographic separations were performed at 30°C with a flow rate of 0.3mL/min, on a Hypersil BDS C18 3μm column (100 x 2.1 mm) with a pre-column (10 × 2.1 mm) with the use of an acidified (0.1% HCOOH) water-acetonitril gradient, from 5 to 60% MeCN over 40 min.

Results

Table A2.1: Recipes in the 2nd and 4th Memoirs in the light of the results of dye-analyses of corresponding samples

Sample	*Ingredients mentioned in the recipe*	*Colorants identified (% at 254 nm)*
p. 47 Chocolate with milk *Chocolat au lait*	Young fustic Cochineal Madder Tartar	Madder (84%) Young fustic (15%) Cochineal (1%) Indigo (traces)
p.53 Crimsoned madder *Garance cramoisillé*	Madder Fernambuco wood	Cochineal (74%) Brazilwood (26%)
p. 56 Festiqui	Weld Soot	Indigo (13%) Flax-leaved daphne (87%)
p. 58 Dark green *Vert brun*	Woad ground of a King's blue Alum *Trentanel* (flax-leaved daphne) Logwood	Indigo (24%) Flax-leaved daphne (76%)
p. 68 Reddish brown *Marron rougeâtre*	Fernambuco wood Red sandalwood *Redoul* Gallnut Copperas	Madder (74%) *Redoul* (20%) Gallnut (gallic acid) (6%)
p. 70 Hazelnut *Noisette*	Gallnut Old fustic Logwood Madder Alum Copperas	Traces of a non-identified compound not corresponding to the composition of any of the colouring sources mentioned in the recipe Indigo (traces)
p. 82 Grass green *Vert d'herbe*	Woad ground of a King's blue Alum Red tartar Weld Flax-leaved daphne	Indigo (17%) Flax-leaved daphne + Weld (83%)
p. 83 Royal wine soup *Soupevin royal*	Alum Red tartar Spoilt cochineal Orchil	Cochineal
p. 86 Chestnut brown *Marron*	Red sandalwood *Redoul* Gallnut Copperas	Red sandalwood (2%) *Redoul* (98%)
p. 89 Lemon *Limon*	Alum Red tartar Weld Flax-leaved daphne	Weld (100%)
p. 90 Festiqui	Flax-leaved daphne Weld Soot	Flax-leaved daphne (100%)
p. 94 Dark purple *Pourpre foncé*	Woad ground of a dark azure blue *Granille* (garbling from cochineal) Larch agaric Red tartar	Indigo (9%) Cochineal (91%)

Appendix 3

Colour measurement and colorimetric data of cloth samples in the *Memoirs on Dyeing*

Iris Brémaud[1] *& Dominique Cardon*

Introduction

In this appendix are presented the colorimetric data obtained for all the samples of Londrins Seconds cloth corresponding to recipes in the second and fourth Memoirs in the manuscript (table A3.1). The colorimetric analysis has consisted in measurements of the samples' spectral reflectance curves as they were illuminated by a standard illuminant. Such measurements aim at providing objective measures correlating with our subjective perception of colours. Several systems have been proposed to represent the double dimension of the appearance of colour, physical (optical) and perceptual (involving the human eye and brain) in the best possible way. The system adopted here is the currently most widely used one, the CIE L*a*b* (also written CIE-LAB) Colour Space, proposed by the *Commission Internationale de l'Éclairage* (International Commission on Illumination). In this system, colour is defined by three co-ordinates: 1) Lightness or L* on the axis of greys (from black, corresponding to L*=0, to white, corresponding to L*=100); 2) on the a* axis, green (negative values of a*) is opposed to red (positive values of a*); 3) along the b* axis, blue (negative values of b*) is opposed to yellow (positive values of b*). The same system can be expressed in terms of chromaticity ($C^* = \sqrt{a^{*2}+b^{*2}}$) and hue angle ($h = \arctan[b^*/a^*]$), C^* describing the chromaticity or colourfulness (i.e. the "amount of colour", based on a given level on the axis of greys L^*), and h defining hue on a circular basis (0°=magenta, 90°=yellow, 180°=green, 270°=blue, 379°= nearly magenta, etc.). Given the human "global" perception of colour, this system (CIE LCh) is easier to intuitively understand than the former one: hue angle is the value best assessed by the human eye, followed by chromaticity, and then lightness.

The purpose of this research was double. It firstly aimed at providing objective data in order to more precisely define the colours of the samples, as a complement to the names given to them by the author of the manuscript and to the colour plates representing them in this book. Even keeping in mind that there may not be a perfect agreement among colour-measuring instruments, the colour coordinates here published should prove helpful guides in future attempts at reproducing any or all of these delightful ancient colours. The issue of reproducibility is also connected with the second aim of this appendix – assessing colour differences between groups of samples in order to answer two related questions: 1) How precisely and consistently does a set of colorimetric data correspond to a particular colour name in the manuscript? To give but one example: to what extent and in what way can a red be different from another red, and still be designated by the same name, scarlet for instance? 2) With what objective degree of exactness did the author manage to dye two pieces of cloth into colours close enough to each other to be given the same name, while using different ingredients, or different proportions of the same ingredients?

The document published here offered a very rare opportunity to answer such questions, since different samples with the same colour name are present in the second and fourth Memoirs, respectively.

Material and methods

All the 171 samples of dyed Londrin Second cloth present in the manuscript of *Memoirs on dyeing* were included into this colorimetric analysis. Measurements were performed in an air-conditioned room protected from direct sunlight.

Measurements were performed using a colorimeter Datacolor® microflash. Data were integrated at between 400 and 710nm (at 10nm intervals), with a diffuse illumination geometry, a 45°/0° measuring head, specular reflectance being excluded. The colorimetric data corresponding to the CIELab and CIELCh systems were obtained according to the CIE 10° standard observer, and using two different standard illuminants: standard illuminant D65, corresponding to day light and commonly used for textiles (this is the illuminant on which all our analyses are based); and standard illuminant A (corresponding to incandescent light, for comparison). The observed surface corresponded to $59mm^2$ (8.7mm diameter aperture). Whenever the size of the sample permitted, each sample was measured at three different points. The values presented here correspond to the average calculated from the data obtained from the three measurements.

Results

The fourth Memoir of annotations describes personal experiments by the author in his attempts to propose more economical recipes as satisfying alternatives to the processes, conform to regulations, which are described in the second Memoir. Only thirty-four colour names are common to recipes described in the two memoirs, but they correspond to 40 samples in the fourth memoir, either because two processes are described as giving a colour with the same name (this is the case for the colours winesoup/*soupevin,* King's colour/*couleur de roy*, grass green/*vert d'herbe* and Saxon blue/*bleu de Prusse*), or because a recipe is illustrated by several samples (golden wax/*cire doré*).

Considering differences in each of the distinct colorimetric values separately (differences in lightness ΔL^*, differences in chroma ΔC^* and differences in hue Δh), it appears that *in average*, samples corresponding to the alternative processes described in the fourth Memoir are generally neither darker nor lighter, and neither more nor less saturated than the standards of the second Memoir. Still *in average*, hue angles do not systematically deviate into any particular direction (more towards yellow or towards red, for instance). Moreover, in about 75% of samples, the difference between the sample resulting from the alternative recipe and the standard of the second Memoir with the same colour name, is inferior or equal to 5 units of L*, C* or h. Of course, a study of the data for each individual pair of samples reveals different natures and magnitudes of differences.

However, an assessment of the global colour difference perceived requires to use colour-difference formulas simultaneously taking into account the different colour coordinates, i.e. lightness, chromaticity and hue. Several different formulas exist to calculate colour-difference (expressed as ΔE). Among them, the quite recent formula ΔE CIEDE2000 is designed to get a better homogeneity, over very different hues, between measured data and visually perceptible colour-differences[2]. Although the calculations involved in this ΔE CIEDE2000 formula are somewhat more complex than for other formulas (which is the main reason why ΔE CIEDE2000 is not more commonly used), the better representativeness of various hues that it allows makes it more suitable in this case, given the wide range of colours represented in the manuscript. One colour-difference unit is generally considered as corresponding to the smallest variation perceptible by the human eye. Limits of acceptability for colour-differences vary according to context, materials concerned, type of production, etc.[3] Two colour-difference units would be considered as acceptable in most industrial branches today.

For the first time in history, the comparison between samples in two memoirs of the exceptional document studied here allows to provide an objective demonstration of the astonishing skill as colourists attained by some dyers of the pre-synthetic colorants era, the author of the *Memoirs on Dyeing* certainly not being an isolated case. Isn't it quite remarkable, that for nearly 40% (13 out of 34) of the samples designated by the same colour name in the two memoirs, the ΔE CIEDE2000 colour difference is inferior or equal to 2 units, although in each pair, the two samples have been obtained by processes differing at least slightly? Furthermore, in 75% of cases, the ΔE CIEDE2000 colour-difference between a sample resulting from one of the recipes conform to regulations of the second memoir and the alternative recipe for the same colour in the fourth memoir, is inferior or equal to 4 colour-difference units.

Conclusion and perspectives

This colorimetric study of the dyed cloth samples in the *Memoirs on Dyeing* therefore contributes much to enrich our vision of the author. He was not only, as we already know, a good manager of a big manufacture, and a good dyer, but he also was an excellent colorist, able to produce a wide range of colours, to invent some, and to reproduce them with an accuracy which is truely impressive, considering the diversity of the mordants and colouring sources he was using, and the variability inherent to natural dyes.

It is hoped that the colorimetric data published in this appendix will soon serve as a basis for comparisons with a wider corpus of data similarly obtained from historical documents including dyed cloth samples and patterns. In this way, the art of ancient European dyers may receive the recognition it deserves as a wonderful source of inspiration.

Notes

1 Scientific Researcher at CNRS (National Centre of Scientific Research, France), Wood Team, Laboratory of Mechanics and Civil Engineering, CNRS/University of Montpellier

2 Luo 2002.

3 Billmeyer and Saltzman 1981, p. 105.

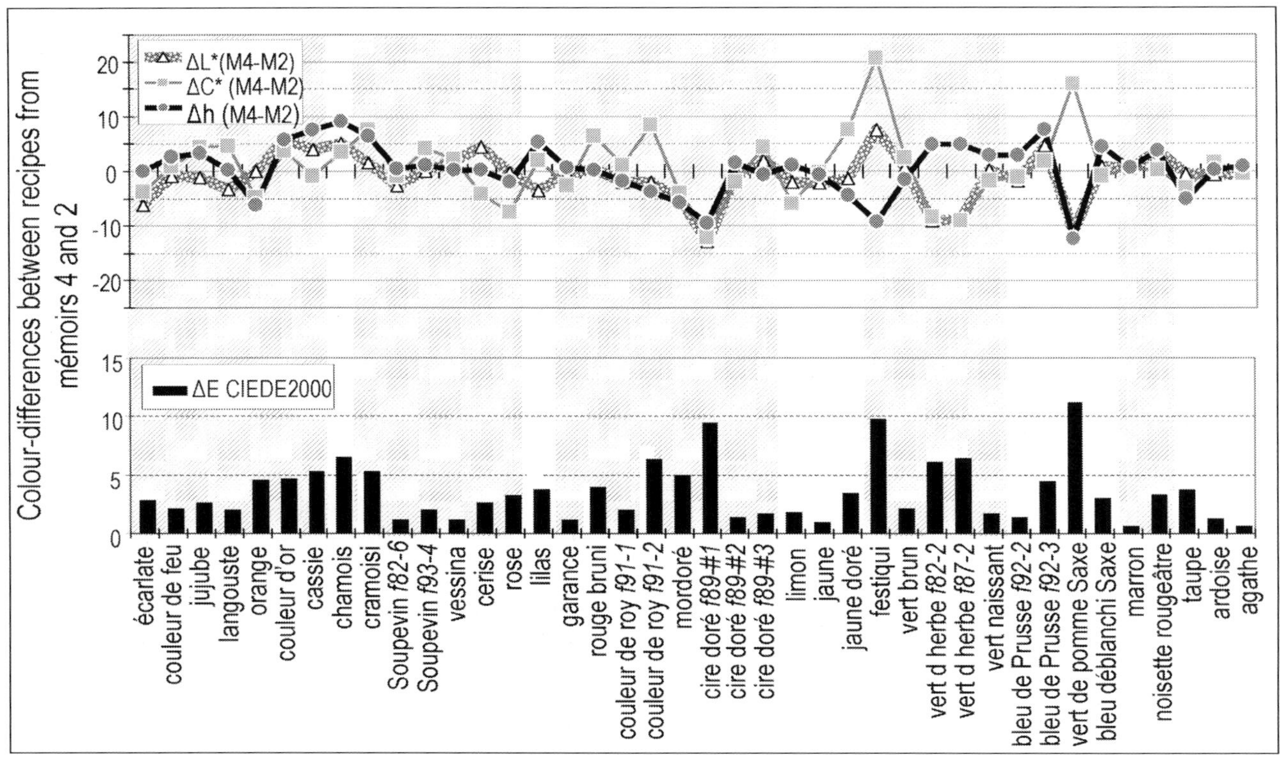

Colour differences between recipes from Memoirs 4 and 2

Table A3.1: Colorimetric data of dyed cloth samples in order of appearance in the Memoirs. Illuminant D65. Horizontal lines separate related groups of recipes in the second Memoir

N° folio-N° sample in page	*Colour name, apparition order in manuscript*	*English translation*	*L**	*a**	*b**	*C**	*h (°)*
Second Memoir							
f41-1	1 *bleu pers*	1 Persian blue	17	2	−10	10	283
f41-2	2 *bleu turquin*	2 Turkish blue	22	1	−11	11	274
f41-3	3 *bleu de Roy*	3 King's blue	21	1	−12	12	277
f41-4	4 *bleu tané*	4 tanned blue	30	−4	−14	14	255
f41-5	5 *bleu d'azur*	5 azure blue	31	−4	−16	16	257
f41-6	6 *bleu céleste*	6 celestial blue	46	−7	−6	9	222
f41-7	7 *bleu mignon*	7 dainty blue	51	−7	−6	9	219
f41-8	8 *bleu de lait*	8 milky blue	56	−7	−1	7	192
f41-9	9 *bleu déblanchi*	9 off-white blue	60	−7	1	7	173
f45-1	10 *écarlate*	10 scarlet	40	54	30	62	29
f45-2	11 *couleur de feu*	11 flame colour	41	53	32	62	31
f45-3	12 *fleur de grenade*	12 pomegranate flower	40	51	35	62	34

(continued)

N° folio-N° sample in page	Colour name, apparition order in manuscript	English translation	L*	a*	b*	C*	h (°)
f45-4	13 *jujube*	13 jujube	45	50	35	61	35
f46-1	14 *langouste*	14 spiny lobster	49	46	40	61	41
f46-2	15 *orange*	15 orange	50	46	50	68	47
f46-3	16 *orange pâle ou abricot*	16 pale orange or apricot	57	40	47	62	49
f46-4	17 *couleur d'or*	17 gold colour	53	32	54	63	59
f46-5	18 *cassie*	18 mimosa flower	57	32	59	67	62
f47-1	19 *jonquille*	19 daffodil	64	23	67	70	71
f47-2	20 *chamois*	20 chamois	66	19	26	32	54
f47-3	21 *biche*	21 doe	69	15	26	30	61
f47-4	22 *café au lait*	22 coffee with milk	69	7	19	20	70
f47-5	23 *chocolat au lait*	23 chocolate with milk	62	14	22	26	57
f48-1	24 *cramoisi*	24 crimson	36	49	19	52	21
f49-1	25 *soupevin*	25 wine soup	29	35	3	35	5
f49-2	26 *vessina*	26 vetch flower	35	42	8	43	11
f49-3	27 *griotte*	27 morello cherry	40	42	8	43	11
f50-1	28 *cerise*	28 cherry	48	48	15	50	18
f50-2	29 *rose*	29 rose	58	37	16	40	24
f50-3	30 *chair*	30 flesh	63	34	18	39	28
f50-4	31 *chair pâle*	31 pale flesh	71	15	15	22	45
f51-1	32 *pourpre*	32 purple	20	15	−6	16	340
f51-2	33 *violet*	33 violet	26	7	−13	15	298
f51-3	34 *lilas*	34 lilac	40	6	−8	10	306
f51-4	35 *gorge de pigeon*	35 dove breast	37	13	−6	14	335
f51-5	36 *mauve*	36 mauve	36	22	−4	22	349
f52-1	37 *gris de lin*	37 linen grey	43	16	−1	16	356
f52-2	38 *fleur de pêcher*	38 peach blossom	52	21	11	24	27
f53-1	39 *garance*	39 madder	29	37	20	42	28
f53-2	40 *garance cramoisillé*	40 crimsoned madder	29	32	7	33	13
f53-3	41 *rouge bruni*	41 saddened red	21	16	8	18	28
f53-4	42 *couleur de Roy*	42 King's colour	24	13	10	16	36
f53-5	43 *couleur de roi sans bruniture*	43 King's colour without saddening	26	26	14	30	28
f54-1	44 *maure doré*	44 golden Moor	30	29	20	36	35
f54-2	45 *cannelle doré*	45 golden cinnamon	41	32	37	49	49
f54-3	46 *canelle rougeâtre*	46 reddish cinnamon	33	35	23	42	34
f54-4	47 *canelle brûlé*	47 burnt cinnamon	32	25	24	35	44
f55-1	48 *cire doré*	48 golden wax	49	24	45	52	62
f56-1	49 *limon*	49 lemon	67	5	65	65	85
f56-2	50 *jaune*	50 yellow	67	10	73	74	83
f56-3	51 *jaune doré*	51 golden yellow	60	11	58	59	79
f56-4	52 *jaune tané*	52 tanned yellow	53	11	47	48	77
f56-5	53 *feuille morte*	53 dead leaf	55	11	48	49	77
f56-6	54 *festiqui*	54 festiqui	42	2	30	30	87

(continued)

N° folio-N° sample in page	*Colour name, apparition order in manuscript*	*English translation*	*L**	*a**	*b**	*C**	*h (°)*
f58-1	55 *vert noir*	55 black green	23	−7	4	8	153
f58-2	56 *vert brun*	56 dark green	20	−4	2	4	146
f58-3	57 *vert de canard*	57 duck green	24	−10	5	11	152
f58-4	58 *vert d'herbe rempli*	58 full-bodied grass green	26	−10	11	15	134
f58-5	59 *vert d'herbe*	59 grass green	33	−14	16	21	131
f59-1	60 *émeraude*	60 emerald	35	−13	18	22	125
f59-2	61 *perroquet*	61 parrot	44	−10	39	40	104
f59-3	62 *vert clair*	62 light green	49	−9	40	41	103
f59-4	63 *vert gay*	63 gay green	54	−11	38	39	107
f59-5	64 *vert naissant*	64 nascent green	56	−3	55	55	93
f60-1	65 *vert pistache*	65 pistachio green	49	3	27	28	83
f60-2	66 *vert pomme*	66 apple green	60	−9	40	40	102
f60-3	67 *vert de chou*	67 cabbage green	32	−8	−5	10	211
f60-4	68 *vert de mer*	68 sea green	39	−9	−8	12	223
f60-5	69 *vert céladon*	69 celadon green	65	−5	26	27	101
f61-1	70 *olive brun*	70 dark olive	24	3	10	10	73
f61-2	71 *olive pourrie*	71 rotten olive	29	5	15	16	71
f61-3	72 *olive*	72 olive	30	−5	17	17	107
f61-4	73 *olive verte*	73 green olive	44	−3	36	36	95
f62-1	74 *olive sèche*	74 dry olive	43	1	31	31	88
f63-1	75 *bleu de Prusse*	75 Saxon blue	54	−7	11	13	124
f64-1	76 *vert de pomme*	76 apple green	57	−14	39	41	110
f65-1	77 *vert de pomme Saxe*	77 Saxon apple green	59	−6	20	21	106
f65-2	78 *bleu de blanchi Saxe*	78 Saxon off-white blue	67	−1	18	18	92
f67-1	79 *marron*	79 brown	19	7	3	8	26
f68-1	80 *marron rougeâtre*	80 reddish brown	26	16	11	19	34
f68-2	81 *girofle rempli*	81 full-bodied clove colour	17	7	3	8	26
f68-3	82 *prune*	82 plum	39	8	10	12	51
f69-1	83 *prune grisâtre*	83 greyish plum	42	0	9	9	88
f70-1	84 *noisette*	84 hazelnut	57	4	18	18	77
f70-2	85 *noisette*	85 hazelnut	47	8	20	21	69
f70-3	86 *noisette rougeâtre*	86 reddish hazelnut	55	12	21	24	60
f70-4	87 *noisette doux*	87 soft hazelnut	58	6	21	22	73
f71-1	88 *noisette biche*	88 doe hazelnut	67	10	25	27	68
f71-2	89 *noisette foncé*	89 dark hazelnut	41	5	17	18	72
f71-3	90 *noisette verdâtre*	90 greenish hazelnut	43	4	17	17	78
f71-4	91 *tabac*	91 tobacco	27	4	14	15	72
f71-5	92 *tabac brun*	92 dark tobacco	28	6	13	14	67
f72-1	93 *tabac d'Espagne*	93 Spanish tobacco	34	28	29	41	46
f72-2	94 *tabac verdâtre*	94 greenish tobacco	28	2	16	16	82
f72-3	95 *café*	95 coffee	21	10	7	12	35
f73-1	96 *noir*	96 black	14	0	−1	1	269

(continued)

N° folio-N° sample in page	Colour name, apparition order in manuscript	English translation	L*	a*	b*	C*	h (°)
f75-1	97 *taupe*	97 mole	29	−4	−4	5	227
f75-2	98 *ardoise*	98 slate	30	−4	−6	8	236
f75-3	99 *gris de plomb*	99 lead grey	37	−4	−2	5	206
f75-4	100 *gris de rat*	100 rat grey	40	−4	3	4	142
f75-5	101 *gris de rat plus foncé*	101 darker rat grey	40	−3	4	5	129
f76-1	102 *gris de rat rougeâtre*	102 reddish rat grey	47	−2	6	6	108
f76-2	103 *agathe*	103 agate	44	−1	−8	8	261
f76-3	104 *agathe clair*	104 light agate	47	−1	−6	7	265
f76-4	105 *rat sur noisette*	105 rat on hazelnut	47	5	18	19	73
f77-1	106 *gris de castor*	106 beaver grey	66	1	15	15	85
f77-2	107 *gris de perle*	107 pearl grey	69	1	15	15	87
f77-3	108 *gris d'argent*	108 silver grey	72	1	16	16	86
Fourth Memoir							
f82-1	109 écarlate	109 scarlet	34	51	28	58	29
f82-2	110 *vert d'herbe*	110 grass green	24	−10	6	12	149
f82-3	111 *cramoisi*	111 crimson	38	53	28	60	28
f82-4	112 *agathe*	112 agate	44	0	−8	8	268
f82-5	113 *couleur de feu*	113 flame colour	40	52	35	63	34
f82-6	114 *soupevin*	114 wine soup	29	39	5	39	7
f83-1	115 *soupevin royal*	115 royal wine soup	30	36	0	36	1
f83-2	116 *soupevin supérieur*	116 superior wine soup	23	28	4	28	8
f83-3	117 *amarante royal*	117 royal amaranth	30	42	7	42	9
f83-4	118 *vessinat*	118 vetch flower	37	44	9	45	11
f83-5	119 *garance*	119 madder	28	34	19	39	29
f84-1	120 *vert bronzé*	120 bronze green	24	−7	−6	9	223
f84-2	121 *langouste*	121 spiny lobster	46	49	43	65	41
f84-3	122 *café verdâtre*	122 greenish coffee	26	7	15	17	66
f84-4	123 *orange*	123 orange	50	47	42	63	42
f84-5	124 *vert de Saxe*	124 Saxon green	36	−2	23	23	95
f84-5	124 *vert de Saxe* 2nd sample	124 Saxon green 2nd sample	38	−7	20	21	110
f85-1	125 *céladon Saxe*	125 Saxon celadon	58	−6	20	20	106
f85-1	125 *céladon Saxe* 2nd sample	125 Saxon celadon 2nd sample	57	−5	23	23	103
f85-2	126 *jonquille vif*	126 bright daffodil	67	20	63	66	72
f85-3	127 *poudre à canon des Anglais*	127 gunpowder of the English	31	−5	−5	8	225
f85-4	128 *rose*	128 rose	58	31	11	33	20
f85-5	129 *lilas*	129 lilac	36	11	−5	12	334
f86-1	130 *gorge de pigeon clair*	130 light dove breast	52	15	7	17	25
f86-2	131 *vert de pomme Saxe*	131 Saxon apple green	48	6	36	37	81
f86-3	132 *maure doré*	132 golden Moor	25	29	13	32	25
f86-4	133 *marron*	133 chestnut brown	19	7	4	8	30

(continued)

N° folio-N° sample in page	Colour name, apparition order in manuscript	English translation	L*	a*	b*	C*	h (°)
f86-5	134 *couleur d'or*	134 gold colour	59	29	60	67	64
f87-1	135 *cassie*	135 mimosa flower	61	24	62	66	68
f87-2	136 *vert d'herbe*	136 grass green	23	−10	6	12	148
f87-3	137 *jaune*	137 yellow	65	10	73	73	82
f88-3	138 *biche anglais*	138 English doe	62	11	23	25	65
f88-2	139 *vert de câpre*	139 caper green	26	−8	−3	8	203
f88-3	140 *pourpre anglais*	140 English purple	19	8	−5	10	331
f89-1	141 *violet anglais*	141 English violet	25	13	−9	16	325
f89-2	142 *limon*	142 lemon	65	4	59	59	86
f89-3	143 *jaune doré*	143 golden yellow	59	18	65	67	75
f89-4one	144 *cire doré*	144 golden wax	48	22	45	50	64
f89-4bis	144 *cire doré* 2nd sample	144 golden wax 2nd sample	37	26	30	40	50
f89-4ter	144 *cire doré* 3rd sample	144 golden wax 3rd sample	51	27	49	56	61
f90-1	145 *canelle*	145 cinnamon	35	30	29	41	44
f90-2	146 *festiqui*	146 *festiqui*	49	15	49	51	73
f90-3	147 *taupe*	147 mole	28	−1	2	2	127
f90-4	148 *ardoise*	148 slate	29	−5	−8	9	237
f90-5	149 *rouge bruni*	149 saddened red	22	21	11	24	28
f91-1	150 *couleur de Roy*	150 King's colour	22	15	9	18	30
f91-2	151 *couleur de Roy*	151 King's colour	22	22	11	25	26
f91-3	152 *noisette rougeâtre*	152 reddish hazelnut	58	9	23	25	69
f91-4	153 *gris cendré*	153 ash grey	67	2	15	15	84
f92-1	154 *jujube*	154 jujube	44	52	40	66	38
f92-2	155 *bleu de Prusse*	155 Saxon blue	52	−9	8	12	137
f92-3	156 *bleu de Prusse*	156 Saxon blue	58	−13	6	15	155
f92-4	157 *bleu de blanchi Saxe*	157 Saxon off-white blue	68	−5	17	17	106
f93-1	158 *violet d'évêque*	158 bishop's violet	22	9	−11	14	309
f93-2	159 *herbe foncé*	159 dark grass	26	−10	11	14	133
f93-3	160 *vert brun*	160 dark green	20	−4	5	7	131
f93-4	161 *soupevin*	161 wine soup	27	34	3	34	6
f93-5	162 *amarante*	162 amaranth	29	35	2	35	4
f94-1	163 *pourpre violet*	163 violet purple	20	6	−11	13	297
f94-2	164 *pourpre foncé*	164 dark purple	20	12	−5	13	337
f95-1	165 *cerise*	165 cherry	52	44	14	46	18
f95-2	166 *vert naissant sans pastel*	166 nascent green without woad	56	−5	53	53	96
f96-1	167 *chamois*	167 chamois	71	12	33	35	69

Bibliography and sources

Abbreviations

AD11 = *Archives départementales de l'Aude*
AD33 = *Archives départementales de la Gironde*
AD34 = *Archives départementales de l'Hérault*
AN = *Archives nationales*
ANOM = *Archives nationales d'Outre Mer*

Manuscripts

An. (1667) *Livret contenant la maniere de bien gouverner le guesde et preparer le pastel pour teindre en bleus – avec autres teintures tant escarlattes rouges, qu'autres telle que je les ay veu faire et pratiquer ches Monsr. Therrieux en la bonne teinture des Gobelins au fauxbourg St Marcel es Paris.* Bibliothèque centrale du Muséum national d'Histoire naturelle. Ms. 2018.

Cardon, D. (1990) *Technologie de la Draperie médiévale d'après la réglementation technique du nord-ouest méditerranéen (Languedoc-Roussillon-Catalogne-Valence-Majorque) – XIIIè-XVè siècles.* Thèse de doctorat d'Histoire, Université Paul Valéry-Montpellier 3.

Delo (1761) *Mémoire sur la teinture du Bleu et du Noir de Sedan par le Sieur Delo, ancien Elève des Manufactures.* AN F/12/737.

Guiraud (s.d.) *Cayer de teinture de Laines, Draps pour le Levant et pour les couleurs mélangées filouzelles et autres* AD 34, 7J7.

Janot, A. (1744) *Mémoire – on trouvera les opérations de la teinture du grand et bon teint des couleurs qui se consomment en Levant avec la quantité et qualité des drogues qui les composent.* AD 34, C 5569.

Maistre, C. (1817) *Livre de Teinture à l'usage de Casimir Maistre.* AD34 11 J 39.

Printed books and articles

An. (1730) *Recueil des Reglemens generaux et particuliers concernant les manufactures et fabriques du Royaume.* Paris: Imprimerie royale.

An. (1743) *Dictionnaire universel françois et latin.... Nouvelle edition.* Paris: Vve Delaune *et al.*

An. (1746) *Manufacture de Carcassonne, Arrêts, édits, règlements.* Carcassonne: imprimerie de Jean-Baptiste Coignet.

Ambrose, G. (1931) "English traders at Aleppo", *Economic History Review* a III (2), pp. 246–267.

Balfour-Paul, J. (2011) *Indigo: Egyptian Mummies to Blue-jeans.* London: Firefly Books.

Batizat (1732) *Mémoire présenté en 1731 par le Sr. Batizat de Carcassonne, à M. de Bernage, Intendant de Languedoc.* Montpellier: François Rochard, imprimeur du Roy.

Berthollet, C. L. and A. B. Berthollet (1804) *Éléments de l'art de la teinture.* 2nd edition. Paris: Firmin Didot.

Billmeyer, F.W. and M. Saltzman (1981) *Principles of Color Technology.* 2nd ed. New York-Chichester: John Wiley & Sons.

Bischoff, J. (1842) *A Comprehensive History of the Woollen and Worsted Manufacturers and the natural and commercial History of Sheep, from the earliest Records to the present period.* London: Smith, Elder and C^o.

Borgard, P., J.-P. Brun and M. Picon (2005) *L'Alun de Méditerranée.* Naples/Aix-en-Provence: Centre Jean Bérard, Centre Camille Julliand.

Bowles, W., G. de Flavigny (trans.) (1776) *Introduction à l'histoire naturelle et à la géographie physique de l'Espagne.* Paris: L. Cellot et Jombert fils.

Buti, G. (2005) "Cochenille mexicaine, négoce marseillais et manufactures languedociennes au XVIII^è siècle" in: Llinares, S. and P. Hrodej (eds) *Techniques et colonies XVIè-XXè siècles.* Paris: Publications de la Société française d'Histoire d'Outre-Mer et de l'Université de Bretagne-Sud, pp. 13–31.

Buti, G. (2008) "Des goûts et des couleurs. Draps du Languedoc pour clientèle levantine au XVIIIè siècle", *Rives méditerranéennes*, 29, pp. 125–140.

Calderan-Giacchetti, H. (1962) "L'exportation de la draperie languedocienne dans les pays méditerranéens d'après les Archives Datini (1380–1410)", *Annales du Midi,* 74(58), pp. 139–176.

Cardon (1989) "Rouge, bleu, blanc, histoire scientifique des trois couleurs" in: D. Cardon, M. Nougarède and C. Potay (éd.) *Rouge, Bleu, Blanc – Teintures à Nîmes.* Nimes: Musée du Vieux Nîmes, pp. 55–78.

Cardon, D. (1991) "Black dyes for wool in Mediterranean textile centres: an example of the chemical relevance of guild regulations", *Dyes in History and Archaeology*, 9, pp. 7–9.

Cardon, D. (1992) "From the medieval woad-vat to the modern indigo-vats: a brief history of the long decline of woad seen through technical sources", *Beiträge zur Waidtagung*, Jargang 4/5, Teil 1. Arnstad: Thüringer Chronik-Verlag, pp. 10–15.

Cardon, D. (1992) "La preuve par l'expérience dans les techniques médiévales de teinturerie", *Cahiers d'Histoire et de Philosophie des Sciences*, 40, pp. 303–310.

Cardon, D. (1992) "New information on the medieval woad vat", *Dyes in History and Archaeology*, 10, pp. 22–31.

Cardon, D. (1995) "Yellow Dyes of historical importance: beginnings of a long-term multi-disciplinary study", *Dyes in History and Archaeology*, 13, pp. 59–73.

Cardon, D. (1995/1998) "The Woad-vat: two previously unpublished sources", in: Cardon, D. and H. E. Müllerott (eds.) *Actes/Papers/Beiträge. 2e Congrès international Pastel, Indigo et autres teintures naturelles: passé, présent, futur* (1995/1998) Arnstadt: Thüringer Chronik-Verlag, pp. 50–60.

Cardon, D. (1999) *La Draperie au Moyen Âge – Essor d'une grande industrie européenne*. Paris: CNRS Editions.

Cardon, D. (2007) *Natural Dyes – Sources, Traditions, Technology and Science*. London: Archetype Publications.

Cardon, D. (2013) *Mémoires de Teinture – Voyage dans le temps chez un maître des couleurs*. Paris: CNRS Editions.

Cardon, D. (2014) *Le Monde des Teintures naturelles*. 2nd ed, updated and augmented. Paris: Belin.

Cardon, D. (2015) "Not Only Red: Cochineal in the Eighteenth-Century Woolen Cloth Industry", in: Padilla, C. and B. Anderson (eds) *A Red Like No Other: How Cochineal Colored the World*. New York: Skira Rizzoli, pp. 134–139.

Cardon, D. and Andary, C. (2001) "Yellow Dyes of historical importance. III. New historical and chemical evidence on a wild Mediterranean dye-plant, *Daphne gnidium*", *Dyes in History and Archaeology* 16/17, pp. 5–9.

Cardon, D. and A. Pinto (2007) "Le redoul, herbe des tanneurs et des teinturiers – Collecte, commercialisation et utilisations d'une plante sauvage dans l'espace méridional (XIIIè-XVè siècles)", *Médiévales* 53, pp. 51–64.

Carrière, Ch. (1980) "La Draperie languedocienne d'exportation", in: Cullen, L. M. and P. Butel (eds.) *Négoce et industrie en France et en Irlande aux XVIIIè et XIXè siècles*. Paris: Editions du CNRS, pp. 92–103.

Cazals, R. and J. Valentin (1984) *Carcassonne, ville industrielle au XVIIIè siècle*. Carcassonne: CDDP de l'Aude.

Charbonnier, P., J.-L. Abbe, B. Brunel, J.-C. Helas, M.-C. Marandet and A. Serpentini (1994) *Les Anciennes Mesures locales du Midi méditerranéen d'après les tables de conversion*. Clermont-Ferrand: Institut d'Etudes du Massif Central.

Chomel, N. (1732) *Dictionnaire oeconomique contenant divers moyens d'augmenter son bien et de conserver sa santé*. Paris: Vve Estienne.

Çolak, M., V. Thirion-Merle, F. Blondé and M. Picon (2005) "Les régions productrices d'alun en Turquie aux époques antique, médiévale et moderne: gisements, produits et transports", in: Borgard, P., J.-P. Brun and M. Picon (2005) *L'Alun de Méditerranée*. Naples/Aix-en-Provence: Centre Jean Bérard, Centre Camille Julliand, pp. 59–68.

Cooper, T. (1815) *A Practical Treatise on Dyeing and Callicoe printing*. Philadelphia: Thomas Dobson.

Cordoba de la Llave, R., A. Franco Silva and G. Navarro Espinach (2005) "L'alun de la Péninsule ibérique durant la période médiévale", in: Borgard, P., J.-P. Brun and M. Picon (2005) *L'Alun de Méditerranée*. Naples/Aix-en-Provence: Centre Jean Bérard, Centre Camille Julliand, pp. 125–137.

Davis, R. (1967) *Aleppo and Devonshire Square*. London/Melbourne/Toronto: Macmillan.

Defoe, D. (1962) *A Tour through the whole Island of Great Britain*. London: J.M. Dent, Everyman's Library.

de Keijzer, M., van Bommel, M., Hofmann-de Keijzer, R. *et al.* (2012) "Indigo carmine: Understanding a problematic blue dye", The International Institute for Conservation of Historic and Artistic Works – Contributions to the Vienna Congress 2012. DOI 10.1179/2047058412Y.0000000058.

de Laborde, A. (1809) *Itinéraire descriptif de l'Espagne, et tableau élémentaire des différentes branches de l'administration et de l'industrie de ce royaume*. 2nd ed. Paris: H. Nicolle. t. V. p. 451.

Delamare, F. (2013) *Blue Pigments. 5000 Years of Art and Industry*. London: Archetype Publications.

Diderot, D. et J. d'Alembert (1751) *Encyclopédie ou dictionnaire raisonné des sciences, des arts et des métiers*. Paris: Briasson, David, Le Breton et Durant, t. 1.

Donkin, R.A. (1977) *Spanish Red – An Ethnogeographical Study of Cochineal and the Opuntia Cactus*. Transactions of the American Philosophical Society, vol. 67, part 5. Philadelphia: The American Philosophical Society.

Doren, A. (1901) *Die Florentiner Wollentuchindustrie vom vierzehten bis zum sechzehnten Jarhundert*. Stuttgart: J. G. Cotta.

Doursther, H. (1965) *Dictionnaire universel des Poids et Mesures anciens et modernes*. Amsterdam: Meridian Publishing Co.

Duhamel du Monceau, H.L. (1765) *L'Art de la draperie, principalement pour ce qui regarde les draps fins*. Paris: H. F. Guérin et L. F. Delatour.

Eldem, E. (1999) *French Trade in Istanbul in the eighteenth century*. Leiden-Boston-Köln.

Fernandez, L. (1778) *Tratado instructivo, y practico sobre el Arte de la Tintura: reglas experimentadas y metodicas para tintar sedas, lanas, hilos de todas clases, y esparto en rama*. Madrid: Blas Roman.

Fougeroux de Bondaroy, A. D. (1768) "Observations sur le lieu appelé Solfatare, situé proche de la ville de Naples", *Histoire de l'Académie royale des Sciences, année 1765*. Paris: Imprimerie royale, pp. 275–285.

Fougeroux de Bondaroy, A. D. (1769) "Mémoire sur les alunières ou alumières de La Tolfa", in: *Histoire de l'Académie royale des Sciences, année 1766*. Paris: Imprimerie royale, pp. 16–21.

Golikov, V. (2001) "The Technology of silk dyeing by Cochineal. III. The experimental investigation of the influences of pH, water quality, cream of tartar and oak galls", *Dyes in History and Archaeology*, 16–17, pp. 21–33.

Grison, T. (1884) *La Teinture industrielle au dix-neuvième siècle en ce qui concerne la laine et les tissus où la laine est prédominante*. Paris: G. Rougier.

Hale, T. (1758) *A Compleat Body of Husbandry*. 2nd ed. London: Osborne, vol. 3.

Hellot, J. (1750) *L'Art de la Teinture des Laines et des Etoffes de Laine en grand et petit teint*. Paris: Vve Pissot.

Hunter, D.M. (1910) *The West of England Woollen Industry under Protection and Free Trade*. London, New York: Cassell & Company, Ltd.

Karpik, L. (1989) "L'économie de la qualité", *Revue française de sociologie*, 30, pp. 187–210.

Kortum, G.M. (1749) *Neue Versuche der Färbekunst betreffend die, bisher unter dem Namen* Sans pareille de Saxe *bekannten blauen und grünen Farben*. Breslau and Leipzig: Johann Jacob Korn.

Lacombe de Prézel, H. (1761) *Dictionnaire du Citoyen, ou abrégé historique, théorique et pratique du commerce: contenant ses principes, le droit public de l'Europe relativement au négoce*. Paris: Grangé.

Luo, R. (2002) "Development of colour-difference formulae" *Review of Progress in Coloration and Related Topics*, 32 (1) pp. 28–39.

Mann, J. de L. (1964) *Documents illustrating the Wiltshire Textile Trade in the eighteenth century*. Devizes: Wiltshire Records Society.

Mann, J. de L. (1971) *The Cloth Industry in the West of England from 1640 to 1880*. Oxford: Clarendon Press.

Marling, W. H. (1913) "The Woollen Trade of Gloucestershire", *Transactions of the Bristol and Gloucestershire Archaeological Society*, vol. 36.

Marquié, C. (1993) *L'Industrie textile carcassonnaise au XVIIIe siècle – Etude d'un groupe social: les marchands-fabricants*. Carcassonne: Société d'Etudes scientifiques de l'Aude.

Prudhomme, L.M. (1804) *Dictionnaire universel, géographique, statistique, historique et politique de la France*. Paris: Baudoin, t. 4, p. 231.

Machabey, A. (1962) *La Métrologie dans les muses de province et sa contribution à l'histoire des poids et mesures en France depuis le treizième siècle*. Paris: CNRS.

McJunkin, D. M. (1991) Logwood: an inquiry into the historical biogeography of Haematoxylum campechianum L. and related dyewoods of the Neotropics. PhD thesis, Los Angeles: University of California.

Masson, P. (1911) *Histoire du Commerce français dans le Levant au XVIIIè siècle*. Paris: Hachette.

Minard, P. (1998) *La Fortune du colbertisme – Etat et industrie dans la France des Lumières*. Paris: Fayard.

Minard, P. (2003) "Réputation, normes et qualité dans l'industrie textile française au XVIIIè siècle", in: A. Stanziani (ed.) *La Qualité des produits en France (XVIIIè-XXè siècles)*, Paris: Belin, pp. 69–89.

Moir, E.A.L. (1955) "The Gentlemen Clothiers – A Study of the organization of the Gloucestershire cloth industry 1750–1835" in: H.P.R. Finberg (ed.) *Gloucestershire Studies*, Leicester: Leicester University Press, pp. 225–266.

Morineau, M. and Ch. Carrière (1968) "Draps du Languedoc et commerce du Levant au XVIIIè siècle", *Revue d'Histoire économique et sociale*, 56, pp. 108–121.

Nowik, W. (2001) "The possibility of differentiation and identification of red and blue 'soluble' dyewoods", *Dyes in History and Archaeology* 16/17, pp. 129–144.

Nowik, W. (2005) "HPLC-PDA characterisation of *Daphne gnidium* L. (Thymelaeaceae) dyeing extracts using two different C-18 stationary phases", *Journal of Separation Science* 28, pp. 1595–1600.

Nowik, W. (2013) "Analyses de colorants d'échantillons de drap teint du second et dernier des *Mémoires de teinture*", in: Cardon 2013, pp. 397–400.

Padilla, C. and Anderson, B. (eds.) (2015) *A Red Like No Other: How Cochineal Colored the World*. New York: Skira Rizzoli.

Partridge, W. (1823, re-ed.1973) *A Practical Treatise on dying of woollen, cotton and skein silk, with the Manufacture of broadcloth and cassimere including the most improved methods in the West of England*. Edington, Wilts: Pasold Research Fund.

Picon, M. (2005) "Mines et aluns de l'Aveyron entre aluns naturels et aluns de synthèse", in: Borgard, P., J.-P. Brun and M. Picon (2005) *L'Alun de Méditerranée*. Naples/Aix-en-Provence: Centre Jean Bérard, Centre Camille Julliand, pp. 139–155.

Playne, A. T. (1915) *A History of the Parishes of Minchinhampton and Avening*. Gloucester: John Bellows.

Ponting, K. G. (1971) *The Woollen Industry of South-West England*. Bath: Adams & Dart.

Porter, J. (1771) *Observations on the Religion, Law, Government and Manners of the Turks. To which is added, the state of the Turkey trade, from its origin to the present time*. 2nd ed. London: J. Nourse.

Pybus, D. (2009) "The Alum Industry in north-east Yorkshire: a preliminary view" in: Sheeran, G. (ed.) *Digging for brass: the impact of the extractive industries on the Yorkshire landscape*, pp.52–59.

Rebora, G. (ed.) (1970) *Un Manuale di Tintoria del Quattrocento*. Milano: Dott. A. Giuffrè.

Richard, E. (ed.) (2011) *Jacques Savary, Le Parfait Négociant*. Genève: Droze.

Richelet, P. (1732) *Dictionnaire de la langue françoise, ancienne et moderne*. New ed. Amsterdam: aux frais de la Compagnie.

Roux, A., F.-A. Aubert de La Chesnaye-Desbois and J. Goulin (1762–1764) *Dictionnaire domestique portatif contenant toutes les connoissances relatives à l'oeconomie domestique et rurale*. Paris: Vincent.

Rudder, S. (1779) *A New History of Gloucestershire*. Cirencester: Samuel Rudder.

Savary des Bruslons, J. and P.-L. Savary (1723) *Dictionnaire universel de commerce*. T. 1 (A-B) Paris: Jacques Estienne.

Savary des Bruslons, J. and P.-L. Savary (1742) *Dictionnaire universel de commerce*. T. 2 (D-O). Genève: Héritiers Cramer et Frères Philibert.

Savary des Bruslons, J. and P.-L. Savary (1741) *Dictionnaire universel de commerce*. T. 3 (L-Z). New ed. Paris: Vve Estienne.

Savary des Bruslons, J. and P.-L. Savary (1748) *Dictionnaire universel de commerce*. T. 2. Paris: Vve Estienne.

Sella, D. (1961) *Commerci e Industrie a Venezia nel secolo XVII*. Venice/Rome: Instituto per la Collaborazione Culturale.

Singer, C. (1948) *The Earliest Chemical Industry: An Essay in the Historical Relations of Economics and Technology Illustrated from the Alum Trade*. London: Folio Society.

Tann, J. (1967) *Gloucestershire Woollen Mills*. Industrial Archaeology. Newton Abbot: David & Charles.

Tann, J. (2012) *Wool and Water – The Gloucestershire Woollen Industry and its Mills*. Stroud: The History Press.

Thomson, J. K. J. (1982) *Clermont-de-Lodève 1633–1789 – Fluctuations in the prosperity of a Languedocian cloth-making town*. Cambridge: Cambridge University Press.

van Laer, G. (1874) *Recueil des principaux procédés de teintures à mordants à l'usage des teinturiers*, III. Verviers: Ch. Vinche.

Wood, A. C. (1935) *A History of the Levant Company*. Oxford: Oxford University Press.

Yvon, T. (2015) *La Production d'indigo en Guadeloupe et Martinique (XVIIe-XIXe siècles)*. Paris: Karthala.

List of tables

Part 1

Part 3

List of figures and diagrams

Part I

Part II – plate section

Part III

Fig. 6.1. *Matrice* (standard) of scarlet.
Fig. 6.2. Passport of cochineal.

Epilogue
Diagram 1. Purples, violets and wine colours of pages 93 and 94 of the *Memoirs on Dyeing*: sequence of operations.
Diagram 2. Greens in pages 93 and 95 of the *Memoirs*: sequence of operations.

Appendix 3
Colour differences between samples corresponding to recipes with same colour names in the 4th and 2nd Memoirs

Illustration acknowledgements

Photos
All the photos of the pages of the manuscript of *Memoirs on Dyeing* with cloth samples have been taken by Pierre-Normann Granier. Their copyright is owned by Dominique Cardon.

All other photos: Dominique Cardon.

Map
Fig. 1.1. Map K. Mercier/D. Cardon, CNRS, CIHAM/UMR 5648.

Diagrams
Fig. 1.7. Trends in the market of Languedocian broadcloths produced for the Levant in the 18th century. M. Morineau and Ch. Carrière, "Draps du Languedoc et commerce du Levant au XVIIIè siècle", *Revue d'Histoire économique et sociale*, 56, 1968, p. 117.

Fig. 1.8. Production of the Royal Manufacture of Bize during the "Gout era" (1757-1789). D. Cardon/I. Brémaud.

Epilogue, diagrams 1-2. K. Mercier/D. Cardon, CNRS, CIHAM/UMR 5648.

Appendix 3, diagrams I. Brémaud.

Other diagrams: D. Cardon.